UNDERSTANDING CONSTRUCTION DRAWINGS

Mark W. Huth

7th edition

UNDERSTANDING CONSTRUCTION DRAWINGS

Mark W. Huth

CENGAGE

Australia • Brazil • Japan • Korea • Mexico • Singapore • Spain • United Kingdom • United States

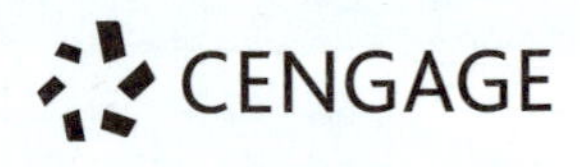

***Understanding Construction Drawings,* Seventh Edition**
Mark W. Huth

SVP, GM Skills & Global Product Management: Jonathan Lau

Product Director: Matt Seeley

Senior Product Manager: Vanessa Myers

Executive Director, Development: Marah Bellegarde

Senior Content Development Manager: Larry Main

Content Developer: Katie Ostler, Lumina Datamatics

Product Assistant: Jason Koumourdas

Vice President, Strategic Marketing Services: Jennifer Ann Baker

Marketing Manager: Scott Chrysler

Senior Content Project Manager: Kara A. DiCaterino

Senior Digital Project Manager: Amanda Ryan

Designer: Angela Sheehan

Cover Designer: Joe Devine

Cover & Interior Design Image(s): Hitdelight/Shutterstock.com istock.com/suprun

Additional Interior Design Image(s): BorisShevchuk/Shutterstock.com Bloomicon/Shutterstock.com

Library of Congress Control Number: 2017950099

Book Only ISBN: 978-1-337-40864-6

Package ISBN: 978-1-337-40863-9

Cengage Learning
20 Channel Center Street
Boston, MA 02210
USA

Notice to the Reader

Publisher does not warrant or guarantee any of the products described herein or perform any independent analysis in connection with any of the product information contained herein. Publisher does not assume, and expressly disclaims, any obligation to obtain and include information other than that provided to it by the manufacturer. The reader is expressly warned to consider and adopt all safety precautions that might be indicated by the activities described herein and to avoid all potential hazards. By following the instructions contained herein, the reader willingly assumes all risks in connection with such instructions. The publisher makes no representations or warranties of any kind, including but not limited to, the warranties of fitness for particular purpose or merchantability, nor are any such representations implied with respect to the material set forth herein, and the publisher takes no responsibility with respect to such material. The publisher shall not be liable for any special, consequential, or exemplary damages resulting, in whole or part, from the readers' use of, or reliance upon, this material.

Printed in the United States of America
Print Number: 03 Print Year: 2018

Contents

Preface

Intended Audience

Understanding Construction Drawings is designed for students in construction programs in two- and four-year colleges and technical institutes, as well as apprentice training. Designed for a course in print reading focused on both residential and commercial construction, the book helps you learn to read the drawings that are used to communicate information about buildings. It includes drawings for buildings that were designed for construction in several parts of North America. The diversity of building classifications and geographic locations ensures that you are ready to work on construction jobs anywhere in the industry. Everyone who works in building construction should be able to read and understand the drawings of the major trades.

How to Use This Book

The book is divided into four major parts and several units within each part. Each part relates to the prints in the separate drawing packet:

- **Part 1, Drawings: Two-Unit Apartment,** introduces you to the basics of print reading by covering views, scales, lines, and symbols, as well as the various plan views, elevations, and sections and details.
- **Part 2, Reading Drawings: Lake House,** provides information on how to interpret drawings for project specifics—everything from footings and foundation walls to room finishing and cabinets.
- **Part 3, Multifamily Construction: Urban Courts,** details more advanced residential print reading and applies the skills learned in Parts 1 and 2 to other types of construction, as well as mechanical and electrical trades.
- **Part 4, Heavy Commercial Construction: School Addition,** presents the need-to-know information on interpreting prints for large commercial construction including structural drawings, mechanical drawings, and electrical drawings.

Features of the Units

The individual units are made up of four elements: Objectives, the main body of the unit, Using What You Learned, and Assignment.

- The ***Objectives*** appear at the beginning of the unit so that you will know what to look for as you study the unit.
- The ***main body*** is the presentation of content with many illustrations and references to the prints for the building being studied in that part.
- ***Using What You Learned*** gives you an opportunity to do a practice exercise that is similar to the exercises found in the assignment questions for that unit. The real-world need to be able to do the exercise is explained first. The exercise is presented and followed by a detailed explanation of how to find the specified information. ***Each unit contains 10 to 20 Assignment*** questions that require you to both understand the content of the unit and apply that understanding to reading the drawings. There are more than 600 questions in all.

The book is divided into four parts, corresponding with the four buildings. At the end of each part there is a test. Additionally units include one or more **Green Notes**, which provide insights and suggestions for green home construction.

The Drawing Packet, Glossary, and Appendix

At the back of the text you will find several helpful aids for studying construction drawings.

- The ***drawing packet*** that is conveniently packaged with the book contains 22 sheets with separate drawings that relate to each of the parts within the book. The drawing packet contains prints for four buildings: a simple two-family duplex that is very easy to understand, a more complex single-family home, one building in an urban development located in an earthquake zone, and an addition to a school.

- The ***Glossary*** defines all the new technical terms introduced throughout the textbook. Each of these terms is defined where it is first used, but if you need to refresh your memory, turn to the Glossary.
- The ***Math Reviews*** in Appendix B are an innovative feature that has helped many construction students through a difficult area. These are concise reviews of the basic math you are likely to encounter throughout the building construction field. As math is required in this textbook, reference is made to the appropriate Math Review. All the math skills needed to complete the end-of-unit assignments in this book are covered in the Math Reviews.
- The **Appendix** also includes a complete list of construction abbreviations commonly used on prints, along with their meaning. There is also a section that explains the most commonly used symbols for materials and small equipment.

New to This Edition

Understanding Construction Drawings, Seventh Edition, represents a major revision of the book. Part 3 is all new, based on a 4-story residential building in an area known to have high seismic activity. One of the new units is entirely devoted to the special considerations that are encountered in areas with seismic activity.

MindTap For Understanding Construction Drawings, 7e

NEW! The MindTap for Understanding Construction Drawings, 7th Edition features an integrated course offering a complete digital experience for the student and teacher. This MindTap is highly customizable and combines assignments, videos, interactivities, and quizzing along with the enhanced ebook to enable students to directly analyze and apply what they are learning and allow teachers to measure skills and outcomes with ease.

- A Guide: Relevant interactivities combined with prescribed readings, featured multimedia, and quizzing to evaluate progress, will guide students from basic knowledge and comprehension to analysis and application.
- Personalized Teaching: Teachers are able to control course content—hiding, rearranging existing content, or adding and creating own content to meet the needs of their specific program.
- Promote Better Outcomes: Through relevant and engaging content, assignments and activities, students are able to build the confidence they need to ultimately lead them to success. Likewise, teachers are able to view analytics and reports that provide a snapshot of class progress, time in course, engagement and completion rates.

Supplements to the Text

Along with the *Understanding Construction Drawings, Seventh Edition* book, we are proud to offer supplemental offerings that will help support classroom instruction and engage students in learning.

The ***Instructor Resources*** available on our Companion website contains *free* helpful tools for the instructor teaching a course on reading and interpreting construction drawings. Each component follows the chapters in the book and is intended to help instructors prepare classroom presentations and student evaluations. To access these helpful tools, please visit www.cengagebrain.com. At the home page, search for this Companion website by typing in the ISBN of the book in the search box at the top of the page. On the page illustrating this book, click on the "Access" button next to "Free Study Tools" and this will direct you to the following resources:

- An ***Instructor's Guide*** provides answers to all the Assignment questions and test questions in the textbook, and explains how the answers were found or calculated. In addition, it contains more than

500 additional questions that can be used for tests, supplemental assignments, and review. The answer to each of these questions is given, along with an explanation of the answer.

- ***PowerPoint Presentations*** include an outline of each chapter along with photos and graphics to help illustrate important points and enhance classroom instruction. These presentations are editable, allowing instructors to include additional notes and photos/graphics from the Image Gallery included on our Companion website.
- ***Cengage Learning Testing Powered by Cognero*** is a flexible, online system that allows you to:
  - author, edit, and manage test bank content from multiple Cengage Learning solutions
  - create multiple test versions in an instant
  - deliver tests from your LMS, your classroom or wherever you want
- ***Image Gallery*** containing graphics and photos from all the chapters in the book provide an additional option for classroom presentations. Instructors may choose to add to the existing PowerPoint, or may wish to create their own presentations based on the book.

Acknowledgments

I am grateful to all who contributed to this textbook. Special thanks are due to Ralph Henderson, School of Science and Technology, Rogue Community College for contributing most of the content of Unit 36 on seismic considerations.

The instructors and their students who have used the previous five editions have given me valuable feedback that has played an instrumental role in shaping this edition. Several companies provided expertise and contributed illustrations—including many of the figures that illustrate this book.

I would especially like to thank the architects and engineers who supplied the drawings for the drawing packet, namely:

Robert Kurzon
Duplex and Lake House

Carl Griffith
Cataldo, Waters, and Griffith Architects, P.C., and HA2F Consultants in Engineering for the School Addition

Chelsea Richardson and Shawn Sidener
Jeffrey DeMure + Associates for architectural drawings of Urban Courts

Karl Freeman
O'Connor, Freeman, & Associates for structural drawings for Urban Courts

Lastly, I would also like to thank the instructors who reviewed the manuscript for the previous editions and for this new edition. They have provided guidance in making it the best print reading textbook it could be.

Ralph Henderson
Department Chair/Instructor for Construction Technology, Rogue Community College, Grants Pass, OR

Steven Peterson
Professor, Weber State University, Ogden, UT

Scott Bretthauer
Teacher, E.M.P.S., College of Lake County, Grayslake, IL.

Joe Dusek
Professor, Construction Management Department, Triton College, River Grove, IL.

Judy Guentzler-Collins
Building Technology Department, Cochise College, Sierra Vista, AZ.

Seongchan Kim
Department of Engineering Technology, Western Illinois University, Macomb, IL

About the Author

The author of this textbook, Mark W. Huth, brings many years of experience in the industry to his writing—first as a carpenter and then as a contractor, building construction teacher, and construction publisher—his career has allowed him to consult with hundreds of construction educators in high schools, colleges, and universities. He has also authored several other successful construction titles, including *Basic Blueprint Reading for Construction*, *Residential Construction Academy: Basic Principles for Construction*, and *Practical Problems in Mathematics for Carpenters.*

A Word about Math

Construction requires the use of mathematics. Whether you are a carpenter planning stairs, a plumber calculating pipe lengths and fitting allowances, or an estimator preparing for a contract bid, you need math to do your job. The math required in this textbook is basic, so you probably have learned enough math to do all of the work required. Most of the math required on a construction job can be done quite easily with a construction calculator, such as the one shown here. Today's construction calculators are preprogrammed to do everything from converting decimals to fractions, calculating the lengths of rafters, figuring cubic yards of concrete, and other standard industry computations.

If you are studying construction, you probably own a construction calculator now or will soon. However, as you progress in your learning and spend more time working on construction sites, you will soon find that you do not always have your trusty calculator handy. If you have learned to do the basic math required with a pencil and paper (or scrap of wood), you will not be hampered by not having your calculator. Also, it is easy to make big mistakes with a calculator—any kind of calculator. With one wrong press of a key, you can add when you meant to multiply or add an extra zero. If you learn the math, you can check your work and ensure that it is close to what the calculator got, so there is less likely to be a catastrophic error.

A typical construction calculator.

For these reasons, you are urged to complete the assignments at the end of each textbook unit by doing the math *without a calculator*, at least until you feel confident in your ability. If you have trouble doing the math, check the Math Reviews in Appendix B at the back of the book. They give easy, step-by-step directions for doing all of the types of math needed in the book.

DRAWINGS: TWO-UNIT APARTMENT

Part 1 helps you develop a foundation upon which to build skills and knowledge in reading the drawings used in the construction industry. The topics of the various units in this section are the basic concepts upon which all construction drawings are read and interpreted. The details of construction are explored in Parts 2, 3, and 4.

Many of the assignment questions in this part refer to the drawings of the Two-Unit Apartment Building (Duplex) included in the drawing packet that accompanies this textbook. The Duplex was designed as income property for a small investor. It was built on a corner lot in a small city in upstate New York. The Duplex is an easy-to-understand building. Its one-story, rectangular design requires only a minimum of views; you can quickly become familiar with the Duplex drawings.

UNIT 1

The Design-Construction Sequence and the Design Professions

Objectives

After completing this unit, you will be able to perform the following tasks:

- Name the professions included in the design and planning of a house or light commercial building.
- List the major functions of each of these professions in the design and planning process.
- Identify the profession or agency that should be contacted for specific information about a building under construction.

The construction industry employs about 15 percent of the working people in the United States and Canada. A large portion of construction workers are involved in constructing new buildings, roads, airports, and industrial facilities. The rest are involved in repairing, remodeling, and maintenance. As the needs of our society change, the demand for different kinds of construction increases. Homeowners and businesses demand more energy-efficient buildings. The shift toward automation in business and industry means that new offices are needed. Our national centers of commerce and industry are shifting. These are only a few of the reasons that new housing starts are considered important indicators of our economic health.

The construction industry is made up of light construction (small buildings) and heavy construction (large buildings, roadways, bridges, industrial installations, etc.). The industry can also be divided into the following four classifications: residential, commercial, industrial, and civil. *Residential construction* includes single-family homes, small apartment buildings, and condominiums (see **Figure 1–1(a)**). *Commercial construction* includes

Figure 1–1(a). Single- and multi-family homes are classified as residential construction.

Figure 1–1(b). Commercial construction.

Courtesy of Bechtel Corporation. Photographer: Terry Lowenthal. Used by permission.

Figure 1–1(c). Industrial construction. Delta Energy Center, water treatment tanks and buildings in the foreground.

office buildings, hotels, stores, shopping centers, and other large buildings (see **Figure 1–1(b)**). *Industrial construction* includes structures other than buildings, such as refineries and paper mills, that are built for industry (see **Figure 1–1(c)**). *Civil construction* (see **Figure 1–1(d)**) is more closely linked with the land and refers to highways, bridges, airports, dams, and the like.

The Design Process

The design process starts with the owner. The owner has definite ideas about what is needed, but may not be expert at describing that need or desire in terms the builder can understand. The owner contacts an architect to help plan the building.

Courtesy of Bechtel Corporation. Photographer: Ray Frayne. Used by permission.

Figure 1–1(d). Civil construction. At 726 feet, Hoover Dam is the highest dam in the United States.

GREEN NOTE

Green construction can be defined in many different ways, and sometimes the definitions sound complicated. In its simplest terms, green construction is the process of designing and constructing a building that minimizes its impact on the environment both during construction, over its useful life, and, ultimately, the recyclability of its materials—or their safe and proper disposal—when that life comes to an end.

A green home is built from environmentally sustainable materials using practices that reduce material use and waste. A durable, long-lasting home has lower maintenance requirements and less overall impact on the environment than a home that needs to be replaced sooner or requires frequent repairs. A green home is also designed to conserve resources such as heating and cooling energy and water.

The architect serves as the owner's agent throughout the design and construction process. Architects combine their knowledge of construction—of both the mechanics and the business—with artistic or aesthetic knowledge and ability. They design buildings for appearance and use.

The architect helps the owner determine how much space is needed, how many rooms are needed for now and in the future, what type of building best suits the owner's lifestyle or business needs, and what the costs will be. As the owner's needs take shape, the architect makes rough sketches to describe the planned building. At first these may be balloon diagrams (see **Figure 1–2**) to show traffic flow and the number of rooms. Eventually, the design of the building begins to take shape (see **Figure 1–3**).

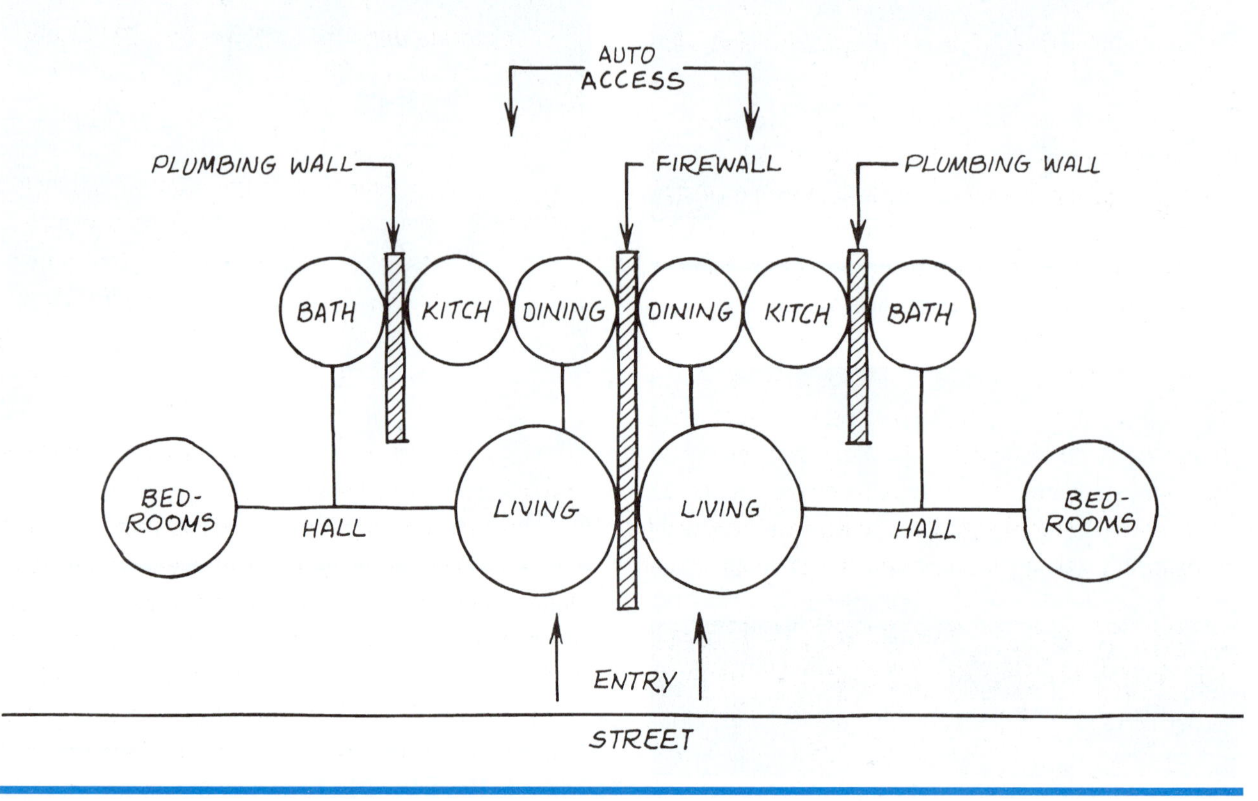

Figure 1–2. Balloon sketch of Duplex.

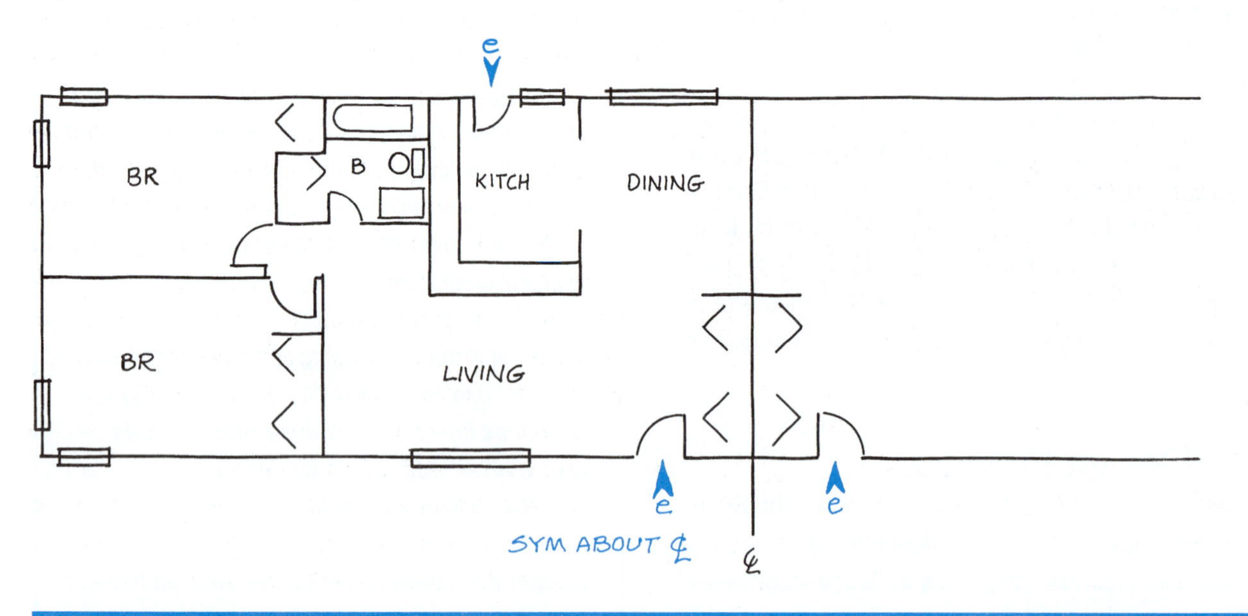

Figure 1–3. Straight line sketch of Duplex.

Before all the details of the design can be finalized, other construction professionals become involved. Building codes specify requirements to ensure that buildings are safe from fire hazards, earthquakes, termites, surface water, and other concerns of the community. Most building codes are based on a model code. For example, the International Code Council (ICC) publishes several model codes, one of which is the *International Building Code®*. It is commonly referred to as the IBC. Another is the *International Residential Code® for One- and Two-Family Dwellings,* the IRC, which includes all of the parts of the IBC that pertain to one- and two-family dwellings plus some additional content that applies to these dwellings (see **Figure 1–4**). The IRC is a model code, because it is a model that may be used by state and local building authorities as a basis for their own local code. A model code has no authority on its own. The government having jurisdiction in a locale must adopt its own building code. Very often the government body having jurisdiction (called the *Authority Having Jurisdiction,* or AHJ) adopts the model code. Sometimes the AHJ adds specific clauses to the model, and, in rare cases, it writes an all-new code. State building codes allow local governments to adopt a local building code, but they require that the local code be at least as stringent as the state code.

GREEN NOTE

The design and planning for a green home involves not only the owner and designer, but the general contractor and key trade contractors as well. The designer's preliminary house plans are reviewed by those who will build the home, each looking for ways to improve energy efficiency, incorporate durable construction details, and simplify utility systems installation. Group meetings are often conducted in which the designer, owner, contractor, and key trade contractors discuss the plans and examine the impact of each recommendation and how the work will be carried out. The general contractor and trade contractors can also recommend green building materials that best suit the project.

Figure 1–4. 2015 *International Residential Code® for One- and Two-Family Dwellings.*

The local building code is administered by a building department of the local government. The building department reviews the architect's plans before construction begins and inspects the construction throughout its progress to ensure that the code is followed.

Most communities also have zoning laws. *A zoning law* divides the community into zones where only certain types of buildings are permitted. Zoning laws prevent such problems as factories and shopping centers being built in the same neighborhood as homes.

Building departments usually require that very specific procedures are followed for each construction project. A building permit is required before construction begins. The building permit notifies the building department about planned construction. Then, the building department can make sure that the building complies with all the local zoning laws and building codes. When the building department approves the completed construction, it issues a *certificate of occupancy.* This certificate is not issued until the building department is satisfied that the construction has been completed according to the local code. The certificate of occupancy is usually issued after a final inspection by the building inspector. The owner is not permitted to move into the new building until the certificate of occupancy has been issued.

If the building is more complex than a home or simple frame building, engineers may be hired to help design the structural, mechanical, electrical, or other aspects of the building. Consulting engineers specialize in certain aspects of construction and are employed by architects to provide specific services. Finally, architects and their consultants prepare construction drawings that show all aspects of the building. These drawings tell the contractor specifically what to build.

Some homes are built from stock plans available from catalogs of house designs, building materials dealers, or magazines (see **Figure 1–5**). However, many states require a registered architect to approve the design and supervise the construction.

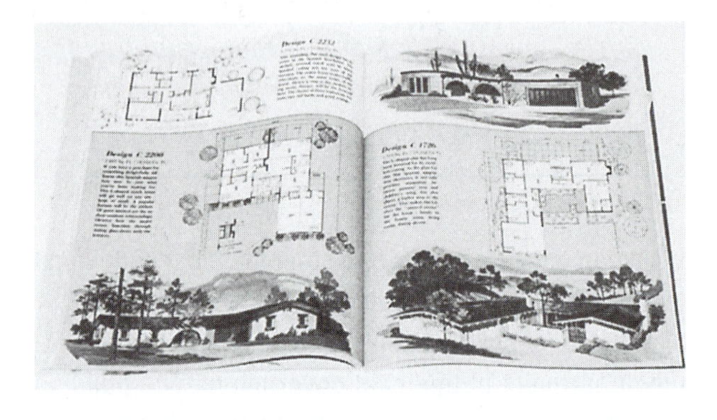

Figure 1–5. Stock plans can be ordered from catalogs.

Starting Construction

After the architect and the owner decide on a final design, the owner obtains financing. The most common way of financing a home is through a mortgage. A *mortgage* is a guarantee that the loan will be paid in installments. If the loan is not paid, the lender has the right to sell the building in order to recover the money owed. In return for the use of the lender's money, the borrower pays interest—a percentage of the outstanding balance of the loan.

When financing has been arranged (sometimes before it is finalized), a contractor is hired. Usually, a general contractor is hired with overall responsibility for completing the project. The general contractor in turn hires subcontractors to complete certain parts of the project. All stages of construction may be subcontracted. The parts of home construction most often subcontracted are excavation, plumbing and heating, electrical, drywall, painting and decorating, and landscaping. The relationships of all the members of the design and construction team are shown in **Figure 1–6**. Utility installers should carefully investigate all the drawings, especially the architectural drawings, in order to determine the installation locations of their equipment.

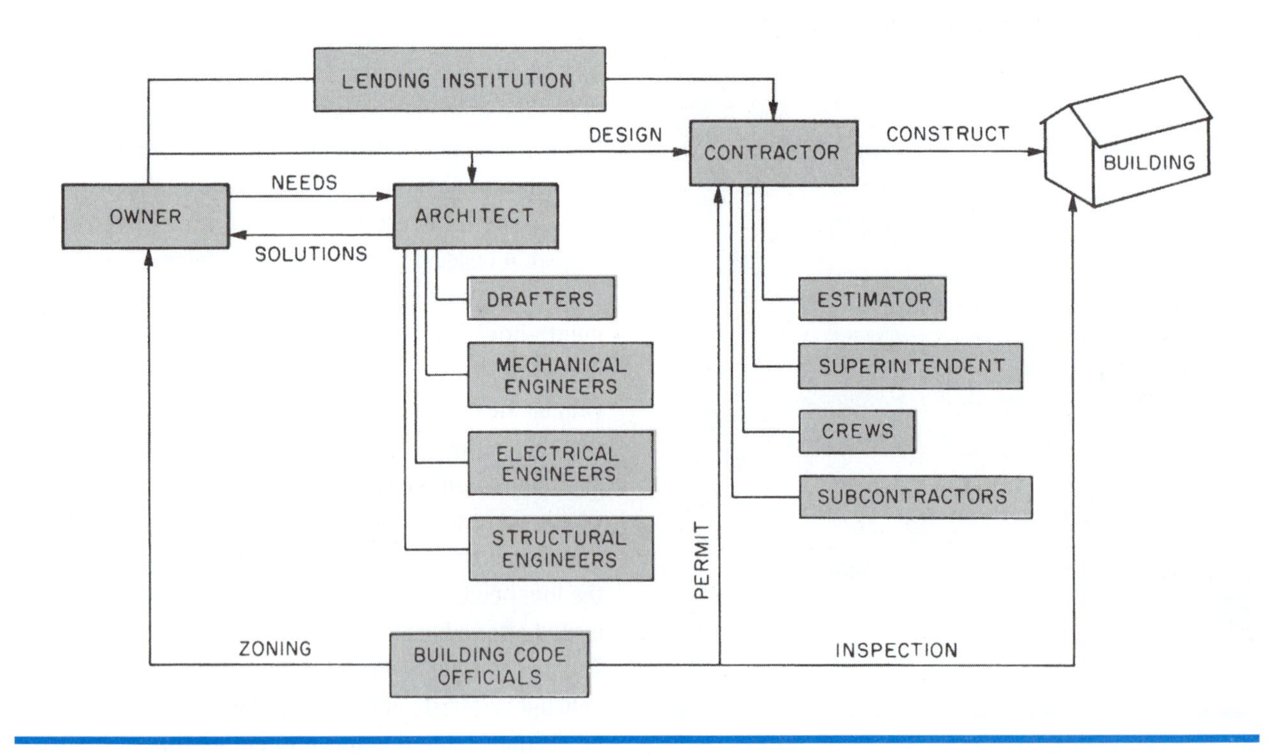

Figure 1–6. Design and construction team.

USING WHAT YOU LEARNED

Everyone involved in the design, construction, and ownership of a building needs to know who the major players are in the process. Only by understanding what role each agency, company, and individual plays in the process can a construction worker know where to go with questions and for information. For example, no work can begin on the site until a building permit has been issued. The owner, contractor, and superintendent all need to know who issues that permit. Building permits are issued by the building department of the city, town, or county where the building is to be constructed.

Assignment

1. Who acts as the owner's agent while the building is being constructed?
2. Who designs the structural aspects of a commercial building?
3. Who would normally hire an electrical engineer for the design of a store?
4. Who is generally responsible for obtaining financing for a small building?
5. To whom would the general contractor go if there were a problem with the foundation design for a home?
6. If local building codes require specific features for earthquake protection, who is responsible for seeing that they are included in a home design?
7. Whom would the owner inform about last-minute changes in the interior trim when the building is under construction?
8. What regulations specify what parts of the community are to be reserved for single-family homes only?
9. Who issues the building permit?
10. What regulations are intended to ensure that all new construction is safe?

UNIT 2 Views

Objectives

After completing this unit, you will be able to perform the following tasks:

- Recognize oblique, isometric, and orthographic drawings.
- Draw simple isometric sketches.
- Identify plan views, elevations, and sections.

Isometric Drawings

A useful type of pictorial drawing for construction purposes is the ***isometric drawing.*** In an isometric drawing, vertical lines are drawn vertically, and horizontal lines are drawn at an angle of 30° from horizontal, as shown in **Figure 2–1**. All lines on one of these isometric axes are drawn in proportion to their actual length. Isometric drawings tend to look out of proportion because we are used to seeing the object appear smaller as it gets farther away.

Isometric drawings are often used to show plumbing layout (see **Figure 2–2**). The ability to draw simple isometric sketches is a useful skill for communicating on the job site. Try sketching a brick in isometric as shown in **Figure 2–3**.

Step 1. Sketch a Y with the top lines about 30° from horizontal.
Step 2. Sketch the bottom edges parallel to the top edges.
Step 3. Mark off the width on the left top and bottom edges. This will be about twice the height.
Step 4. Mark off the length on the right top and bottom edges. The length will be about twice the width.
Step 5. Sketch the two remaining vertical lines and the back edges.

Other isometric shapes can be sketched by adding to or subtracting from this basic isometric brick (see **Figure 2–4**). Angled surfaces are sketched by locating their edges and then connecting them.

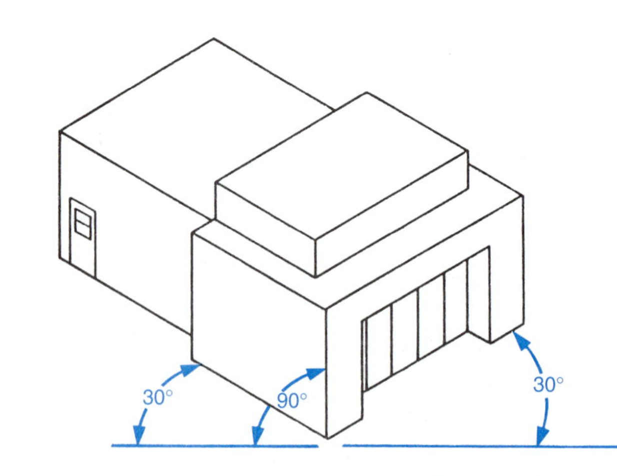

Figure 2–1. Isometric of building.

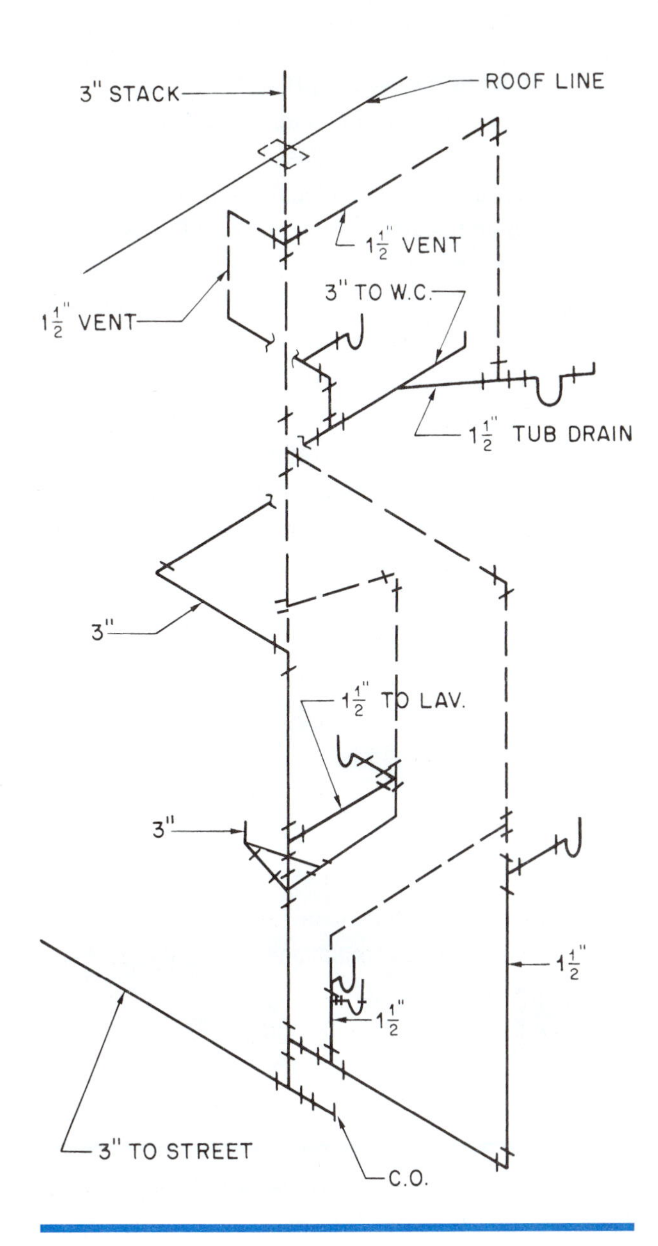

Figure 2–2. Single-line plumbing isometric.

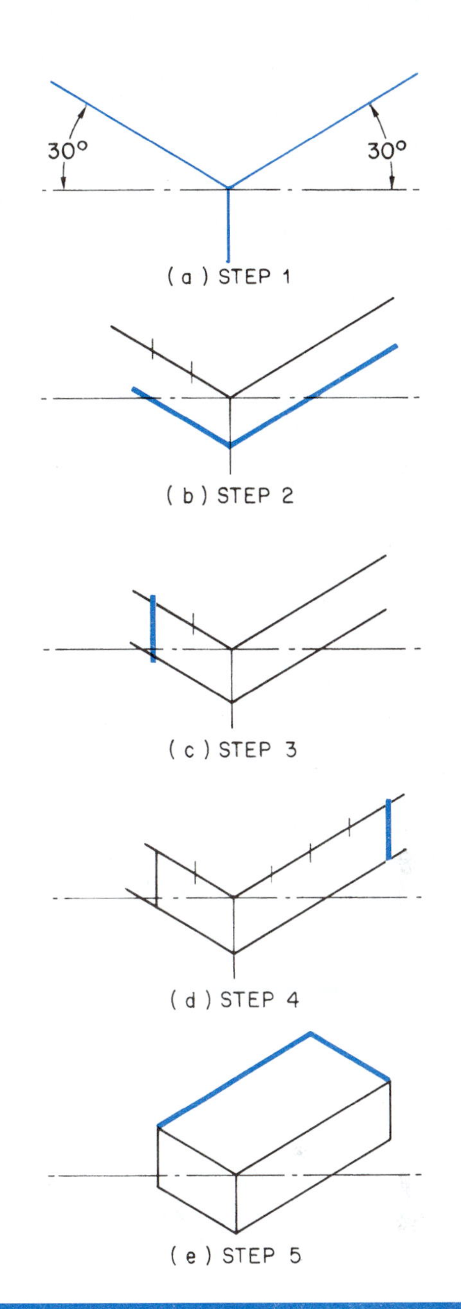

Figure 2–3. Sketching an isometric brick.

GREEN NOTE

Waste water is classified as either black water or gray water. Black water is that which is discharged from toilets and kitchens, where the water can be very contaminated and must be extensively treated before it can be released into the environment or reused. Gray water is from sinks, laundry areas, and storm water and is less dangerous to the environment. It is becoming increasingly common for green homes to treat gray water and reuse it for irrigation and toilets.

Oblique Drawings

When an irregular shape is to be shown in a pictorial drawing, an ***oblique drawing*** may be best. In oblique drawings, the most irregular surface is drawn in proportion as though it were flat against the drawing surface. Parallel lines are added to show the depth of the drawing as shown in **Figure 2–5**.

Orthographic Projection

To show all information accurately and to keep all lines and angles in proportion, most construction drawings are drawn by ***orthographic projection.*** Orthographic projection is most often explained by imagining the object to be drawn inside a glass box. The corners and the lines representing the edges of the object are then projected onto the sides of the box (see **Figure 2–6**). If the box is unfolded, the images projected onto its sides will be on a single plane, as on a sheet of paper (see **Figure 2–7**). In other words, in orthographic projection, each view of an object shows only one side (or top or bottom) of the object.

All surfaces that are parallel to the plane of projection (the surface of the box) are shown in proportion to their actual size and shape. However, surfaces

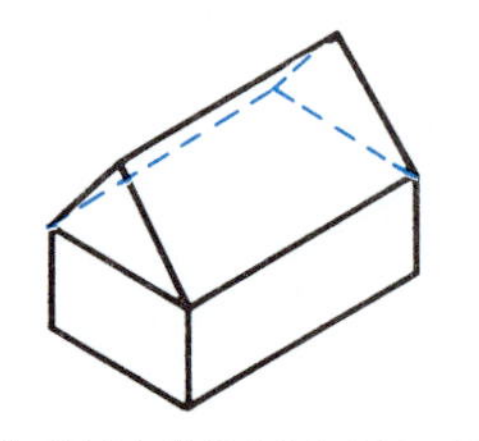

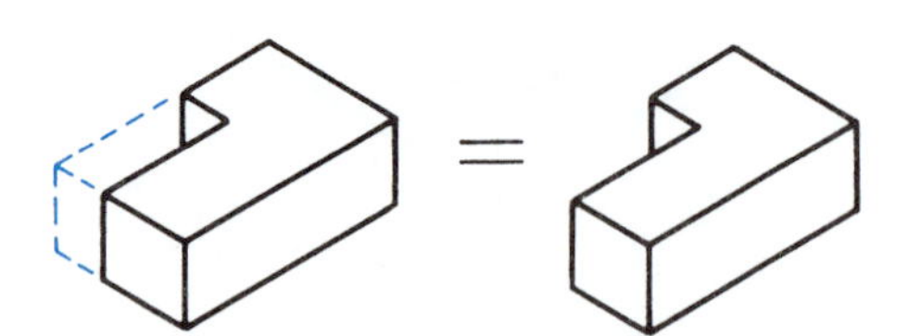

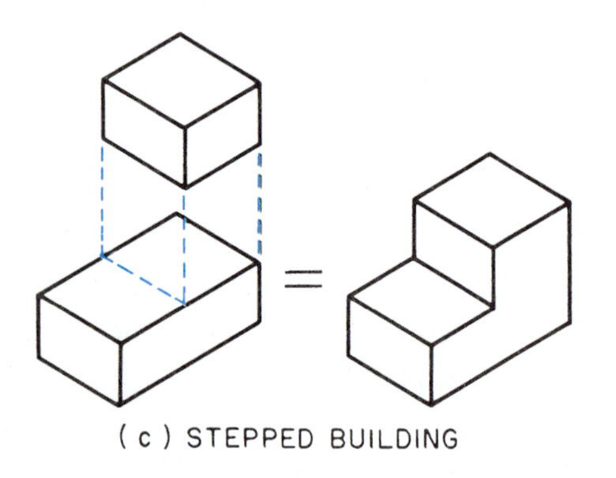

Figure 2–4. Variations on the isometric brick.

(a) END VIEW OF CROWN MOLDING

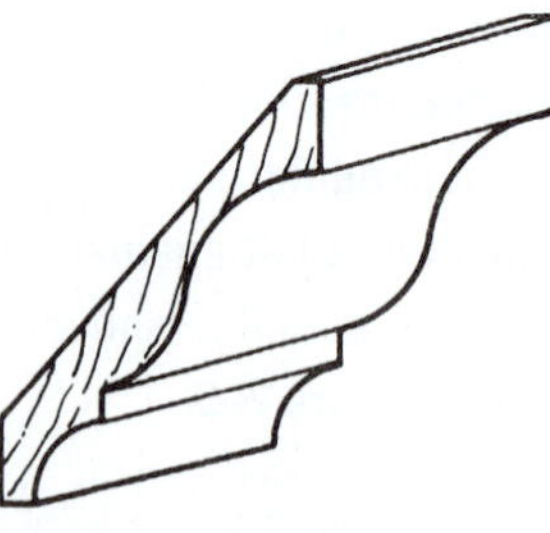

Figure 2–5. Oblique drawing.

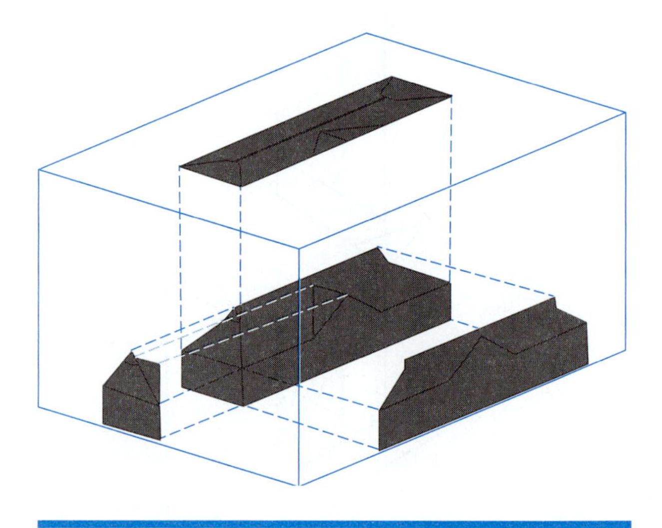

Figure 2–6. Duplex inside a glass box; method of orthographic projection of roof, front side, and end.

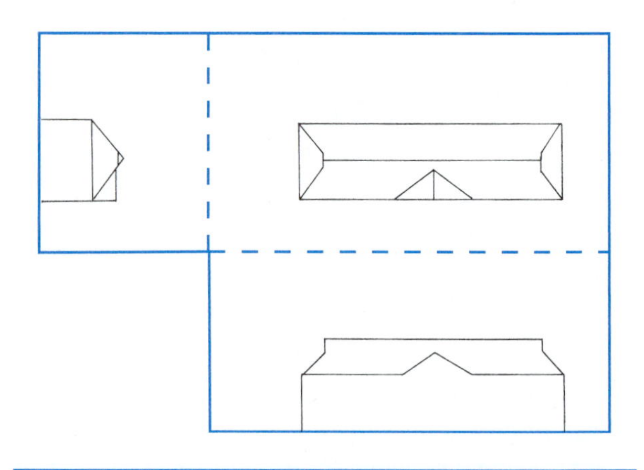

Figure 2–7. Orthographic projection unfolded on a flat sheet of paper.

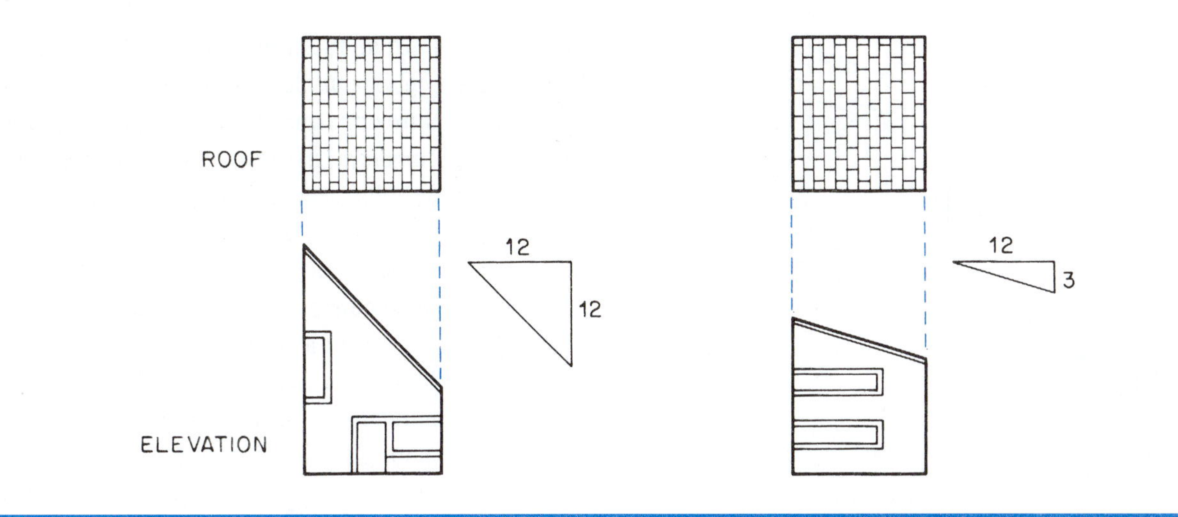

Figure 2–8. Views of two shed roofs.

that are not parallel to the plane of projection are not shown in proportion. For example, both of the roofs in the top views of **Figure 2–8** appear to be the same size and shape, but they are quite different. To find the actual shape of the roof, you must look at the end view.

In construction drawings, the views are called plans and elevations. A *plan view* shows the layout of the object as viewed from above (see **Figure 2–9**). A set of drawings for a building usually includes plan views of the site (lot), the floor layout, and the foundation. ***Elevations*** are drawings that show height. For example, a drawing that shows what would be seen standing in front of a house is a building elevation (see **Figure 2–10**). Elevations are also used to show cabinets and interior features.

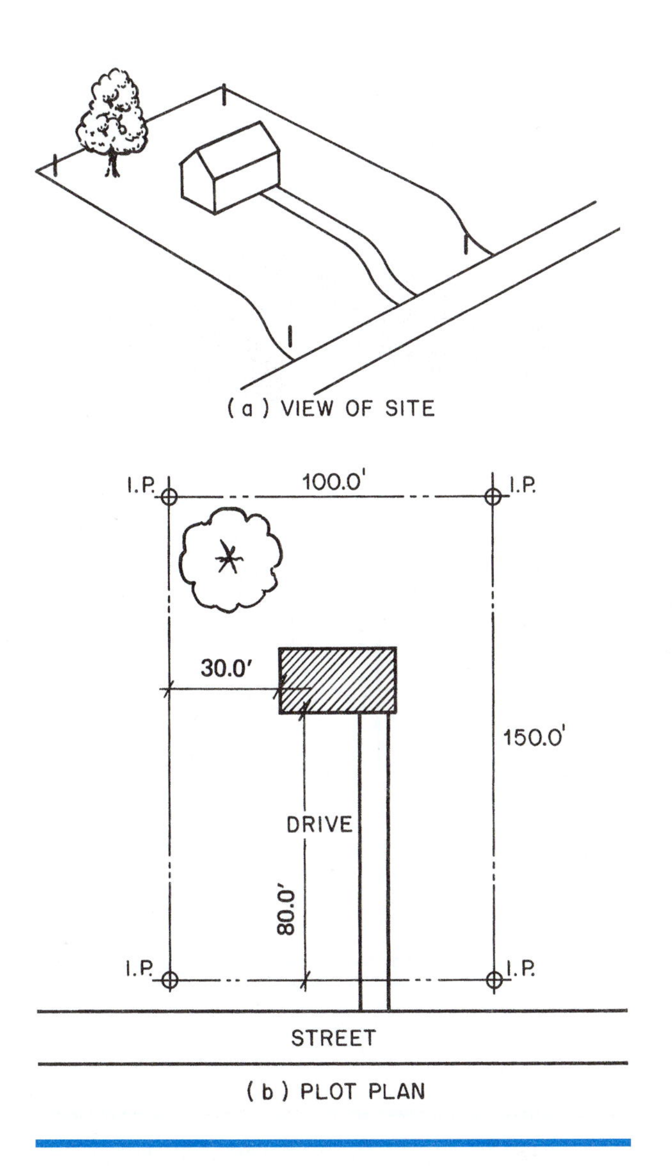

Figure 2–9. Plan view.

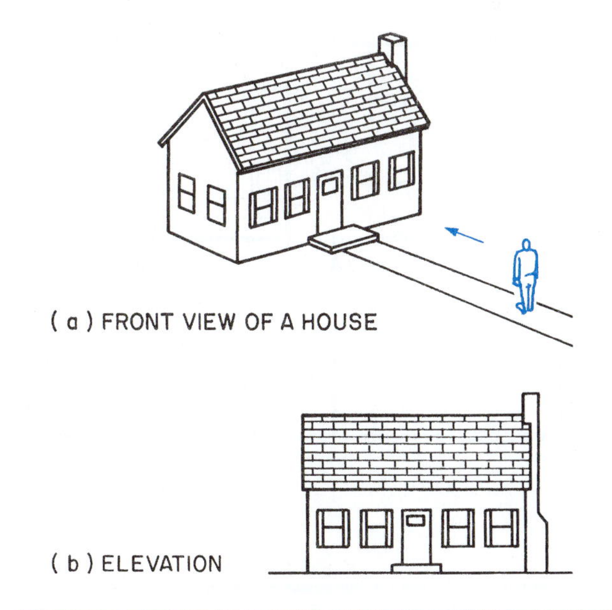

Figure 2–10. Building elevation.

Because not all features of construction can be seen in plan views and elevations from the outside of a building, many construction drawings are **section views.** A section view, usually referred to simply as a *section,* shows what would be exposed if a cut were made through the object (see **Figure 2–11**). Actually, a *floor plan* is a type of section view (see **Figure 2–12**). It is called a *plan* because it is in that position—viewed from above—but it is a type of *section* because it shows what would be exposed if a cut were made through the building. Most section views are called sections, but floor plans are customarily referred to as plans or floor plans.

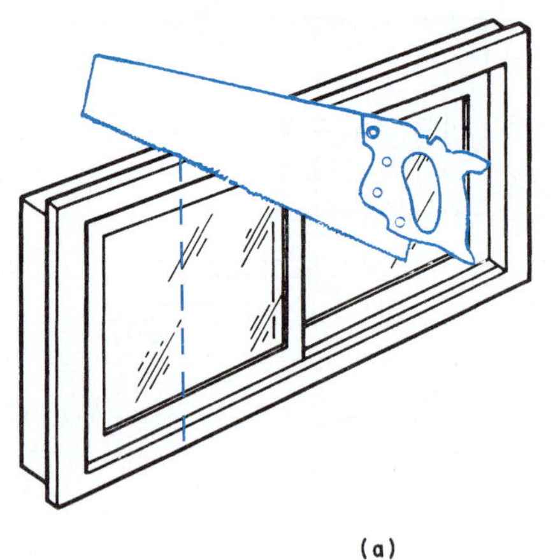

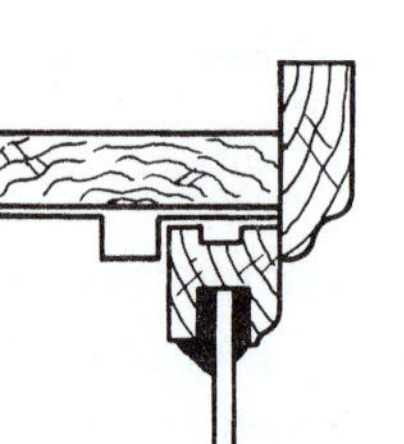

Figure 2–11. Section of a window sash.

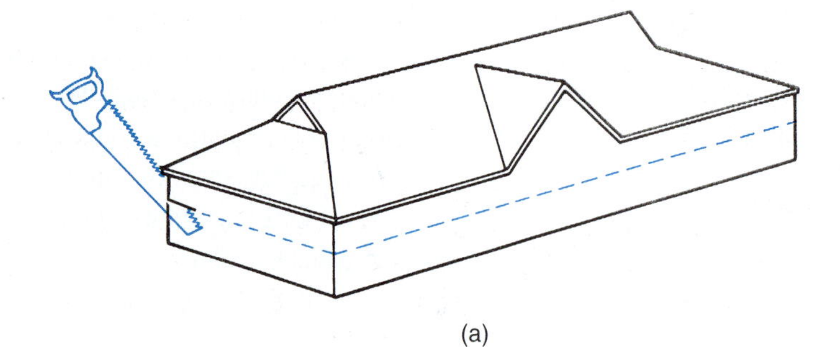

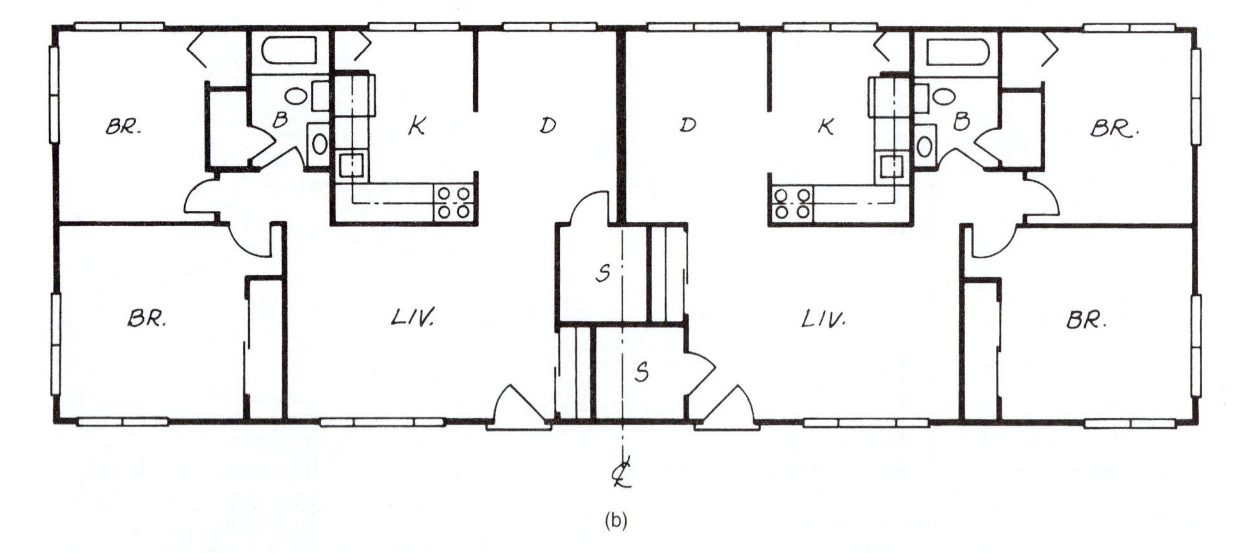

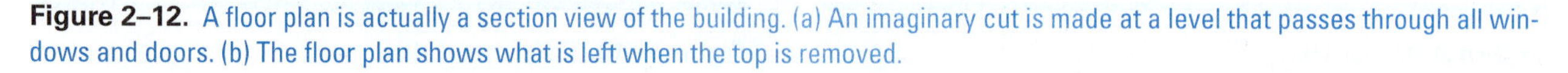

Figure 2–12. A floor plan is actually a section view of the building. (a) An imaginary cut is made at a level that passes through all windows and doors. (b) The floor plan shows what is left when the top is removed.

USING WHAT YOU LEARNED

As you look for specific information on a set of construction prints, it is helpful to know what type of drawing you are looking at. For example, if it is an orthographic projection, the lines you see will be drawn in true proportion to their actual sizes. However, if it is oblique or isometric, they may not be in proportion. The Assignment questions in this unit require you to identify various drawing types. Take a look at the door frame types on Sheet 2 of the Two-Unit Apartment in the drawing packet accompanying this textbook. It is a section view because it shows parts as though a cut were made through the door jamb, revealing the interior construction. It is a plan view because it shows what would be seen looking straight down from above. It is an orthographic projection drawing because what we see is what would have been projected onto the top of a glass box placed over the cut door jamb.

Assignment

1. Identify each of the drawings in **Figure 2–13** as oblique, isometric, or orthographic.
2. Identify each of the drawings in **Figure 2–14** as elevation, plan, or section.
3. In the view of the house shown in **Figure 2–15**, which lines are true length?
4. What type of pictorial drawing is easiest to draw on the job site?
5. What type of drawing is used for working drawings?

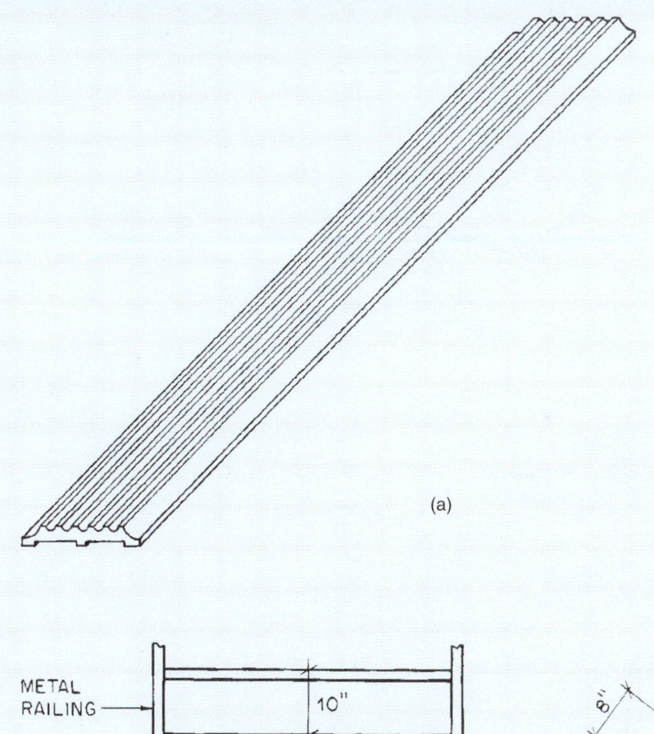

(a)

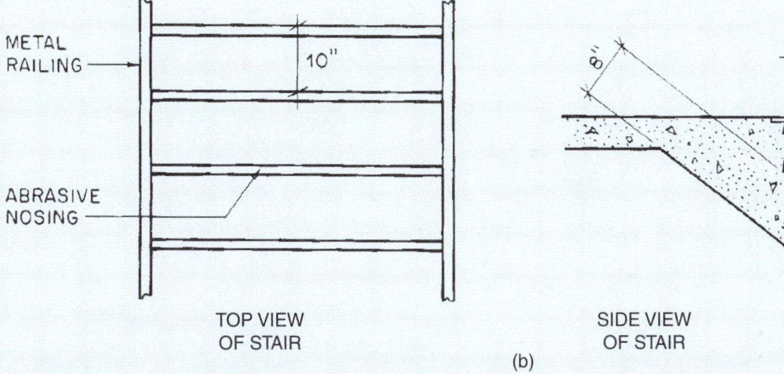

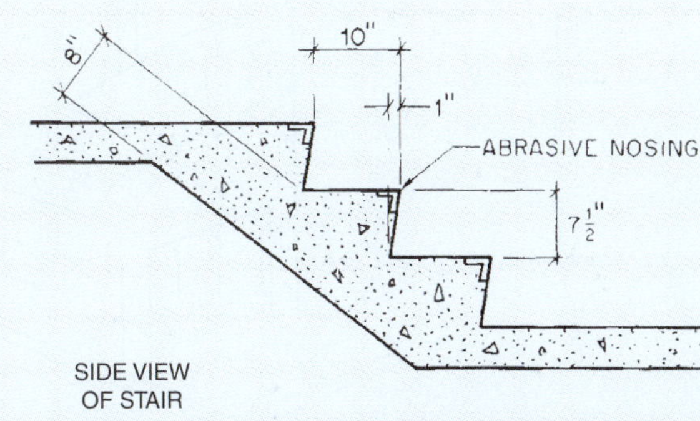

(b)

Figure 2–13.

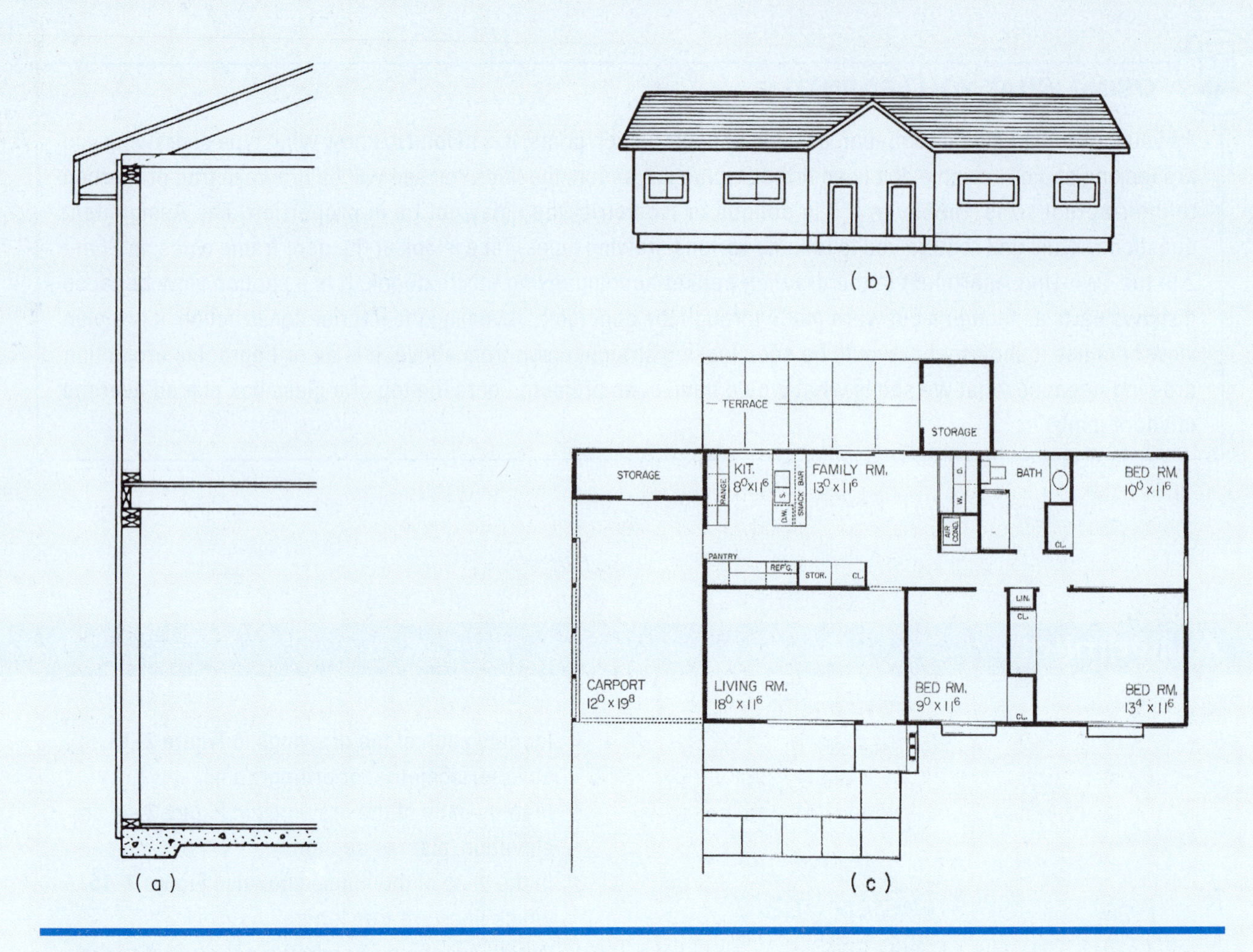

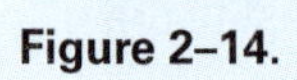

Figure 2–14.

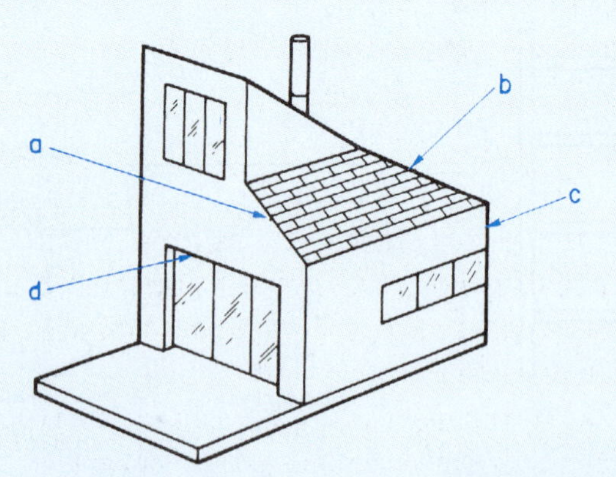

Figure 2–15.

UNIT 3

Scales

Scale Drawings

Because construction projects are too large to be drawn full size on a sheet of paper, everything must be drawn proportionately smaller than it really is. For example, ***floor plans*** for a house are frequently drawn 1/48th of the actual size. This is called *drawing* to scale. At a scale of ¼″ = 1′-0″, ¼ inch on the drawing represents 1 foot on the actual building. When it is necessary to fit a large object on a drawing, a small scale is used. Smaller objects and drawings that must show more detail are drawn to a larger scale. The floor plan in **Figure 3–1** was drawn to a scale of ¼″ = 1′-0″. The detail drawing in **Figure 3–2** was drawn to a scale of 3″ = 1′-0″ to show the construction of one of the walls on the floor plan.

The scale to which a drawing is made is noted on the drawing. The scale is usually indicated alongside or beneath the title of the view.

Reading an Architect's Scale

All necessary dimensions should be shown on the drawings. The instrument used to make drawings to scale is called an ***architect's scale*** (see **Figure 3–3**). Measuring a drawing with an architect's or engineer's scale is a poor practice. At small scales it is especially difficult to be precise. The following discussion of how to read an architect's scale is presented only to ensure an understanding of the scales used on drawings. The triangular architect's scale includes eleven scales frequently used on drawings.

Full Scale	
3/32″ = 1′- 0″	3/16″ = 1′- 0″
1/8″ = 1′- 0″	1/4″ = 1′- 0″
3/8″ = 1′- 0″	3/4″ = 1′- 0″
1/2″ = 1′- 0″	1″ = 1′- 0″
11/2″ = 1′- 0″	3″ = 1′- 0″

Two scales are combined on each face, except for the full-size scale, which is fully divided into sixteenths (see **Figure 3–4**). The combined scales work together because one is twice as large as the other, and their zero points and extra divided units **are on** opposite ends of the scale.

The fraction, or number, near the zero at each end of the scale indicates the unit length in inches that is used on the drawing to represent 1 foot of the actual building. The extra unit near the zero end of the scale is subdivided into twelfths of a foot (inches) as well as fractions of inches on the larger scales.

Objectives

After completing this unit, you will be able to perform the following tasks:

- Identify the scale used on a construction drawing.
- Read an architect's scale.

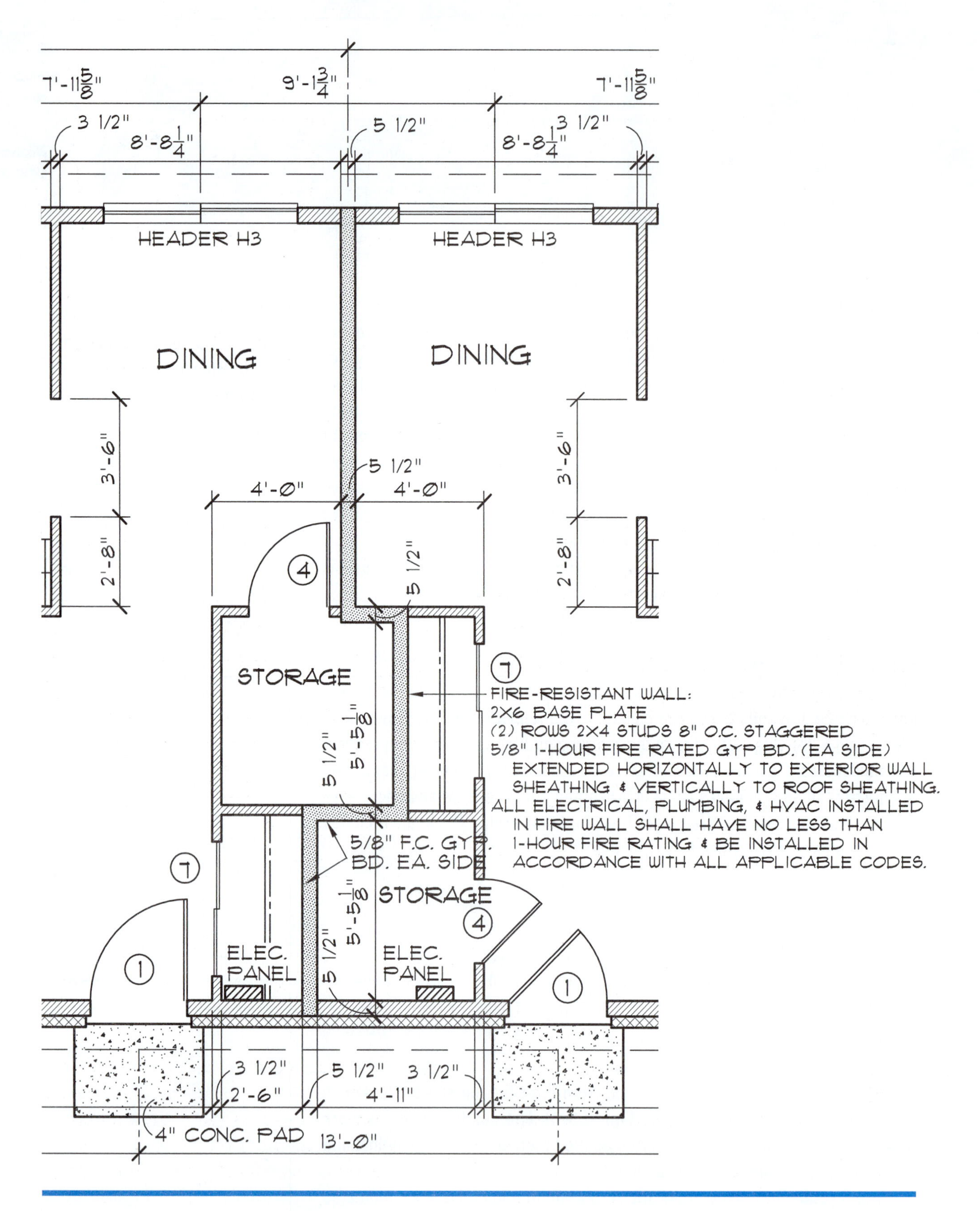

Figure 3–1. Portion of a plan view with a firewall. ¼" = 1'–0".

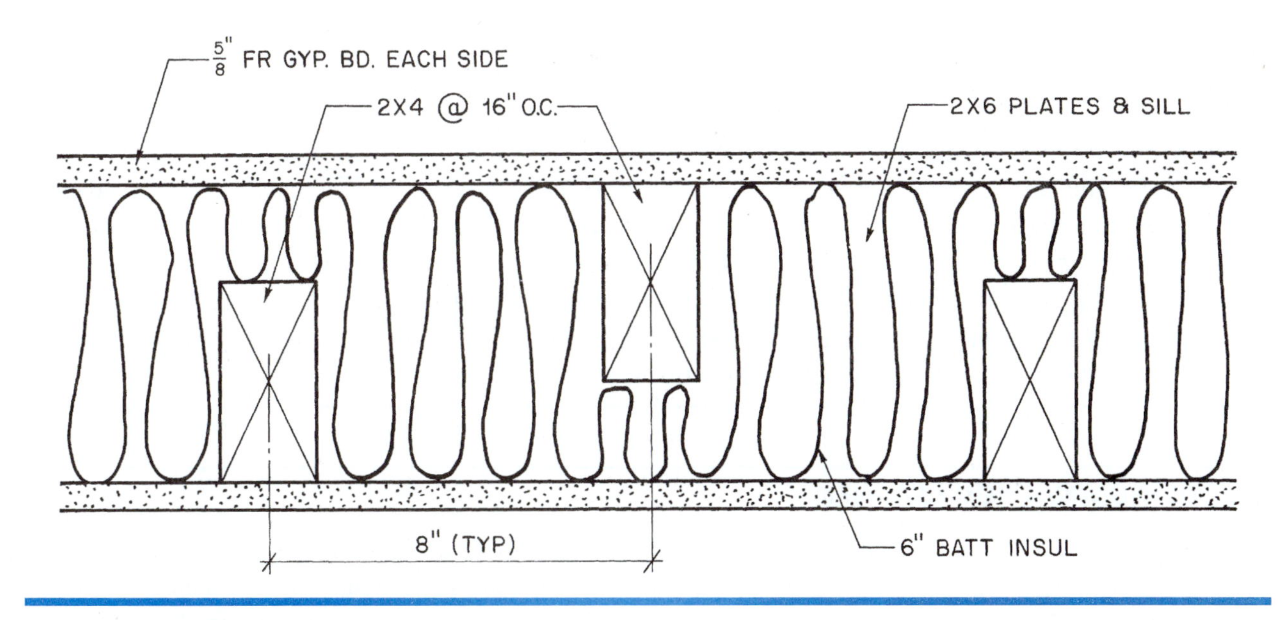

Figure 3–2. Detail (plan at firewall). 3" = 1'–0".

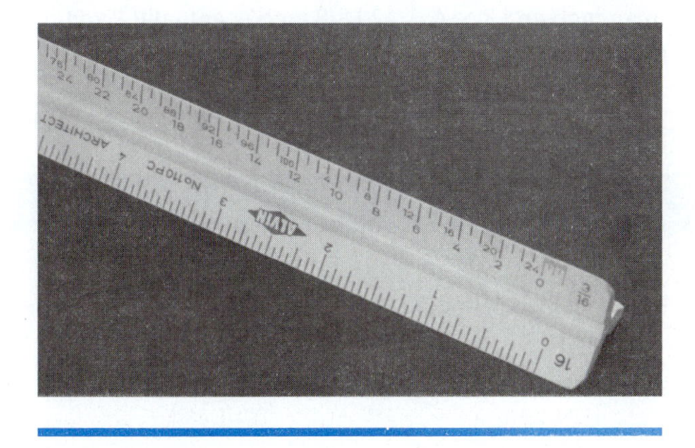

Figure 3–3. Architect's scale.

To read the architect's scale, turn it to the ¼-inch scale. The scale is divided on the left from the zero toward the ¼ mark so that each line represents 1 inch. Counting the marks from the zero toward the ¼ mark, there are 12 lines marked on the scale. Each one of these lines is 1 inch on the ¼" = 1'-0" scale.

The fraction 1/8 is on the opposite end of the same scale. This is the 1/8-inch scale and is read in the opposite direction. Notice that the divided unit is only half as large as the one on the ¼-inch end of the scale. Counting the lines from zero toward the 1/8 mark, there are only six lines. This means that each line represents 2 inches at the 1/8-inch scale.

Now look at the 1½-inch scale. The divided unit is broken into twelfths of a foot (inches) and also fractional parts of an inch. Reading from the zero toward the number 1½, notice the figures 3, 6, and 9. These figures

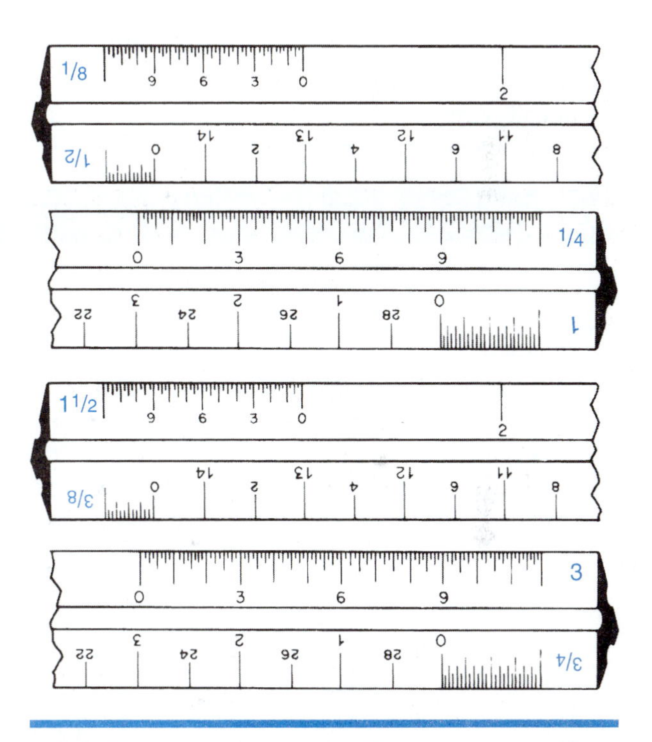

Figure 3–4. Architect's triangular scales.

represent the measurements of 3 inches, 6 inches, and 9 inches at the 1½" = 1'-0" scale. From the zero to the first long mark, that represents 1 inch (which is the same length as the mark shown at 3) and four lines. This means that each line on the scale is equal to ¼ of an inch. Reading from the zero to the 3, read each line as follows: ¼, ½, ¾, 1, 1¼, 1½, 1¾, 2, 2¼, 2½, 2¾, and 3 inches. Do not confuse the engineer's scale with the architect's scale. The engineer's scale uses feet and decimal parts of a foot.

GREEN NOTE

Many homes have green or sustainable features such as solar panels to provide energy or extra insulation to conserve energy, but what makes a home a green home? There are programs to certify homes as green. A green certification program is a set of standards for green building practices that are followed by the building team and verified by an independent third party. These programs usually have a system for awarding points meeting certification standards that, when added up, determine whether the home can be certified as a green home. The two best known national green certification programs are LEED (Leadership in Energy and Environmental Design) and the National Green Building Standard developed by Home Innovation Laboratories and the National Association of Home Builders (NAHB).

USING WHAT YOU LEARNED

Construction drawings are rarely drawn the actual size of what they depict. They are almost always smaller than the actual object. Drawings for buildings are drawn to one of the scales found on an architect's scale. For this reason it is important to understand how to read an architects scale. If a drawing is made to a scale of ¼″ = 1′, what would be the dimension represented by a line 3³/₈ inches long? Each ¼ inch represents 1 foot, so 3 inches represents 12 feet. (There are four ¼s in an inch and 3 × 4 = 12.) 3/8 inch is actually 1 and ½ quarters of an inch, so 3/8″ represents 1½′ or 1′ foot 6″. 12 feet plus 1 foot six inches is 13 feet 6 inches, normally written as 13′-6″.

Assignment

1. What are the dimensions indicated on the scale in **Figure 3–5**?
2. What scales are used for the following views of the duplex? (Refer to the duplex drawings in your textbook packet.)
   a. Floor plan
   b. Site plan
   c. Front elevation
   d. Typical wall section

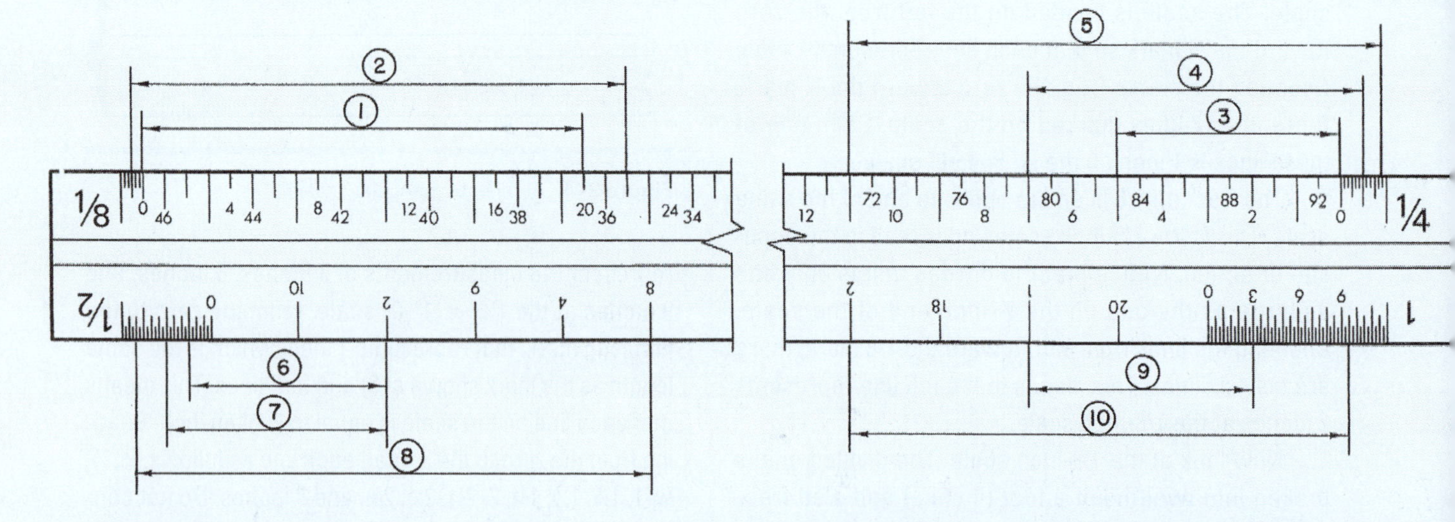

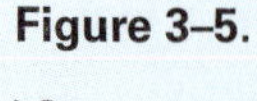

Figure 3–5.

Alphabet of Lines

UNIT 4

That drawings are used in construction for the communication of information has already been discussed in Unit 2. Indeed, drawings serve as a language for the construction industry. The basis for any language is its alphabet. The English language uses an alphabet made up of twenty-six letters. Construction drawings use an *alphabet of lines* (see **Figure 4–1**).

The weight or thickness of lines is sometimes varied to show their relative importance. For example, in **Figure 4–2** notice that the basic outline of the building is heavier than the lines used for the smaller architectural details. This difference in line weight sometimes helps distinguish the basic shape of an object from surface details.

Objectives

After completing this unit, you will be able to identify and understand the meaning of the listed lines:

- Object lines
- Dashed lines (hidden and phantom)
- Extension lines and dimension lines
- Centerlines
- Leaders
- Cutting-plane lines

Object Lines

Object lines are used to show the shape of an object. All visible edges are represented by object lines. All the lines in **Figure 4–2** are object lines. Drawings usually include many solid lines that are not object lines, however. Some of these other solid lines are discussed here. Others are discussed later.

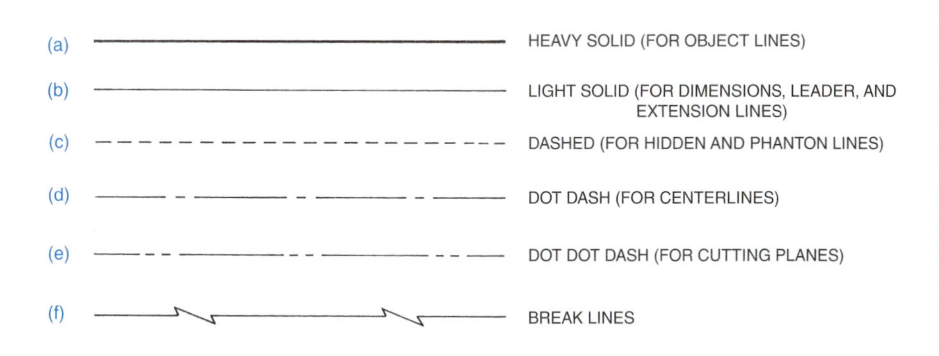

Figure 4–1. Alphabet of lines.

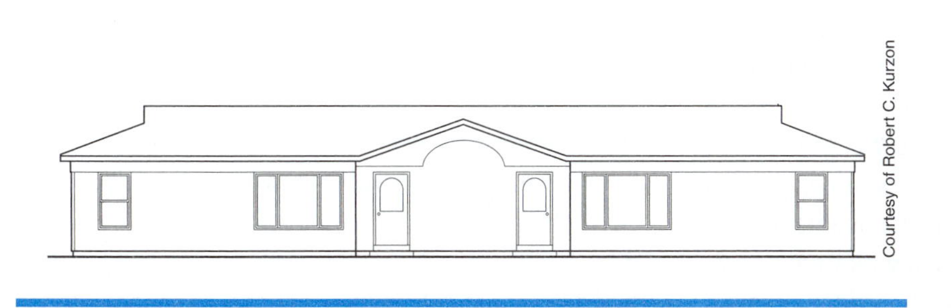

Figure 4–2. Elevation outlined.

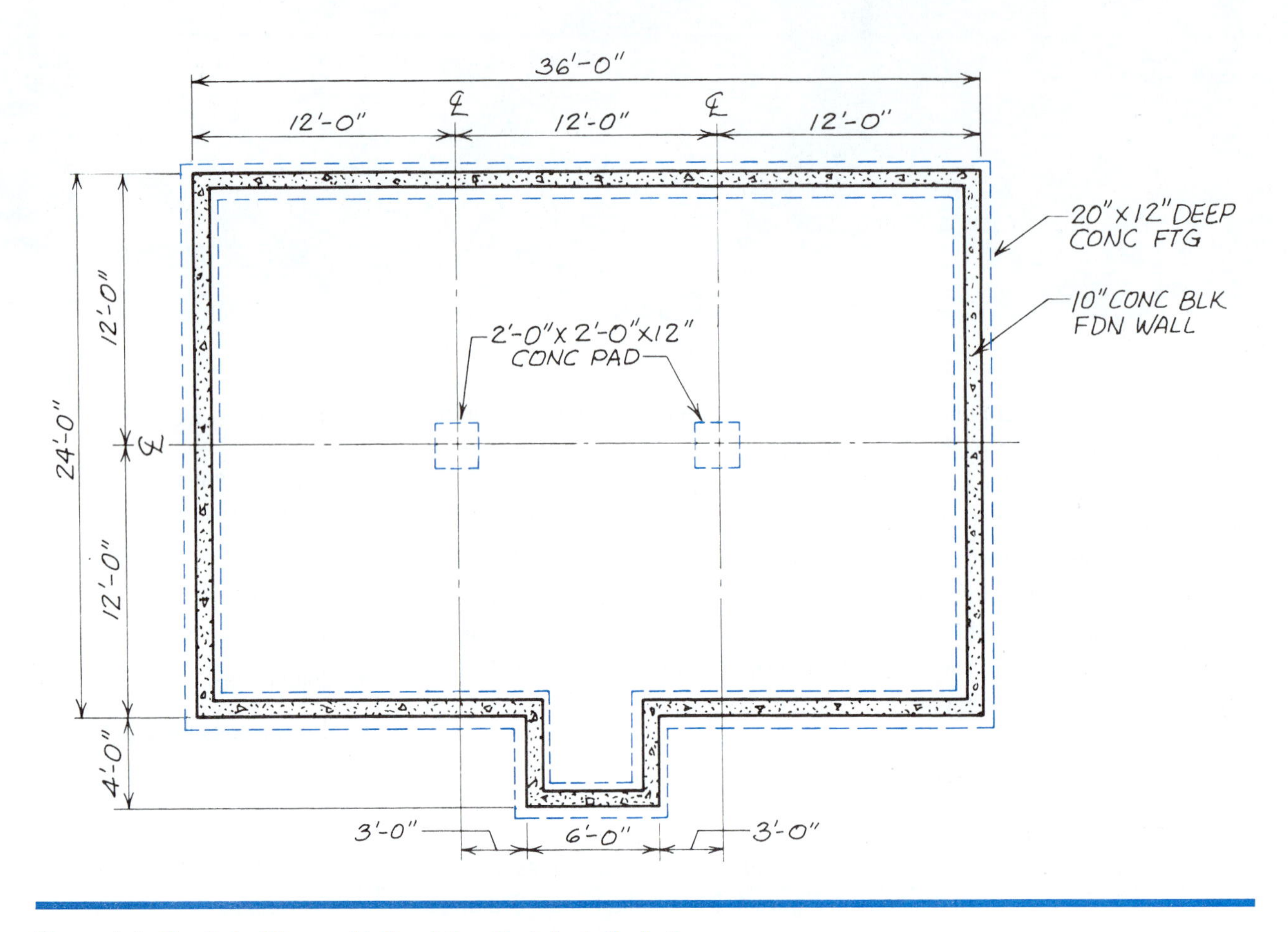

Figure 4–3. The dashed lines on this foundation plan indicate the footing.

Dashed Lines

Dashed lines have more than one purpose in construction drawings. One type of dashed line, the *hidden line*, is used to show the edges of objects that would not otherwise be visible in the view shown. Hidden lines are drawn as a series of evenly sized short dashes (see **Figure 4–3**). If a construction drawing were to include hidden lines for all concealed edges, the drawing would be cluttered and hard to read. Therefore, only the most important features are shown by hidden lines.

Another type of dashed line is used to show important overhead construction (see **Figure 4–4**). These dashed lines are called *phantom lines*. The objects they show are not hidden in the view—they are simply not in the view. For example, the most practical way to show exposed beams on a living room ceiling may be to show them on the floor plan with phantom lines. Phantom lines are also used to show alternate positions of objects (see **Figure 4–5**). To avoid confusion, the dashed lines may be made up of different weights and different length dashes, depending on the purpose (see **Figure 4–6**).

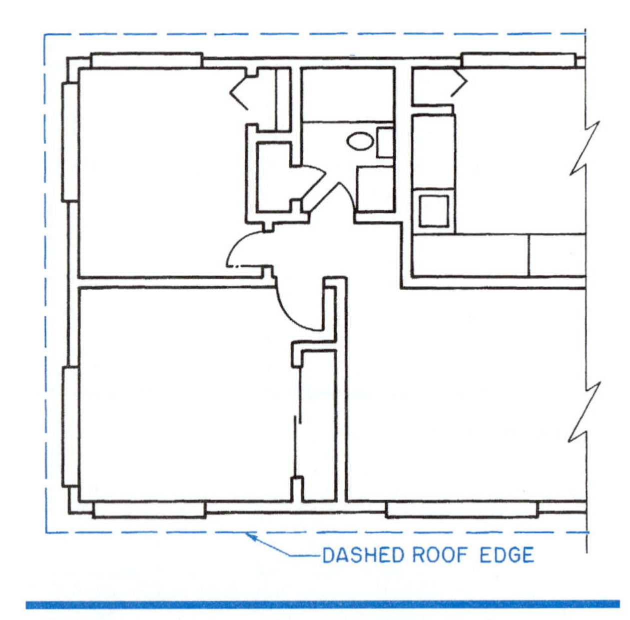

Figure 4–4. The dashed lines on this floor plan indicate the edge of the roof overhang.

Extension Lines and Dimension Lines

Extension lines are thin, solid lines that project from an object to show the extent or limits of a dimension. Extension lines do not quite touch the object they indicate (see **Figure 4–7**).

Dimension lines are solid lines of the same weight as extension lines. A dimension line is drawn from one extension line to the next. The dimension (distance between the extension lines) is lettered above the dimension line. On construction drawings, dimensions are expressed in feet and inches. The ends of dimension lines are drawn in one of three ways, as shown in **Figure 4–8**.

Dimensions that can be added together to come up with one overall dimension are called *chain dimensions.*

Figure 4–5. The dashed lines here are phantom lines to show alternate positions of the double-acting door and the door of the dishwasher.

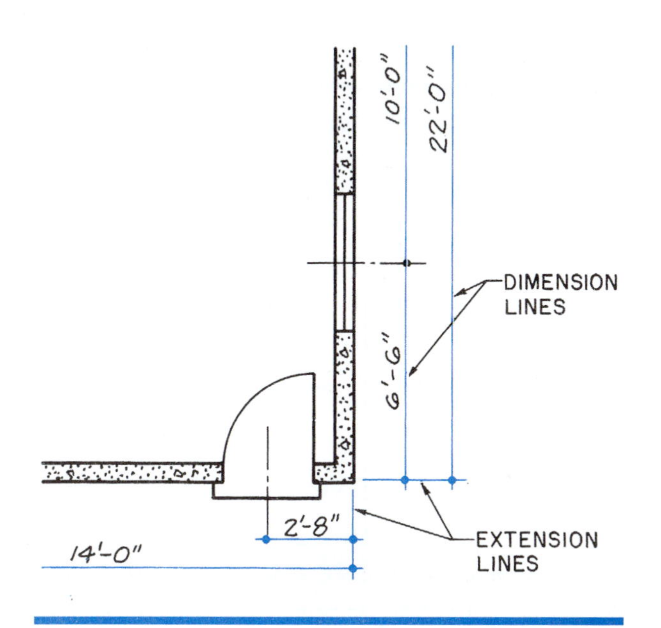

Figure 4–7. Dimension and extension lines.

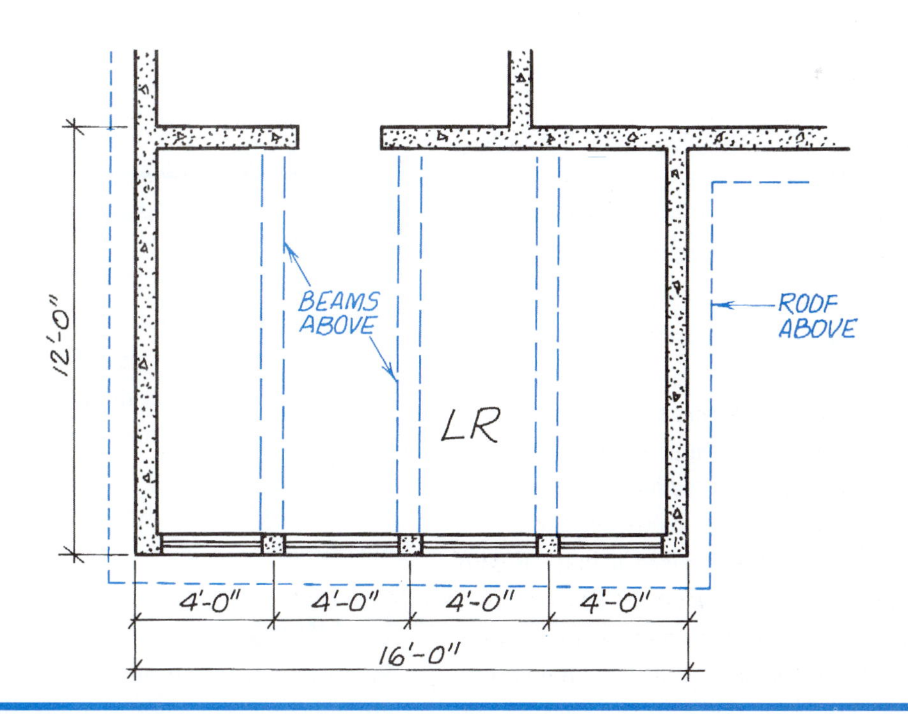

Figure 4–6. Different types of dashed lines are used to show different features.

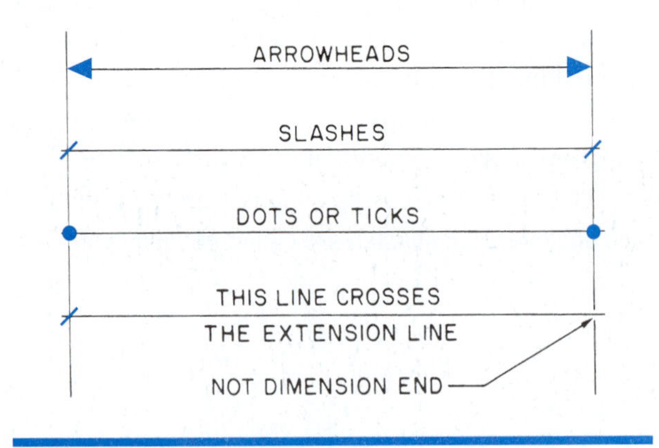

Figure 4–8. Dimension line ends.

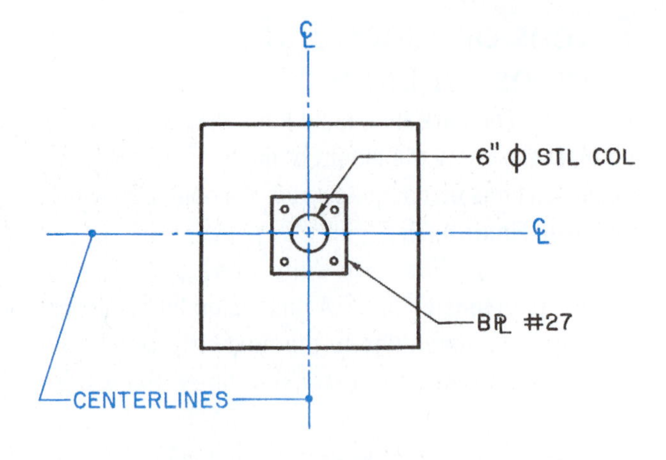

Figure 4–10. When centerlines show the center of a round object, the short dashes of two centerlines cross.

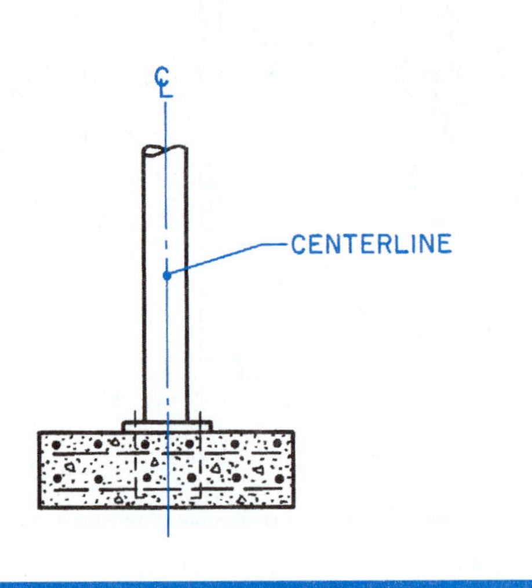

Figure 4–9. This centerline indicates that the column is symmetrical, or the same, on both sides of the centerline.

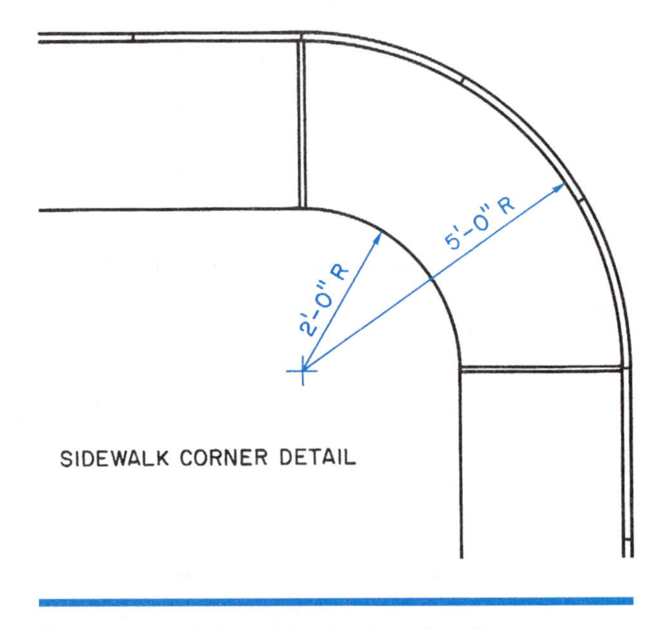

Figure 4–11. Method of showing the radius of an arc.

The dimension lines for chain dimensions are kept in line as much as possible. This makes it easier to find the dimensions that must be added to find the overall dimension.

Centerlines

Centerlines are made up of long and short dashes. They are used to show the centers of round or cylindrical objects. Centerlines are also used to indicate that an object is *symmetrical,* or the same on both sides of the center (see **Figure 4–9**). To show the center of a round object, two centerlines are used so that the short dashes cross in the center (see **Figure 4–10**).

To lay out an arc or part of a circle, the radius must be known. The *radius* of an arc is the distance from the center to the edge of the arc. On construction drawings, the center of an arc is shown by crossing centerlines. The radius is dimensioned on a thin line from the center to the edge of the arc (see **Figure 4–11**).

Rather than clutter the drawing with unnecessary lines, only the short, crossing dashes of the centerlines are shown. If the centerlines are needed to dimension the location of the center, only the needed centerlines are extended.

Leaders

Some construction details are too small to allow enough room for clear dimensioning by the methods described earlier. To overcome this problem, the

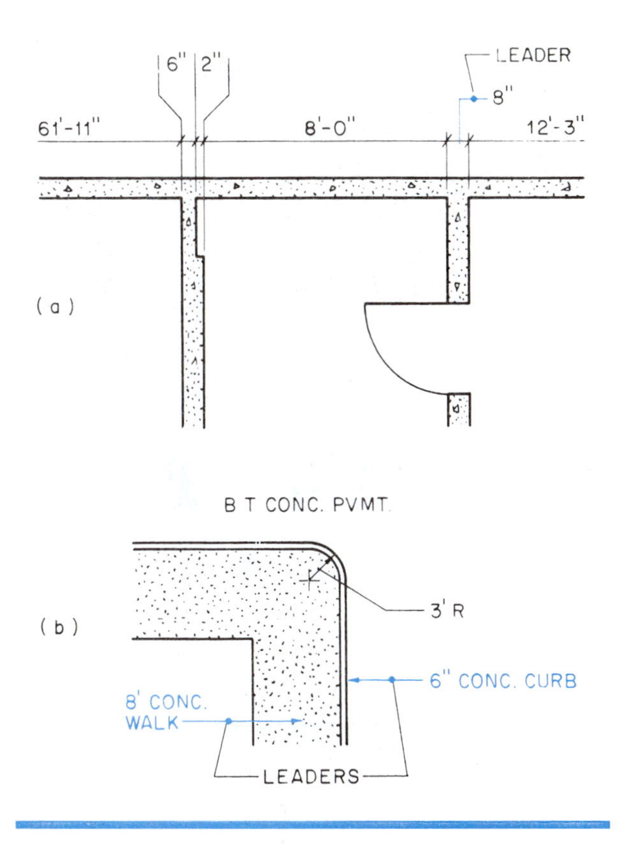

Figure 4–12. Leaders used for dimensioning.

dimension is shown in a clear area of the drawing. A thin line called a *leader* shows where the dimension belongs (see **Figure 4–12**).

GREEN NOTE

There are many ways to wire a house, plumb a bathroom, or frame a house that meet the minimum requirements of building codes. Green building goes beyond the building codes and raises the bar for construction practices to the highest level of quality. Materials that are installed just well enough to meet a building code might not perform efficiently or last as long as those that are installed perfectly. For example, fiberglass insulation that is overstuffed into a wall cavity and has air spaces along the top and bottom may only perform at half the expected insulating value due to the installation defects. Best practices require insulation to be properly installed so that it fits every space snugly without over packing it.

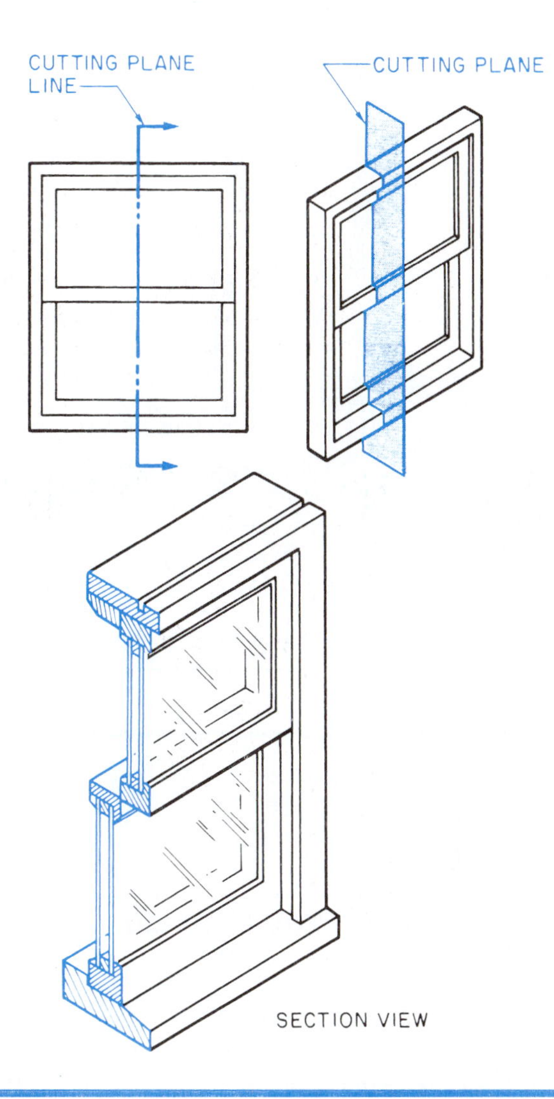

Figure 4–13. A cutting-plane line indicates where the imaginary cut is made and how it is viewed.

Cutting-Plane Lines

It was established earlier that section views are needed to show interior detail. In order to show where the imaginary cut was made, a *cutting-plane line* is drawn on the view through which the cut was made (see **Figure 4–13**). A cutting-plane line is usually a heavy line with long dashes and pairs of short dashes. Some drafters, however, use a solid, heavy line. In either case, cutting-plane lines always have some identification at their ends and arrowheads to indicate the direction from which the section is viewed. Cutting-plane-line identification symbols are discussed in the next unit.

Some section views may not be referenced by a cutting-plane line on any other view. These are *typical sections* that would be the same if drawn from an imaginary cut in any part of the building (see **Figure 4–14**).

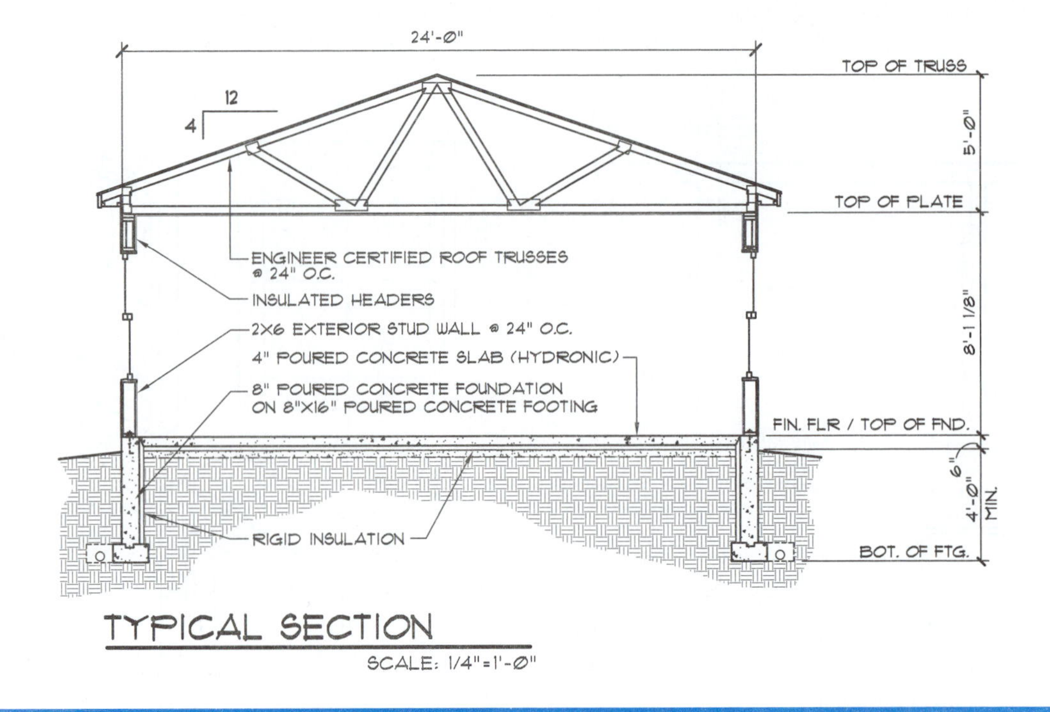

Figure 4–14. Building section.

USING WHAT YOU LEARNED

On very simple drawings, it is usually easy to understand what each line represents, but on complex drawings, conveying a lot of information, there can be many types of lines, each with a different meaning. Look at the Site Plan for the Two-Unit Apartment. Why would some of the lines representing the building be much heavier than the lines across the middle of the building? What kinds of lines are these? The bold lines are object lines showing the basic shape and location of the building as viewed from above. The thinner lines are not part of the basic building outline. However, because this building is two dwelling units, the drafter has used these thinner lines to show how the building is divided. A better practice might have been to have used dashed hidden lines to show the division, because they cannot be seen from above the building.

Assignment

Refer to the drawings of the Two-Unit Apartment in your textbook packet. For each of the lines numbered A5.1 through A5.10, identify the kind of line and briefly describe its purpose on these drawings.

The broad arrows with A5 numbers are for use in this assignment.

Example: A5.E, object line, shows the end of the building.

Use of Symbols

5 UNIT

An alphabet of lines allows for clear communication through drawings; the use of standard symbols makes for even better communication. Many features of construction cannot be drawn exactly as they appear on the building. Therefore, standard symbols are used to show various materials, plumbing fixtures and fittings, electrical devices, windows, doors, and other common objects. Notes are added to drawings to give additional explanations.

It is not important to memorize all the symbols and abbreviations used in construction before you learn to read drawings. There are commonly accepted standards for architectural symbols, but many architects and drafters use their own variations of standard symbols. Even so, with very little practice, you can develop the ability to interpret the symbols that are commonly used on construction drawings, whether standard or a variation. Typically, an architectural symbol is a simplified picture of the material or item it represents. In many cases, the material represented by a symbol is also labeled with words or abbreviations. Some of the most common symbols are shown in this chapter and additional symbols are shown in the Appendix.

Objectives

After completing this unit, you will be able to identify and understand the meaning of the listed symbols:

- Door and window symbols
- Materials symbols
- Electrical and mechanical symbols
- Reference marks for coordinating drawings
- Abbreviations

Door and Window Symbols

Door and window symbols show the type of door or window used and the direction the door or window opens. There are three basic ways for household doors to open—swing, slide, or fold (see **Figure 5–1**). Within each of these basic types

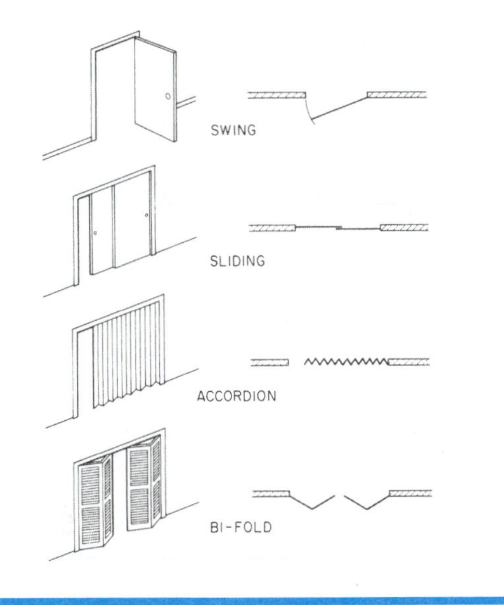

Figure 5–1. Types of doors and their plan symbols.

PLAN

DOUBLE HUNG

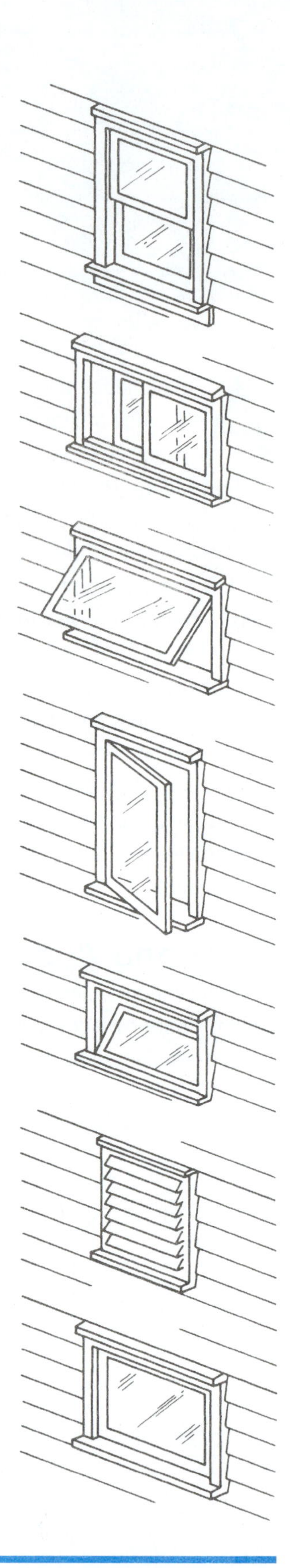

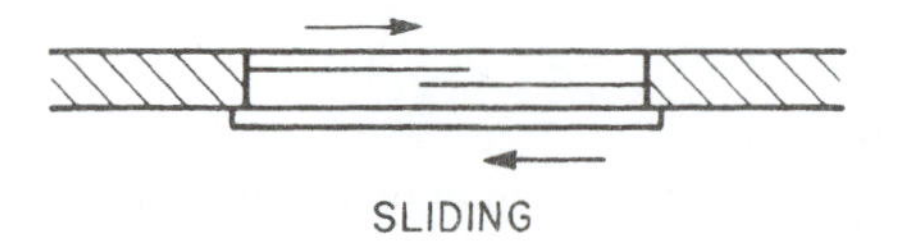

SLIDING

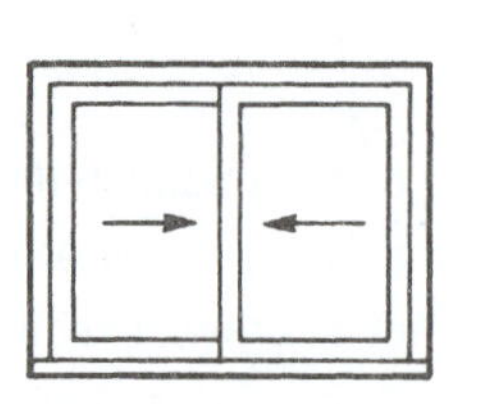

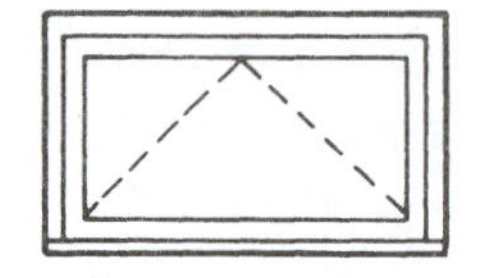

AWNING

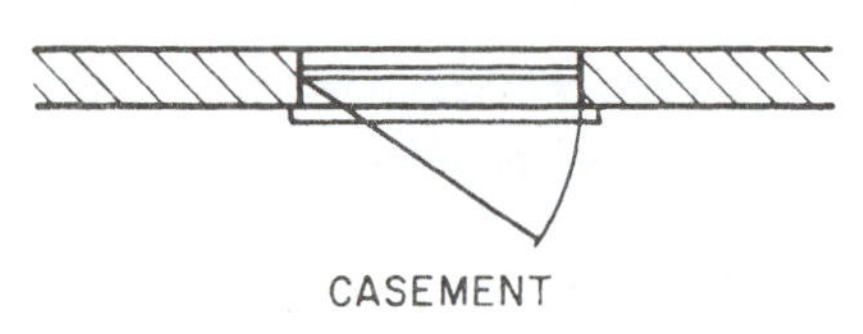

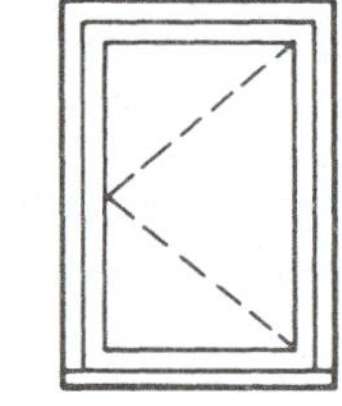

CASEMENT

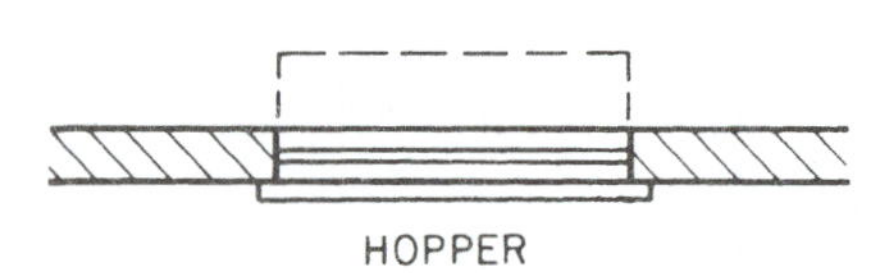

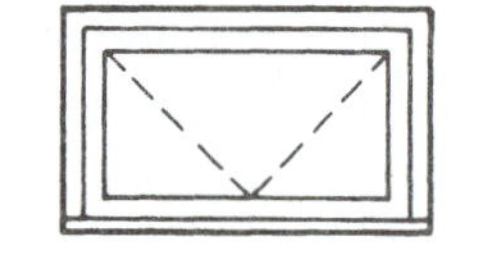

HOPPER

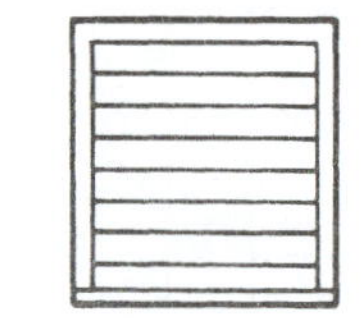

JALOUSIE

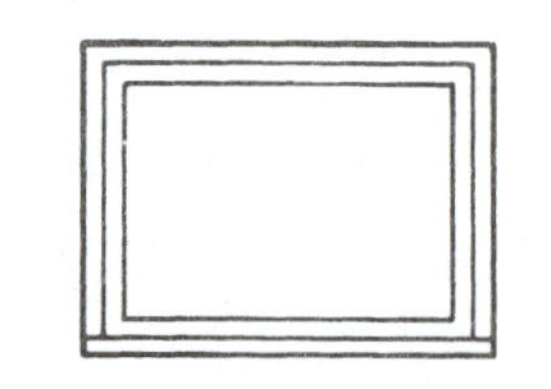

FIXED

Figure 5–2. Window symbols.

there are variations that can be readily understood from their symbols. The direction a swing-type door opens is shown by an arc representing the path of the door.

There are seven basic types of windows. They are named according to how they open (see **Figure 5–2**). The symbols for hinged windows—***awning, casement,*** and ***hopper***—indicate the direction they open. In elevation, the symbols include dashed lines that come to a point at the hinged side, as viewed from the exterior.

The sizes of windows and doors are usually shown on a special window schedule or door schedule, but they might also be indicated by notes on the plans near their symbols. Door and window schedules are explained later. The notations of size show width first and height second. Manufacturers' usually list several sets of dimensions for every window model (see **Figure 5–3**). The glass size indicates the area that will actually allow light to pass. The rough opening size is important for the carpenter, who will frame the wall into which the window will be installed. The masonry opening is important to masons. The notations on plans and schedules usually indicate nominal dimensions. A *nominal dimension* is an approximate size and may not represent any of the actual dimensions of the unit. Nominal dimensions are usually rounded off to whole inches or feet and inches and are used only as a convenient way to refer to the window or door size. The actual dimensions should be obtained from the manufacturer before construction begins.

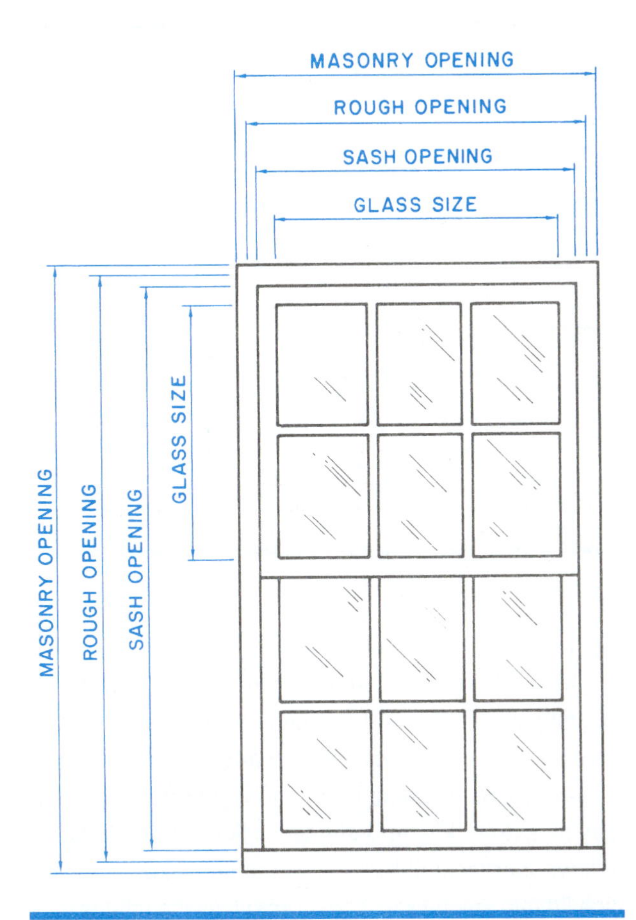

Figure 5–3. Windows and doors can be measured in several ways.

Material Symbols

The drawing of an object shows its shape and location. The outline of the drawing may be filled in with a material symbol to show what the object is made of (see **Figure 5–4**). Many materials are represented by one symbol in elevations and another symbol in sections. Examples of such symbols are concrete block and brick. Other materials look pretty much the same when viewed from any direction, so their symbols are drawn the same in sections and elevations.

GREEN NOTE

Life Cycle Assessment (LCA), also called cradle-to-grave assessment, is a technique to evaluate or assess all of the environmental impacts involved with the harvesting, mining, or manufacture; transportation; use; repair, and maintenance; and eventual disposal of a product. This analysis can be used to help determine which materials are most advantageous for a green home project.

When a large area is made up of one material, it is common to only draw the symbol in a part of the area (see **Figure 5–5**). Some drafters simplify this even further by using a note to indicate what material is used and omitting the symbol altogether.

Electrical and Mechanical Symbols

The electrical and mechanical systems in a building include wiring, electrical devices, piping, pipe fittings, plumbing fixtures, registers, and heating and air conditioning ducts. It is not practical to draw these items as they would actually appear, so standard symbols have been devised to indicate them.

The electrical system in a house includes wiring as well as devices such as switches, receptacles,

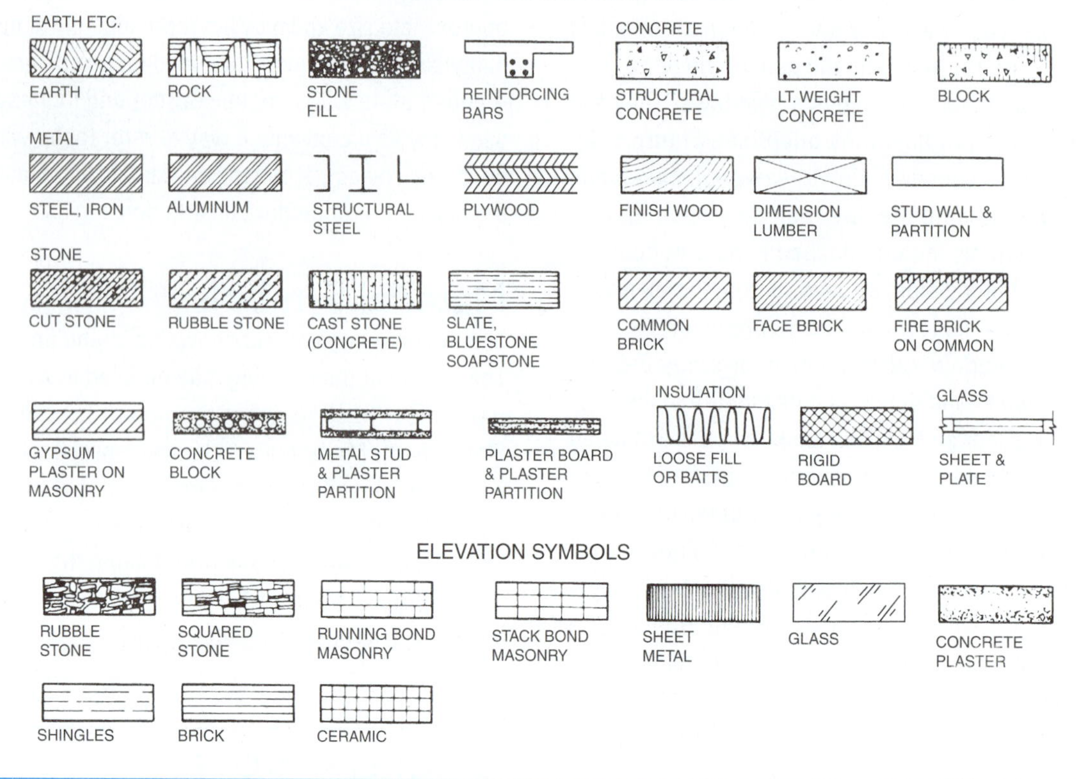

Figure 5–4. Material symbols.

Figure 5–5. Only part of the area is covered by the brick symbol, although the entire building will be brick.

light fixtures, and appliances. Wiring is indicated by lines that show how devices are connected. These lines are not shown in their actual position. They simply indicate which switches control which lights, for example. Outlets (receptacles) and switches are usually shown in their approximate positions. Major fixtures and appliances are shown in their actual positions. A few of the most common electrical symbols are shown in **Figure 5–6**.

Mechanical systems—plumbing and HVAC (heating, ventilating, and air conditioning)—are not usually shown in much detail on drawings for single-family homes. However, some of the most important features may be shown. Piping is shown by lines; different types of lines represent different kinds of piping. Symbols for pipe fittings are the same basic shape as the fittings they represent. A short line, or *hash mark*, represents the joint between the pipe and the fitting. Plumbing fixtures are

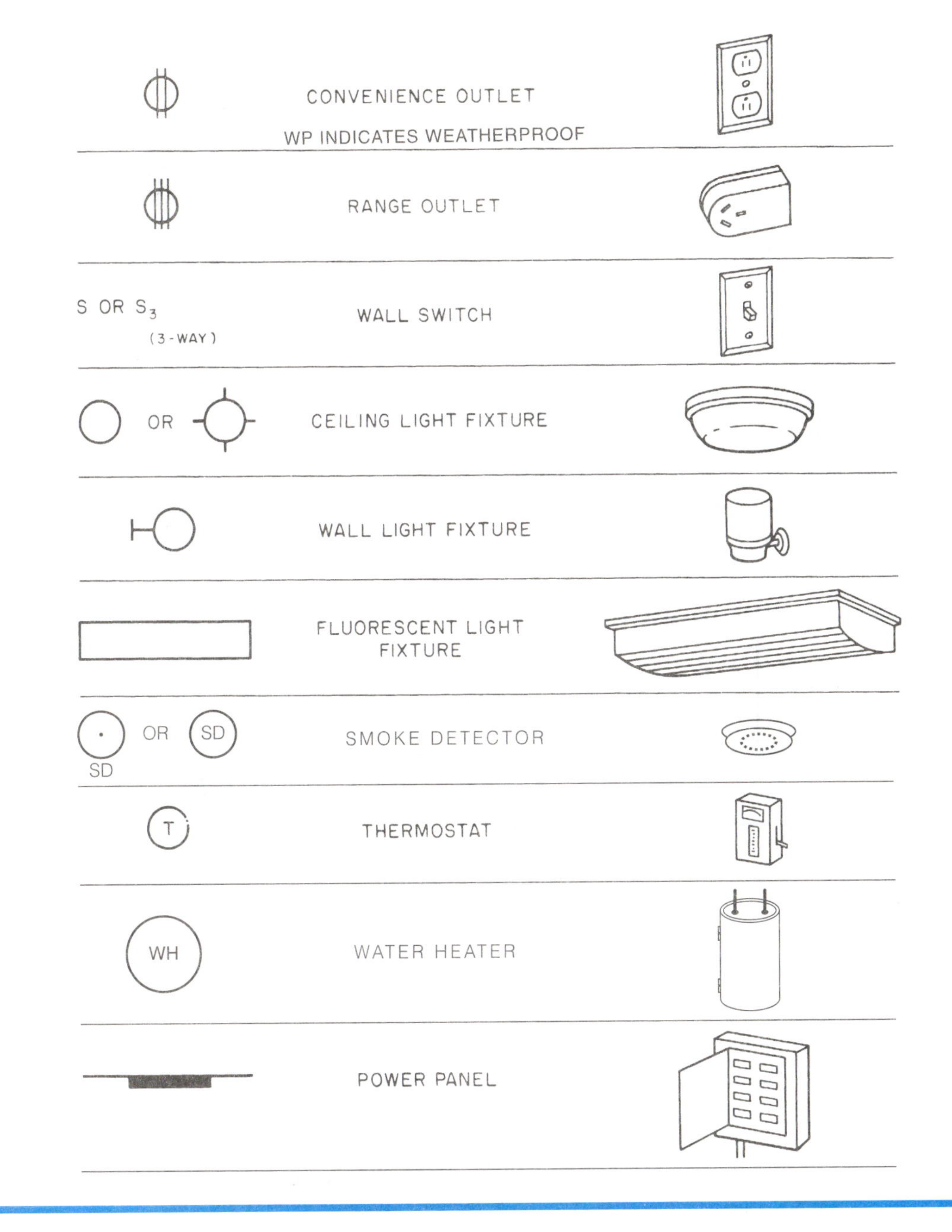

Figure 5–6. Some common electrical symbols.

drawn pretty much as the actual fixture appears. A few plumbing symbols are shown in **Figure 5–7.**

Reference Marks

A set of drawings for a complex building may include several sheets of section and detail drawings. These sections and details do not have much meaning without some way of knowing what part of the building they are meant to show. Callouts, called *reference marks,* on plans and elevations indicate where details or sections of important features have been drawn. To be able to use these reference marks for coordinating drawings, you must first understand the numbering system used on the drawings. The simplest numbering system for drawings consists of numbering the drawing sheets and naming each of the views. For example, Sheet 1 might include a site plan and foundation plan; Sheet 2, floor plans; and Sheet 3, elevations.

On large, complex sets of drawings, the sheets are numbered according to the kind of drawings shown. Architectural drawing sheets are numbered A-1, A-2, and so on for all the sheets. Electrical drawings are

Assignment

1. What is represented by each of these symbols?

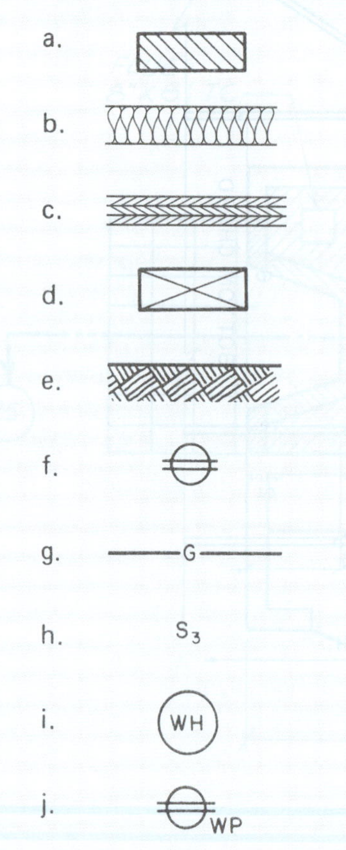

2. What is meant by each of these abbreviations?
   a. GYP. BD.
   b. FOUND.
   c. FIN. FL.
   d. O.C.
   e. REINF.
   f. EXT.
   g. COL.
   h. DIA.
   i. ELEV.
   j. CONC.
3. Where in a set of drawings would you find a detail numbered 6.4?
4. Where in a set of drawings would you find a detail numbered 5/M–3?

Plan Views

UNIT 6

Objectives

After completing this unit, you will be able to describe the general kinds of information shown on the listed plans and how that information applies to a building project:

- Site plans
- Foundation plans
- Floor plans

You learned earlier in Unit 2 that plans are drawings that show an object as viewed from above. Many of the detail and section drawings in a set show parts of the building from above. Some of the plan views that show an entire building are discussed here. This brief explanation will help you feel more comfortable with plans, although it does not cover plans in depth. You will use plans frequently throughout your study of the remainder of this textbook. Each of the remaining units helps you understand plan views more thoroughly.

Site Plans

A ***site plan*** gives information about the site on which the building is to be constructed. The boundaries of the site (property lines) are shown. The property line is usually a heavy line with one or two short dashes between longer line segments. The lengths of the boundaries are noted next to the line symbol. Property descriptions are usually the result of a survey by a surveyor. Surveyors and engineers usually work with decimal parts of feet, rather than feet and inches. Therefore, site dimensions are usually stated in tenths or hundredths of feet (see **Figure 6–1**).

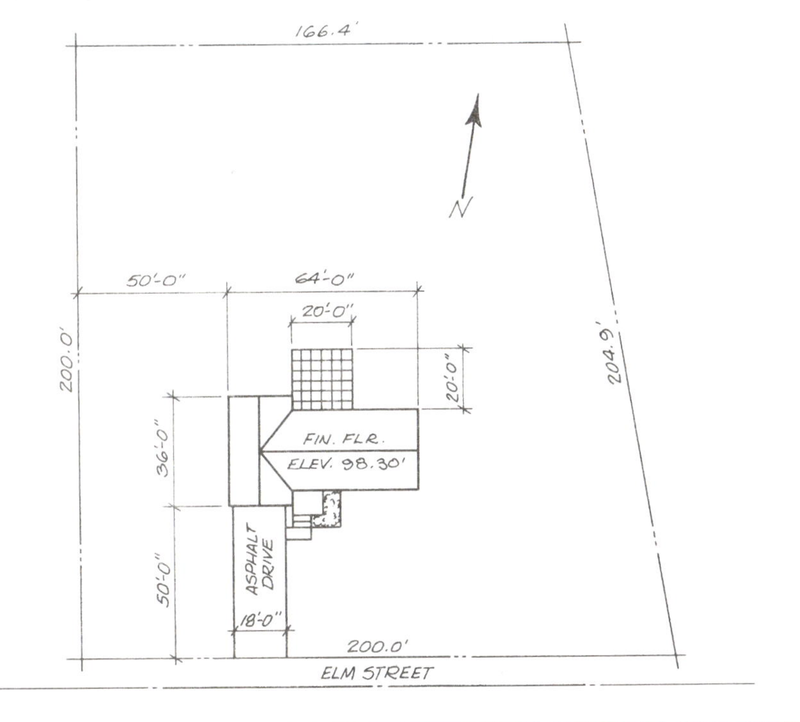

Figure 6–1. Minimum information shown on a site plan.

Assignment

Refer to the drawings for the Two-Unit Apartment (which are included in your textbook packet) to complete this assignment.

1. In what direction does the apartment face?
2. What is the length and width of the apartment site?
3. How far is the front of the apartment from the front property line?
4. What is the overall length and width of the apartment?
5. What are the inside dimensions of the living room?
6. What is the thickness of the partitions between the two bedrooms?
7. What is the thickness of the interior wall between the two dining rooms?
8. With two exceptions, the units in the apartment are exactly reversed. What are the two exceptions?
9. What is the distance from the west end of the apartment to the centerline of the west front entrance?
10. What is indicated by the small rectangle on the floor plan outside each main entrance?
11. What is the distance from the ends of the apartment to the centerlines of the $6^0/6^8$ sliding glass doors?
12. What is indicated by the dashed line just outside the front and back walls on the floor plan of the apartment?

Drawings that show the height of objects are called *elevations.* However, when builders and architects refer to building elevations, they mean the exterior elevation drawings of the building (see **Figure 7–1**). A set of working drawings usually includes an elevation of each of the four sides of the building. If the building is very complex, there may be more than four elevations. If the building is simple, there may be only two elevations—the front and one side.

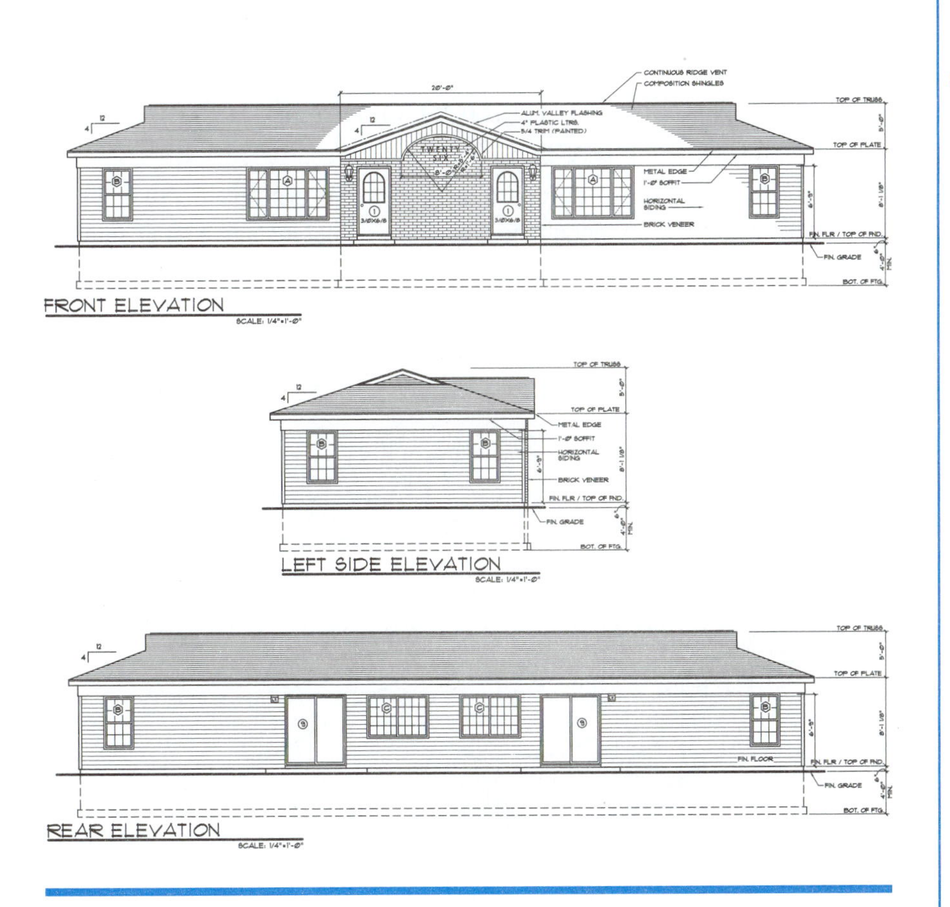

Figure 7–1. Building elevations.

Objectives

After completing this unit, you will be able to perform the following tasks:

- Orient building elevations to building plans.
- Describe the kinds of information shown on elevations.

Orienting Elevations

It is important to determine the relationship of one drawing to another. This is called *orienting* the drawings. For example, if you know which elevation is the front, you must be able to picture how it relates to the front of the floor plan.

Elevations are sometimes named according to compass directions (see **Figure 7–2**). The side of the house that faces north is the north elevation, and the side that faces south is the south elevation, for example. When the elevations are named according to compass direction, they can be oriented to the floor plan, foundation plan, and site plan by the north arrow on those plans. It might help to label the edge of the plans according to the north arrow (see **Figure 7–3**).

Labeling elevations according to compass direction is not always possible. When drawings are prepared to be sold through a catalog, or when they are for use on several sites, the compass directions cannot be included. In this case, the elevations are named according to their position as you face the building (see **Figure 7–4**). To orient these elevations to the plans, find the front on the plans. The front is usually at the bottom of the floor plan, but it can be checked by the location of the main entrance.

> **GREEN NOTE**
>
> *One of the earliest green practices designers and builders discovered was that by facing the building in the best direction, they could take advantage of natural heating and cooling. If the side with the most window area faces west, the afternoon sun will add considerable heat to the building. This can help heat the building in cold weather—but it can also add to the cooling load during the warm months. Designers have learned to both take advantage of these factors as well as adapt to them when planning buildings that use less energy.*

Information on Building Elevations

Building elevations are normally quite simple. Although the elevations do not include a lot of detailed

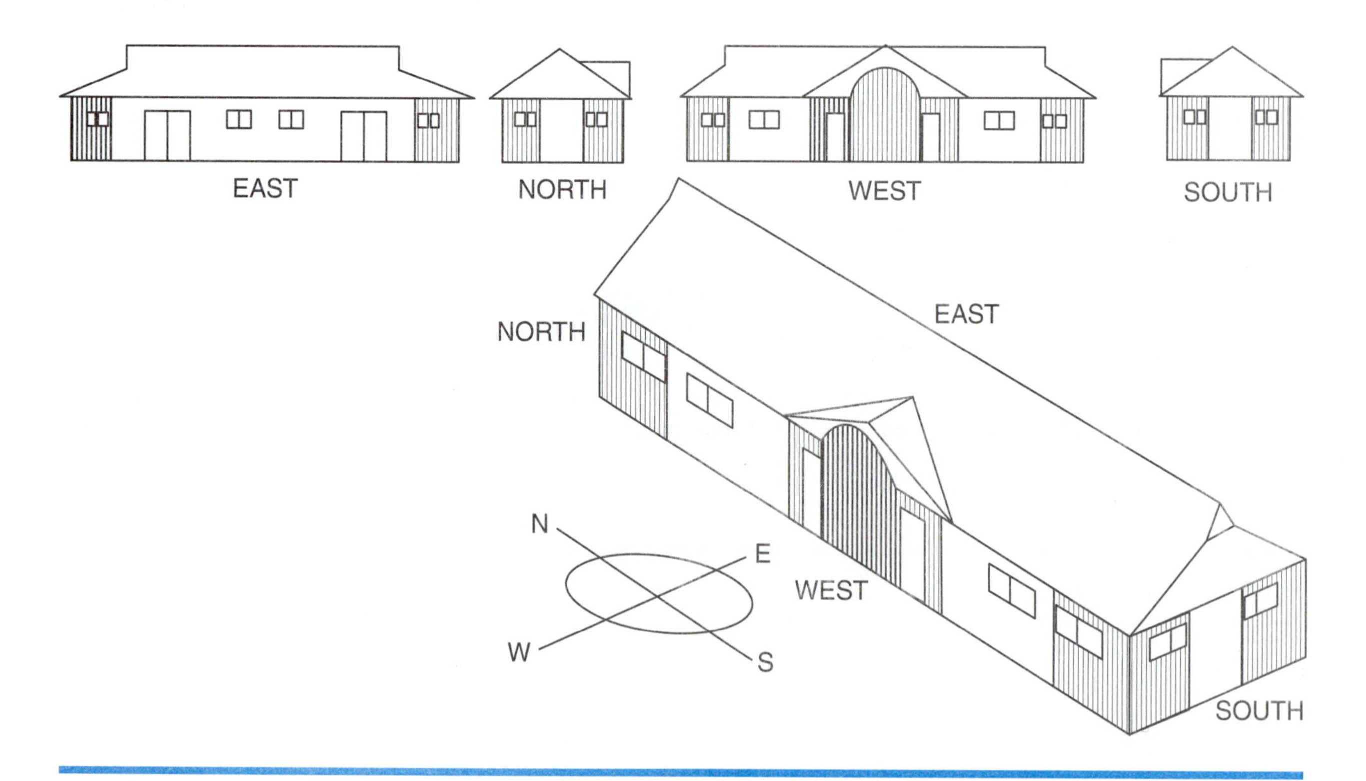

Figure 7–2. Elevations are usually named according to their compass directions.

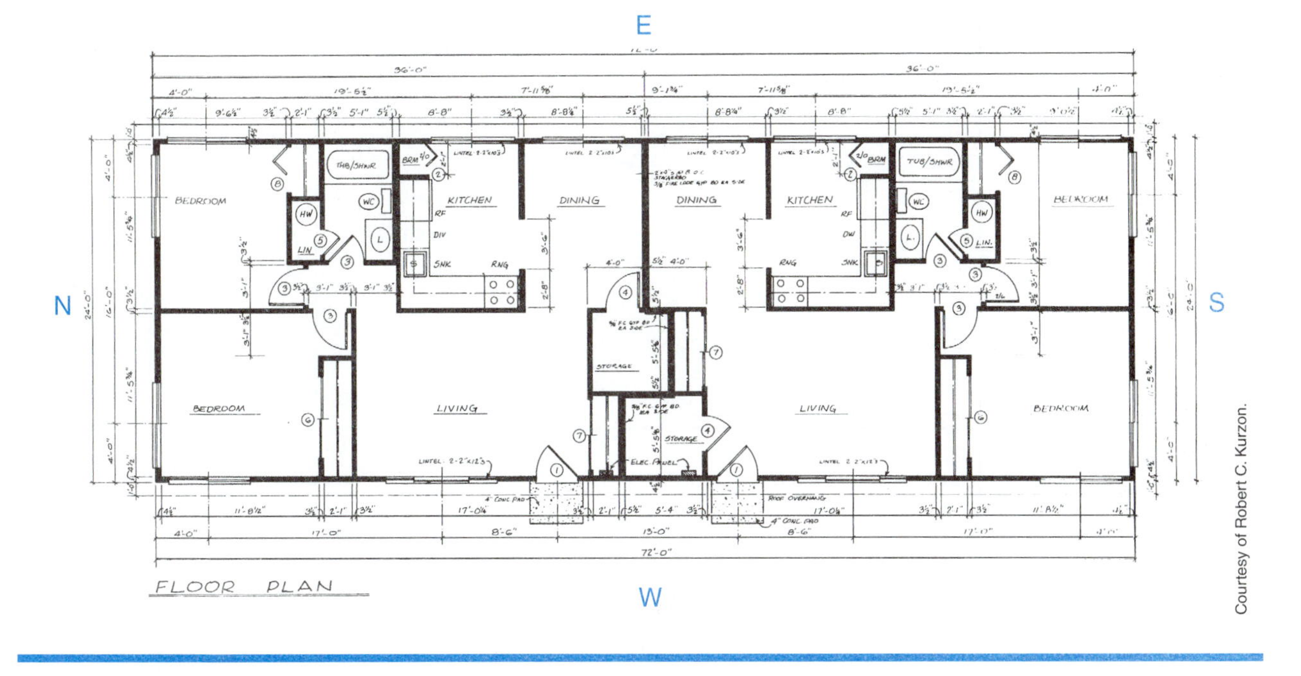

Courtesy of Robert C. Kurzon.

Figure 7–3. Plan labeled to help orientation to north arrow.

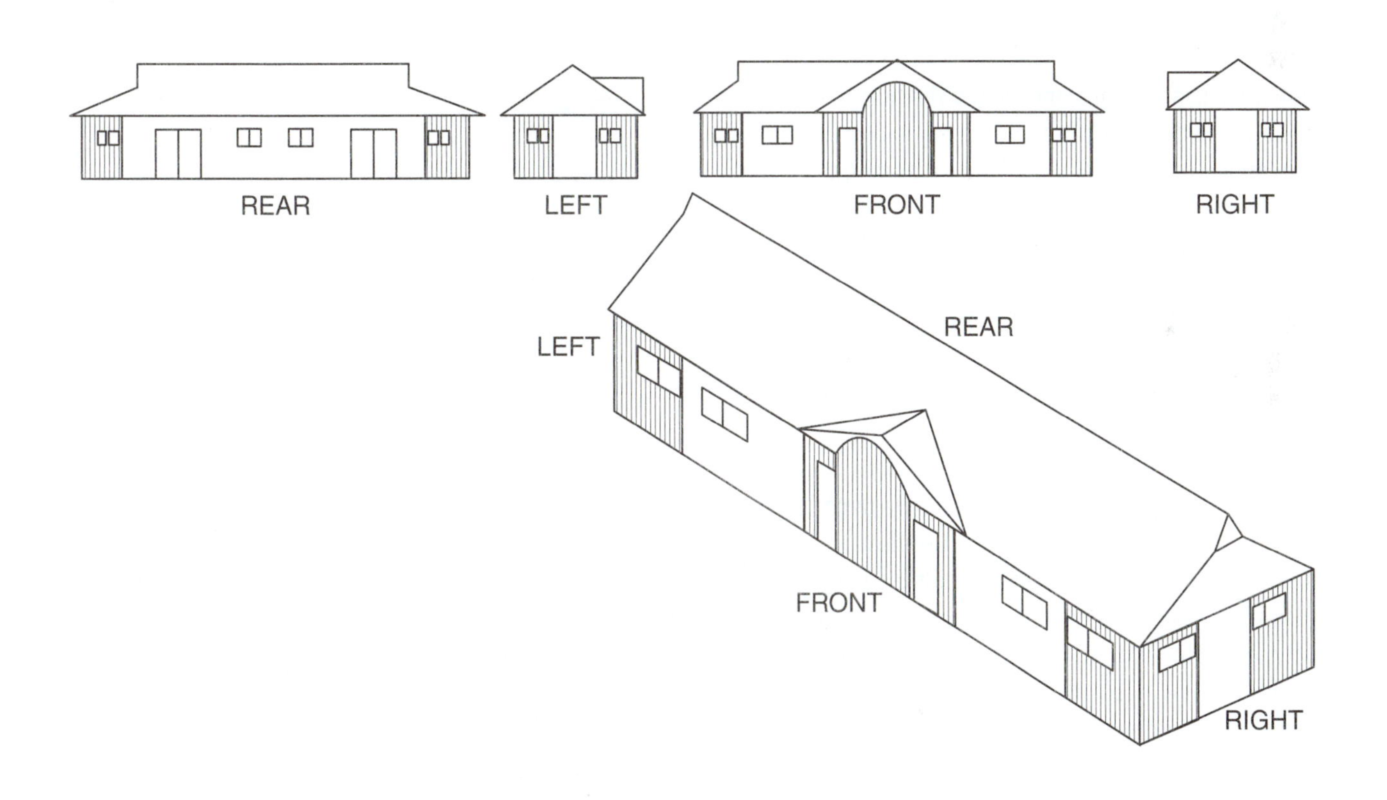

Figure 7–4. Elevations can be named according to their relative positions.

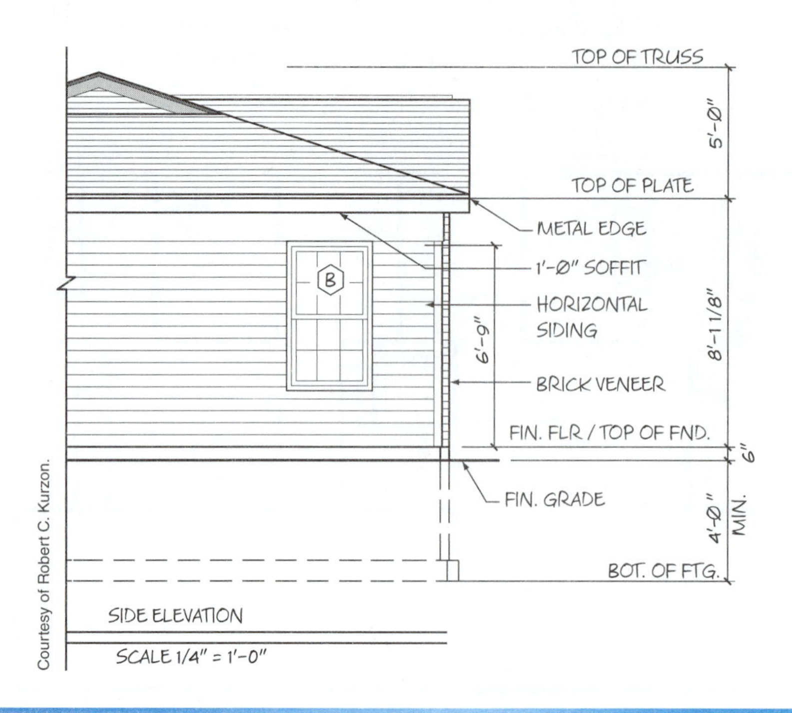

Figure 7–5. Underground portion of the building is shown with dashed lines.

dimensions and notes, they show the finished appearance of the building better than other views. Therefore, elevations are a great aid in understanding the rest of the drawing set.

The elevations show most of the building, as it will actually appear, with solid lines. However, the underground portion of the foundation is shown as hidden lines (see **Figure 7–5**). The footing is shown as a rectangle of dashed lines at the bottom of the foundation walls.

The surface of the ground is shown by a heavy solid line, called a grade line. The *grade line* might include one or more notes to indicate the elevation above sea level or another reference point (see **Figure 7–6**). Elevation used in this sense is altitude, or height—not a type of drawing. All references to the height of the ground or the level of key parts of the building are in terms of elevation. Methods for measuring site elevations are discussed in Unit 9.

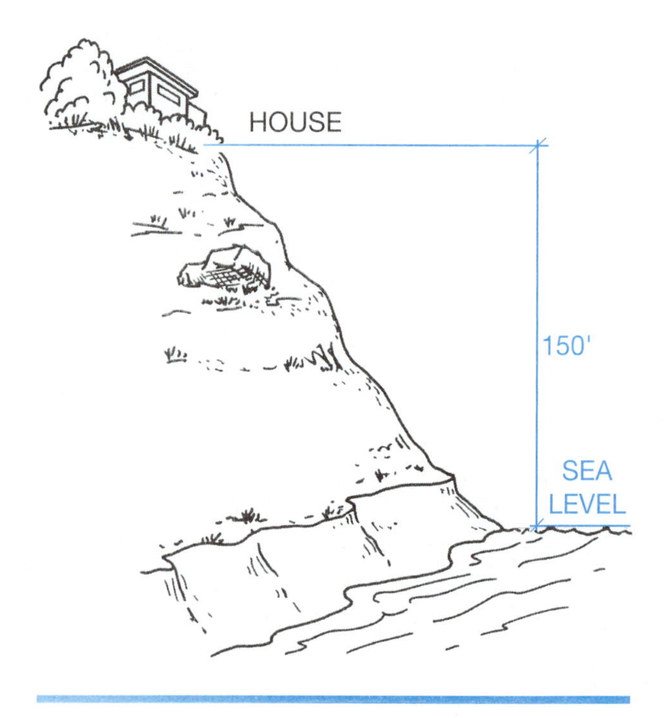

Figure 7–6. The elevation of this site is 150 feet.

Some important dimensions are included on the building elevations. Most of them are given in a string at the end of one or more elevations (see **Figure 7–7**). The dimensions most often included are listed here:

- Thickness of footing
- Height of foundation walls
- Top of foundation to finished first floor
- Finished floor to ceiling or top of plate (The ***plate*** is the uppermost framing member in the wall.)
- Finished floor to bottom of window headers (The ***headers*** are the framing across the top of a window opening.)
- Roof overhang at eaves

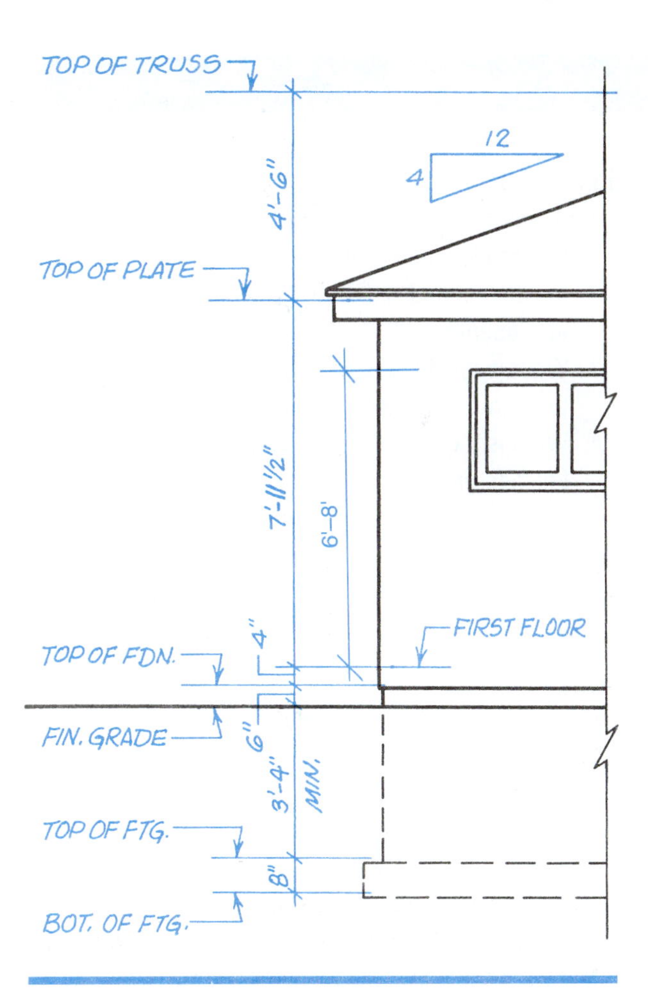

Figure 7–7. Dimensions on an elevation.

USING WHAT YOU LEARNED

The building elevations are often the best drawings for a quick picture of what the finished building will look like. They also convey a lot of useful information in the construction of a building.

One requirement sometimes found in zoning ordinances is a limit on the overall height of a building. What is the height of the ridge of the two-unit apartment above the finished grade? This information is shown at the right end of each of the building elevations. The finished floor is 6 inches above the finished grade. The top of the wall plate is 7′–11 ½″ above the finished floor. The top of the roof (i.e., top of truss) is 4′–6″ above the top of the plate. All of these added together, 12′–11 ½″ is the overall height of the building.

Assignment

Refer to the drawings of the Two-Unit Apartment in your textbook packet to complete this assignment.

1. Which elevation is the north elevation?
2. In what compass direction does the left end of the apartment face?
3. What is the dimension from the surface of the floor to the top of the wall framing?
4. What is the overall height of the building above the finished grade?
5. How far does the foundation project above the ground?
6. How far below the surface of the ground does the foundation wall extend?
7. What is the total height of the foundation walls?
8. What is the minimum depth of the bottom of the footings?

8 UNIT

Sections and Detail

It is not possible to show all the details of construction on foundation plans, floor plans, and building elevations. Those drawings are meant to show the relationships of the major building elements to one another. To show how individual pieces fit together, it is necessary to use larger-scale drawings and section views. These drawings are usually grouped together in the drawing set. They are referred to as ***sections*** and ***details*** (see **Figure 8–1**).

Objectives

After completing this unit, you will be able to perform the following tasks:

- Find information shown on section views.
- Find information shown on large-scale details.
- Orient sections and details to the other plans and elevations.

GREEN NOTE

The fill under the concrete slab in the Two-Unit Apartment in your textbook packet is run-of-bank (ROB) gravel. Many slabs are placed on crushed stone, but by using ROB gravel, the apartment reduces the impact on the environment. The gravel does not require running a stone crusher and grader, which consumes energy and emits stone dust into the atmosphere. The gravel may be available more locally than crushed stone, thereby reducing the need to transport it from any great distance.

Sections

Nearly all sets of drawings include, at least, a typical wall section. The typical section may be a section view of one wall, or it may be a full section of the building. Full sections are named by the direction in which the imaginary cut is made. **Figure 8–2** shows a transverse section. A *transverse section* is taken from an imaginary cut across the width of the building. Transverse sections are sometimes called *cross sections.* A full section taken from a lengthwise cut through the building is called a *longitudinal section* (see **Figure 8–3**).

Full sections and wall sections normally have only a few dimensions but have many notes with leaders to identify the parts of the wall. The following is a list of the kinds of information that are included on typical wall sections with most sets of drawings:

- Footing size and material (This is specified by building codes.)
- Foundation wall thickness, height, and material
- Insulation, waterproofing, and interior finish for foundation walls
- Fill and waterproofing under concrete floors
- Concrete floor thickness, material, and reinforcement

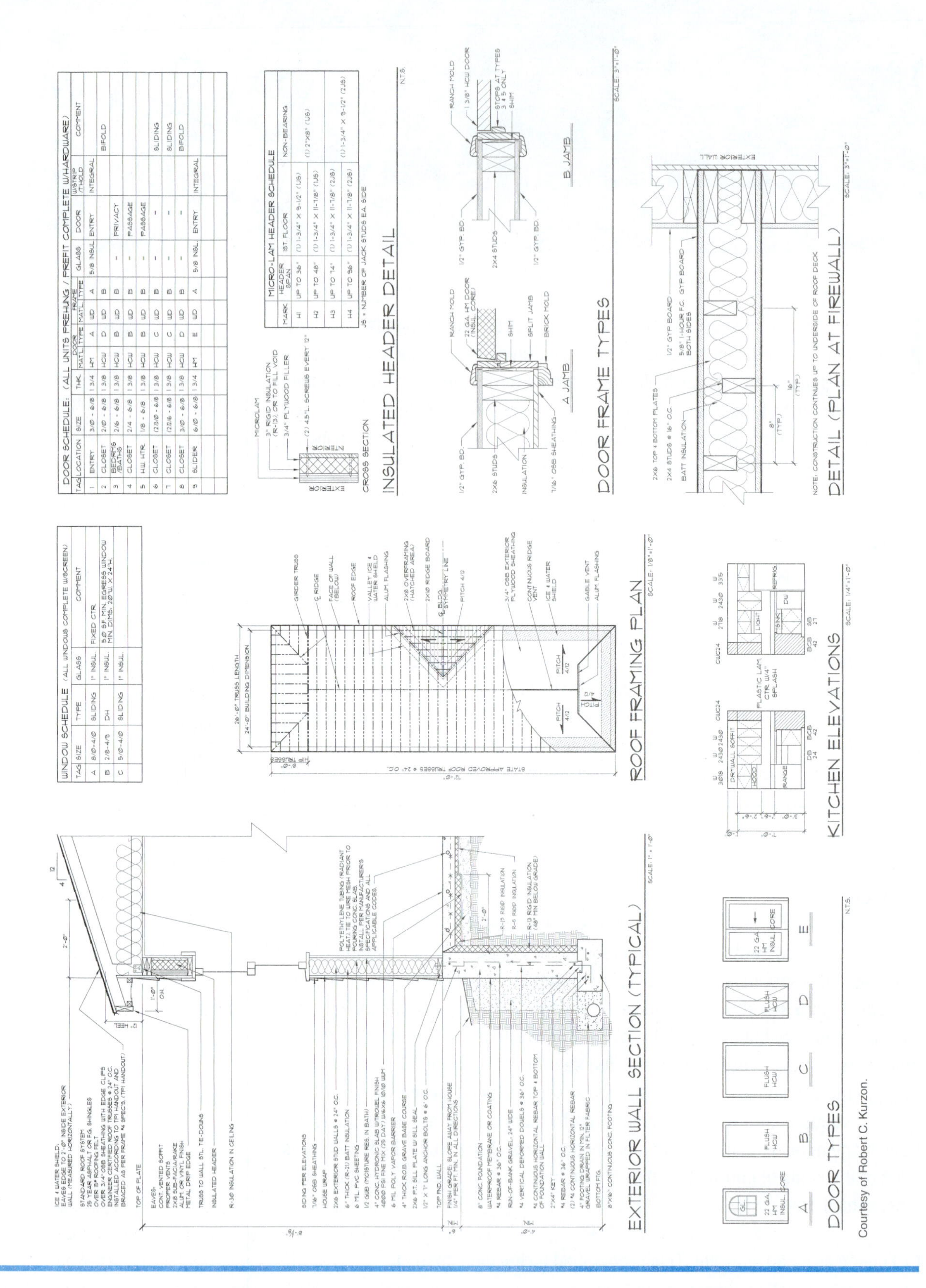

Figure 8–1. Typical sheet of sections and details for a small building.

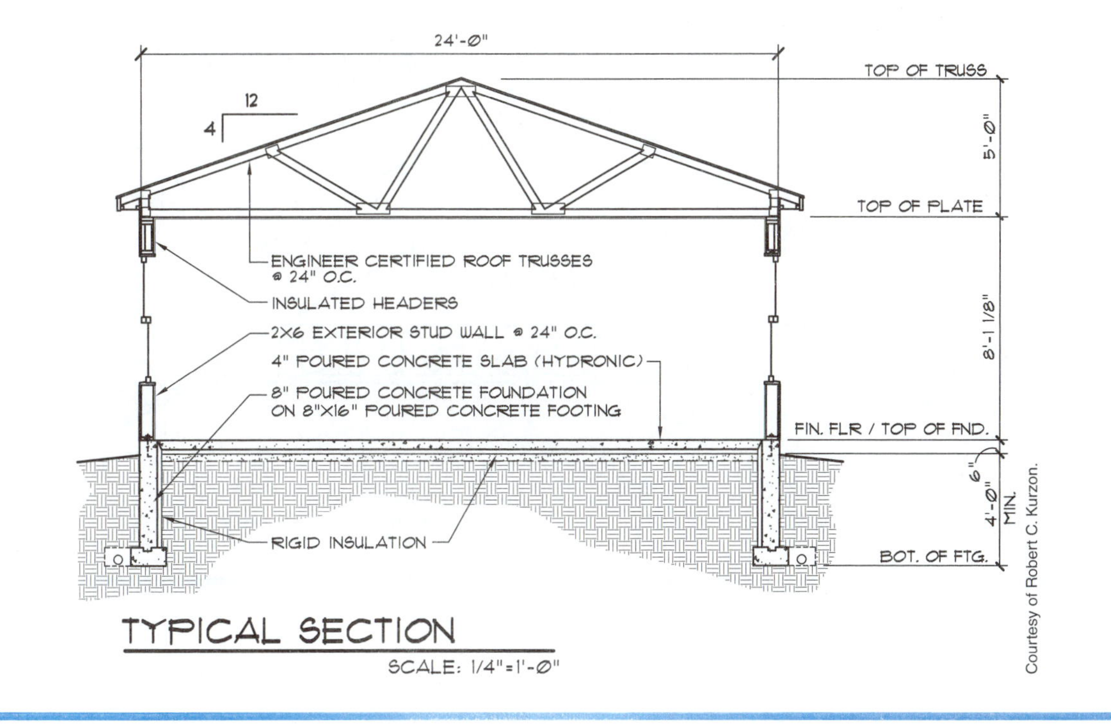

Figure 8–2. Transverse section.

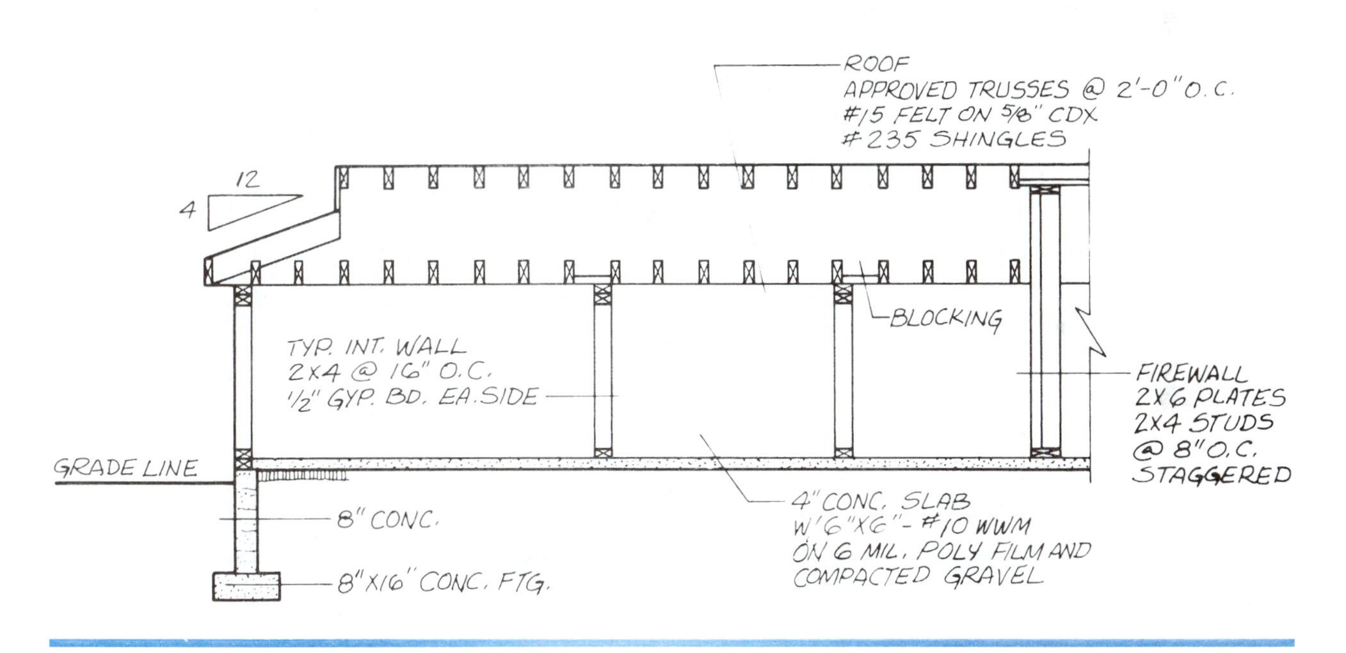

Figure 8–3. Longitudinal section.

- Sizes of floor framing materials
- Sizes of wall framing materials
- Wall covering (sheathing, siding, stucco, masonry, and interior wall finish) and insulation
- Cornice construction—materials and sizes (The ***cornice*** is the construction at the roof eaves.)
- Ceiling construction and insulation

Other section drawings are included as necessary to explain special features of construction. Wherever wall construction varies from the typical wall section, another wall section should be included. Section views are used to show any special construction that cannot be shown on normal plans and elevations. **Figure 8–4** is an example of a special section in elevation. This section view is said

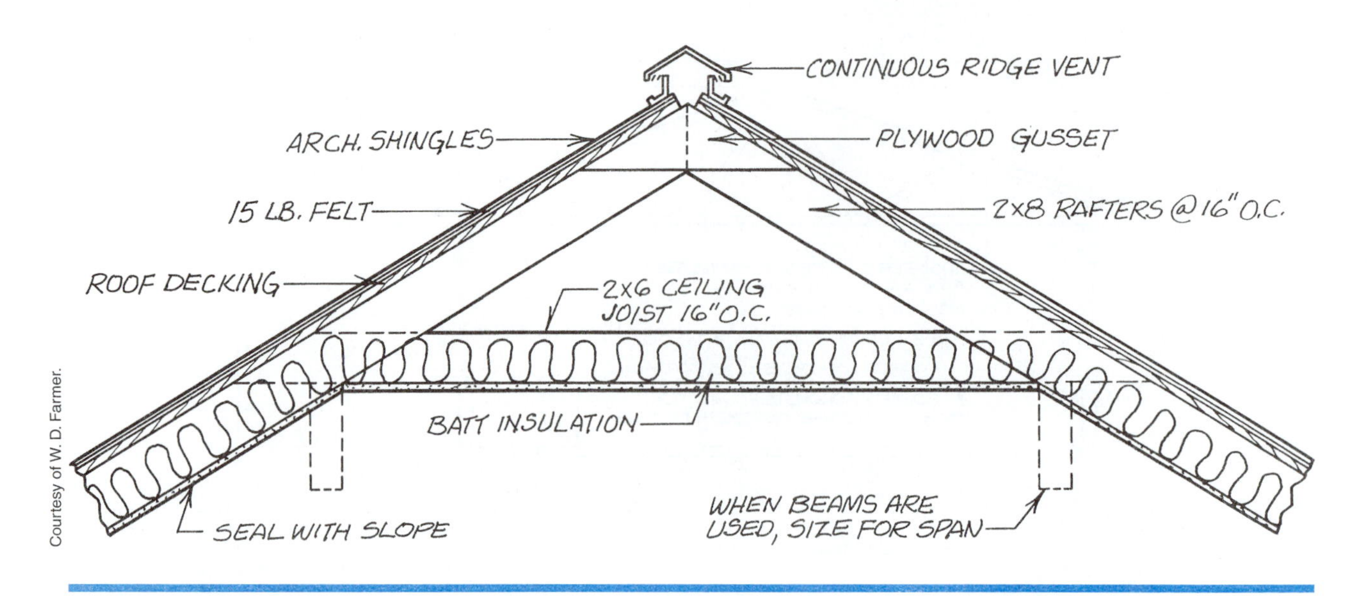

Courtesy of W. D. Farmer.

Figure 8–4. Special section of ventilated ridge.

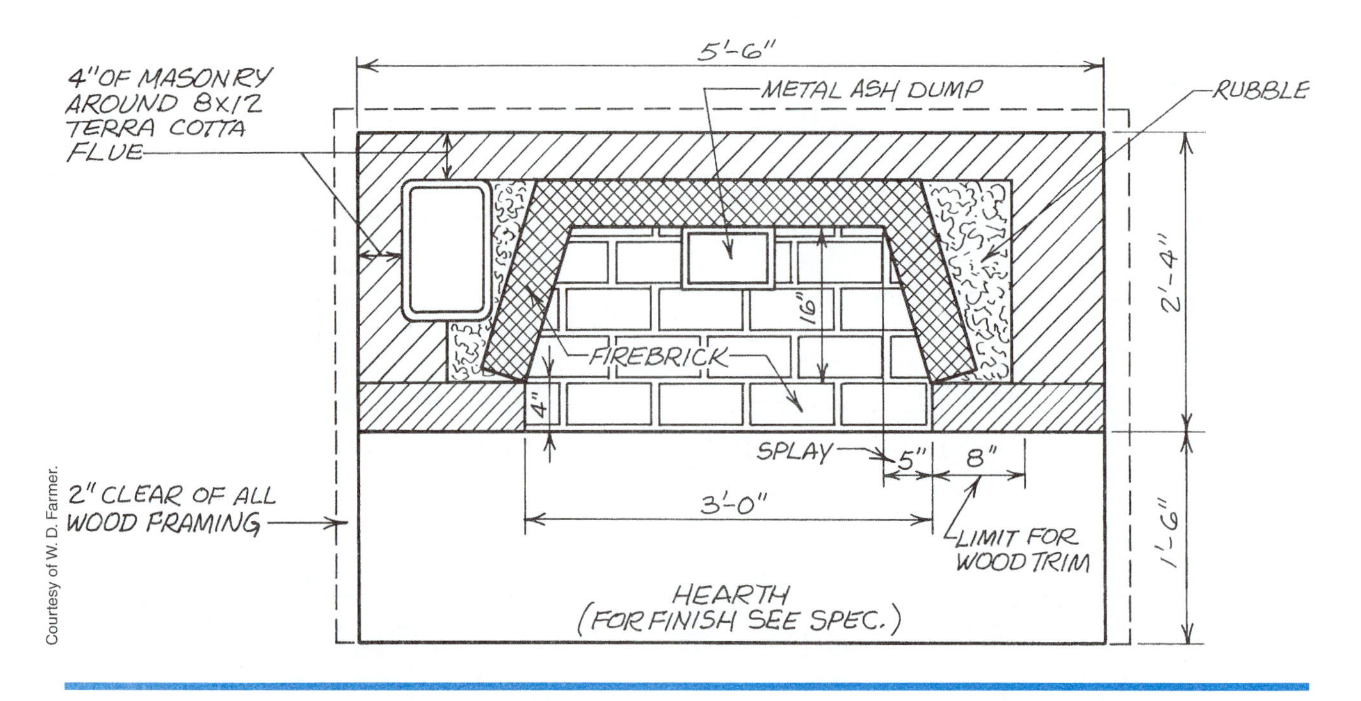

Courtesy of W. D. Farmer.

Figure 8–5. A section in plan.

to be in elevation because it shows the height of the ridge construction. **Figure 8–5** is in plan because it shows the interior of the fireplace as viewed from above.

Other Large-Scale Details

Sometimes necessary information can be conveyed without showing the interior construction. A large scale may be all that is needed to show the necessary details. The most common examples of this are on cabinet installation drawings (see **Figure 8–6**). Cabinet elevations show how the cabinets are located, without showing the interior construction.

Many details are best shown by combining elevations and sections or by using isometric drawings. **Figure 8–7** shows an example of an elevation and a section used together to explain the construction of a fireplace. **Figure 8–8** shows an isometric detail drawing that includes sections to show interior construction.

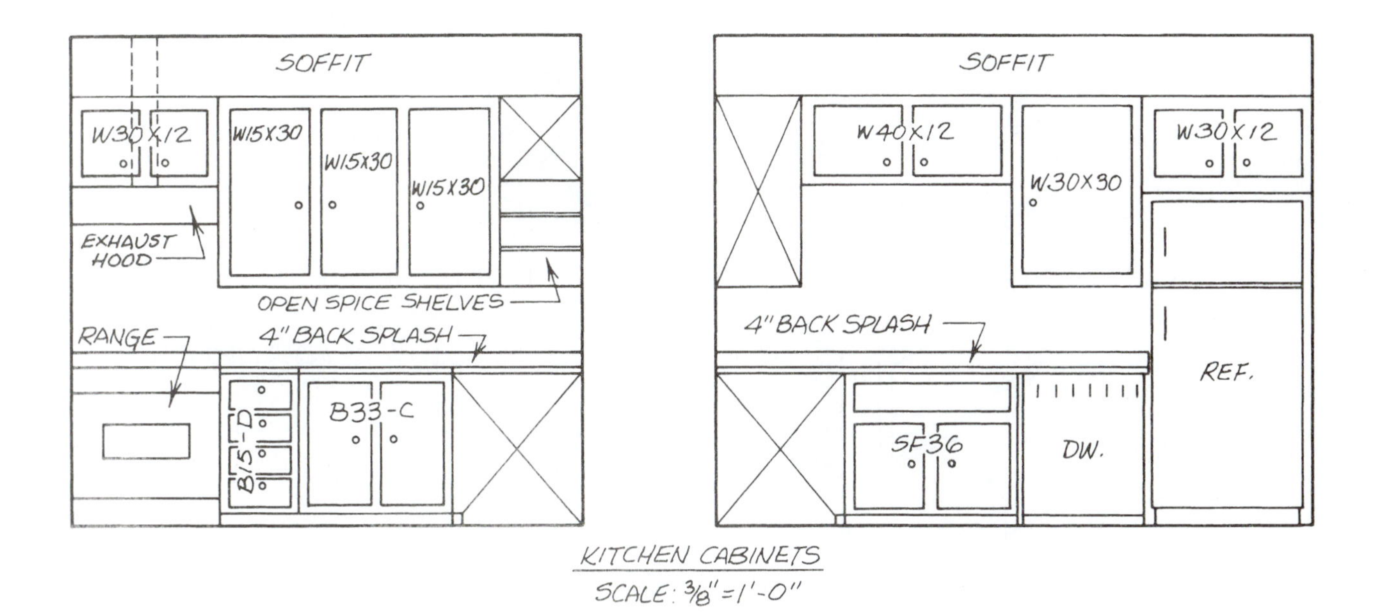

Figure 8–6. Cabinet elevations.

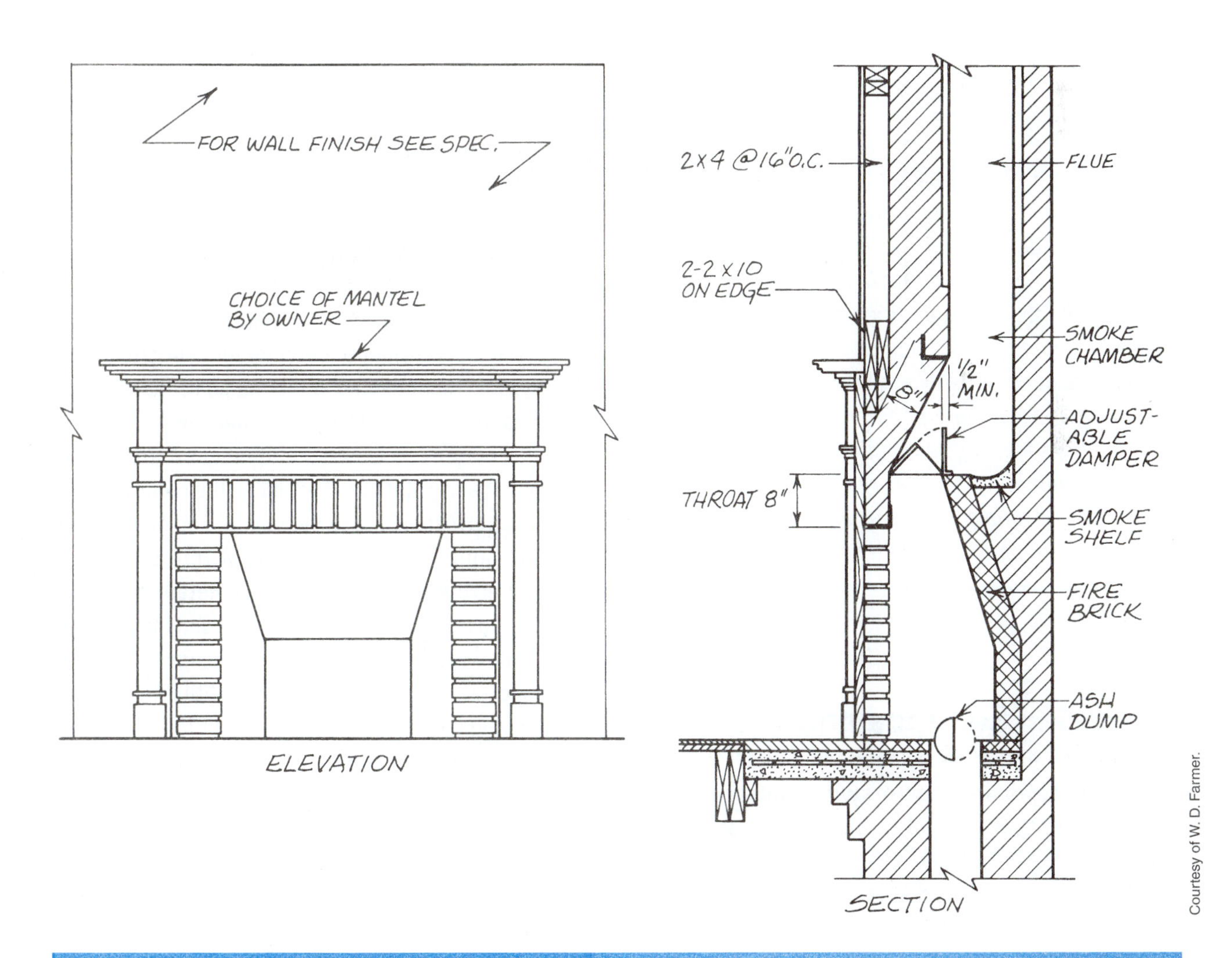

Figure 8–7. Fireplace details.

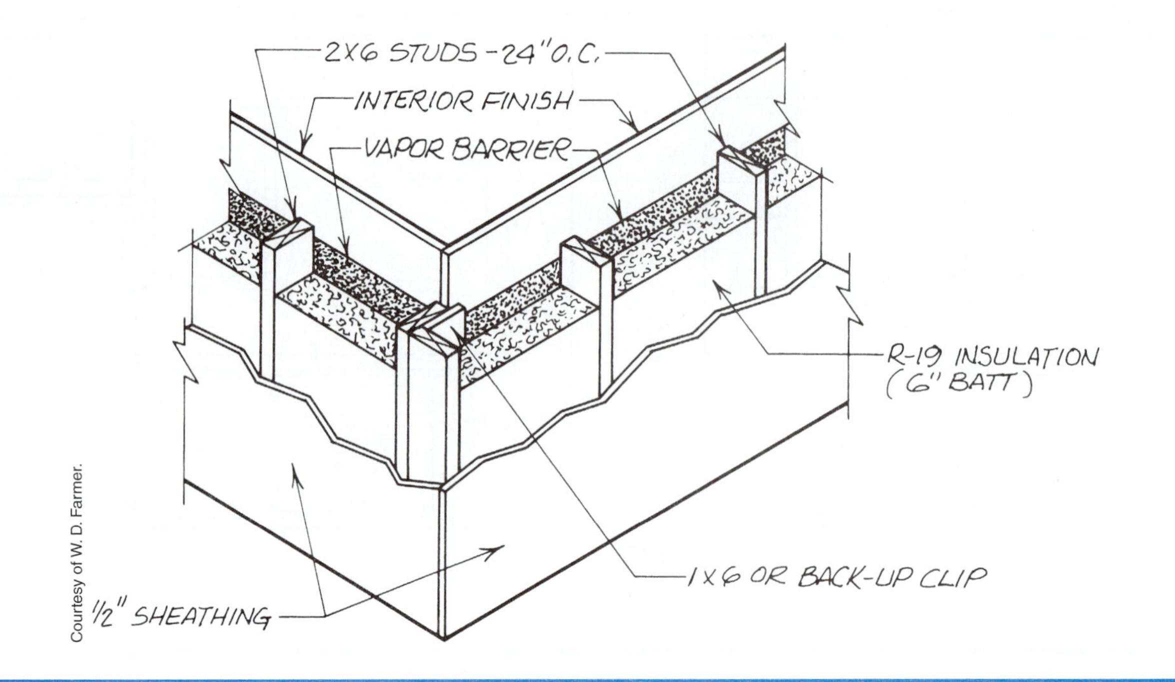

Figure 8–8. Isometric section.

GREEN NOTE

When frame construction was first developed, horizontal or diagonal boards were applied as sheathing. After World War II, plywood replaced board sheathing. Today Oriented Strand Board (OSB) is the most commonly used sheathing material, but other engineered products are becoming increasingly popular. Gypsum-based structural sheathing and magnesium oxide structural sheathing are water resistant, decay resistant, and form a vapor barrier, eliminating the need for a separate moisture barrier.

Orienting Sections and Details

As explained earlier, some sections and details are labeled as typical. These drawings describe the construction that is used throughout most of the building.

Sections and details that refer to a specific location in the building include a reference that indicates where the section or detail came from. That larger source drawing has a cutting-plane line to show exactly where the section cut or detail is taken from. The cutting-plane line has an arrowhead or some other indication of the direction from which the detail is viewed. The top drawing in **Figure 8–9** shows that there is a section view or detail of the construction at the skylight. The little flag at the top points to the right, so that is the direction from which the detail at the bottom of the figure is viewed. The label on the bottom drawing, the skylight detail, includes the number of the drawing, corresponding to the number indicated at the cutting-plane line in the top view. A reference mark near the arrow indicates where the detail drawing is shown. The reference marks that are used for orienting details may vary from one set of drawings to another. It is important, although not usually difficult, to study the drawings and learn how the architect references details. Usually a system of sheet numbers and view numbers is used. One such numbering system was explained earlier.

Some basic principles of details and sections have been discussed here. You will gain more practice later in reading details and sections.

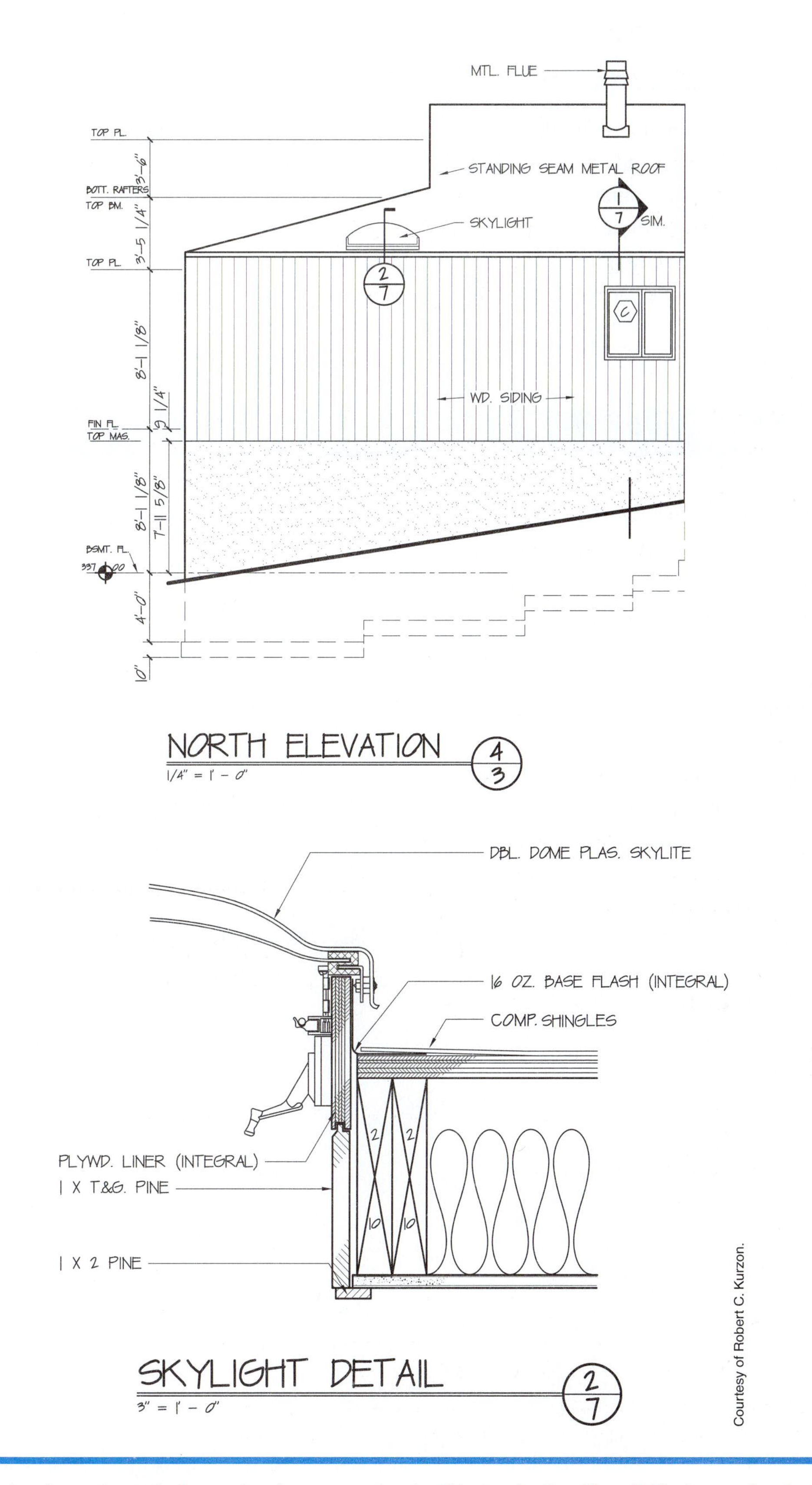

Figure 8–9. The elevation at the top indicates that the construction detail is drawing 2 on Sheet 7. The bottom drawing is that detail.

USING WHAT YOU LEARNED

A lot of information can be found on the various section views and detail drawings. Some of this is shown simply by drawing a symbol and adding a note or callout. Any material that is to be placed beneath a concrete slab or imbedded in the slab must be in place before the concrete is delivered. Beneath the concrete slab in the Two-Unit Apartment in your textbook packet there is rigid foam insulation. What is immediately beneath the insulation? The exterior wall section on Sheet 2 has a callout and a leader pointing to a line of long dashes. The callout identifies this as 6 mil poly vapor barrier.

Assignment

Refer to the drawings of the Two-Unit Apartment in your textbook packet to complete the assignment.

1. What is used to show the detail of a complex design, installation, or product?
2. What kind of section drawing is the Typical Section on Sheet 1?
3. What kind and size material is to be used for the foundation walls?
4. What material is immediately beneath the outer two feet of the concrete slab?
5. What kind and size of insulation is used around the foundation? Is this insulation used on the inside or outside of the foundation?
6. What kind and size of material is to be used on the inside face of the frame walls?
7. Sheet 2 includes a firewall detail. Where in the apartment is this firewall?
8. What is the distance between the centerlines of the studs in the firewall?
9. What is the total thickness of the firewall? (Remember that a 2×6 is actually 5½″ wide.)
10. Were the cabinet elevations drawn of the kitchen on the east side or the west side of the Two-Unit Apartment?
11. How would the kitchen elevations be different if they were drawn from the other kitchen?
12. What is the distance from the kitchen counter top to the bottom of the wall cabinets?
13. How far does the roof overhang project beyond the exterior walls?
14. Where is the electrical panel located for the west apartment?

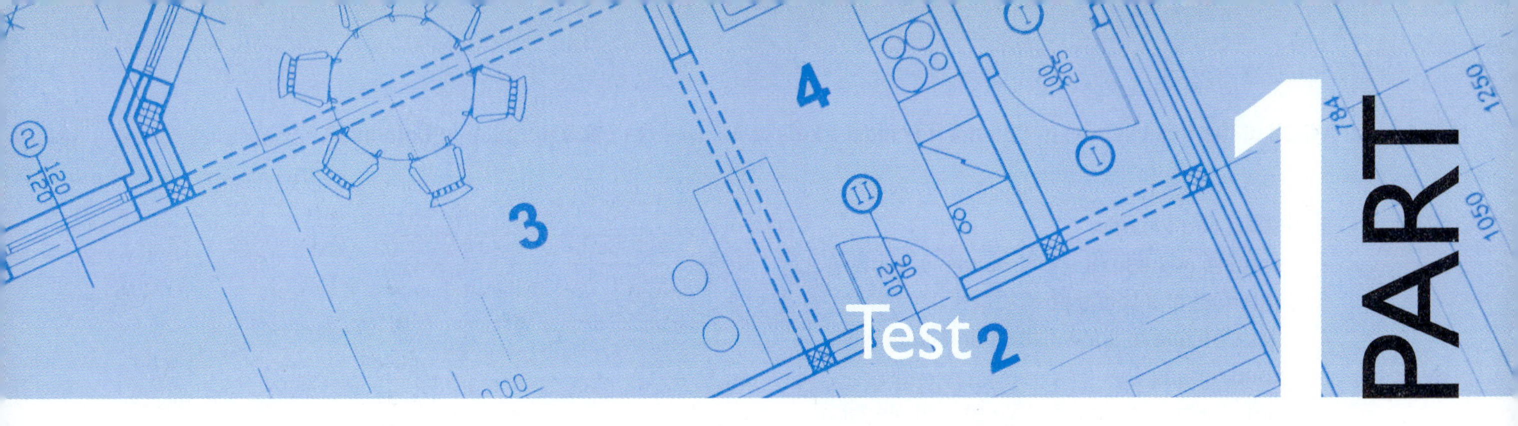

A. Identify each of the dimensions indicated on the illustrated scale.

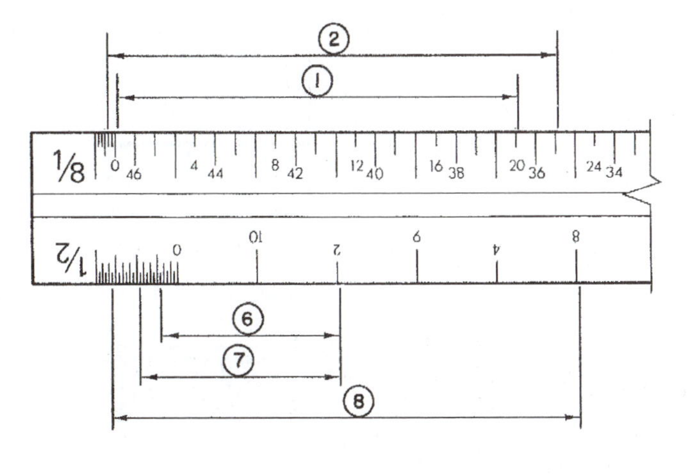

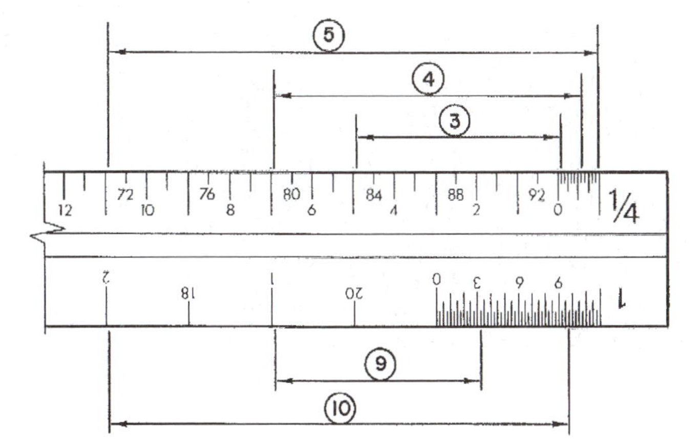

B. Which occupation or individual listed in Column II performs the task listed in Column I?

I	II
1. Obtains a building permit	a. Architect
2. Issues the building permit	b. Building inspector
3. Acts as the owner's representative	c. Owner
4. Issues certificate of occupancy	d. Mechanical engineer
5. Lays out rooms for efficient use	e. Municipal building department
6. Designs plumbing in large buildings	f. General contractor
7. Hires and supervises carpenters	
8. Checks to see that codes are observed	

C. Which of the lines shown in Column II is most likely to be used for each purpose in Column I?

I	II
1. Outline of a window	a.
2. Alternate position of a fold-down countertop	b.
3. Centerline of a round post	c.
4. Extension line to show extent of a dimension	d.
5. Buried footing	e.
6. Point at which an imaginary cut is made for a section view	f.

D. Which of the symbols shown in Column II is used for each of the objects or materials in Column I?

I

1. Awning window in elevation
2. Bifold door in plan
3. Earth
4. Dimension lumber
5. Batt insulation
6. Concrete
7. Ceiling light fixture
8. Finish wood
9. Shutoff valve (plumbing)
10. Hopper window in elevation

II

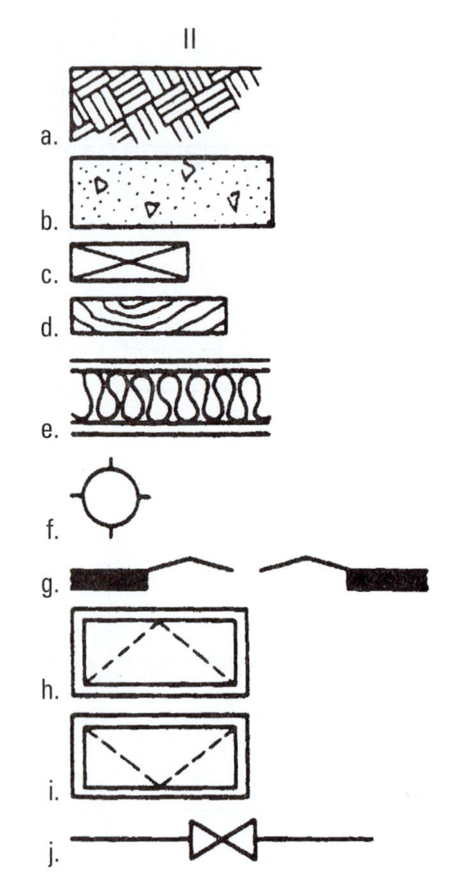

E. Select the one best answer for each question.

1. What kind of regulation controls the types of buildings allowed in each part of a community?
   a. building code
   b. zoning law
   c. specification
   d. certificate of occupancy
2. Which of the listed kinds of information can be clearly shown on construction drawings?
   a. size of parts
   b. location of parts
   c. shape of parts
   d. all of these

3. What type of drawing is the 2 × 4 shown in Illustration 3?
   a. isometric
   b. oblique
   c. perspective
   d. none of these

3

4. What type of drawing is the 2 × 4 shown in Illustration 4?
   a. isometric
   b. oblique
   c. perspective
   d. none of these

4

5. What type of drawing is the 2 × 4 shown in Illustration 5?
   a. isometric
   b. oblique
   c. perspective
   d. none of these

5

6

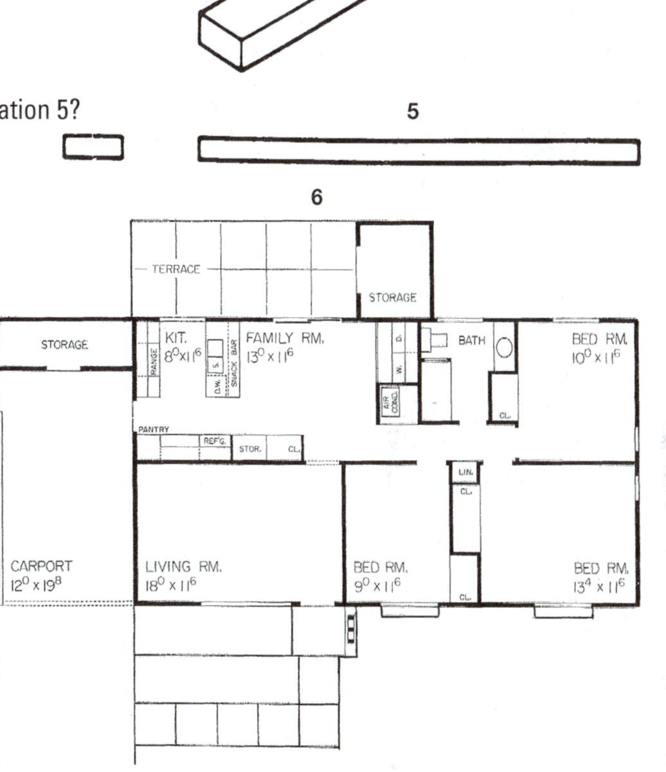

6. What kind of drawing is shown in Illustration 6?
   a. elevation
   b. detail
   c. rendering
   d. plan

7. What kind of drawing is shown in Illustration 7?
   a. elevation
   b. detail
   c. rendering
   d. plan

7

8. What kind of drawing is shown in Illustration 8?
   a. elevation
   b. detail
   c. rendering
   d. plan

8

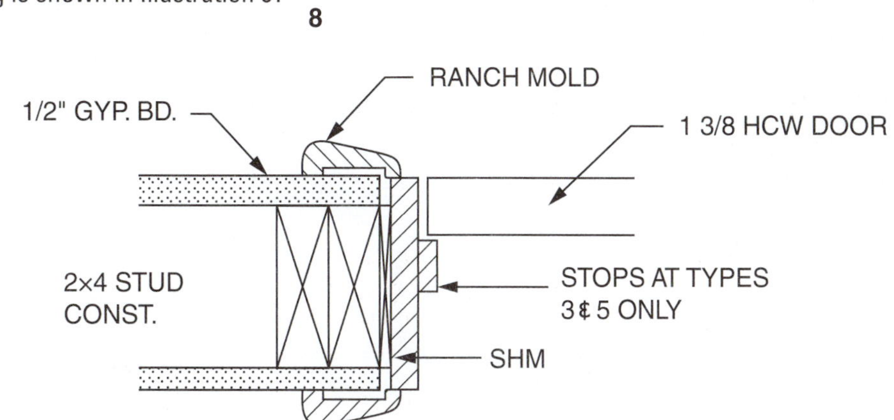

9. If an object 12 feet long is drawn at a scale of ¼″= 1′-0″, how long is the drawing?
   a. 48 inches
   b. 3 inches
   c. 3 feet
   d. none of these
10. If an object 1′-6″ long is drawn at a scale of 1½″= 1′-0″, how long is the drawing?
   a. 2¼ inches
   b. 2 inches
   c. 27 inches
   d. none of these
11. Where in a set of drawings would you find detail number 9.6?
   a. sheet 6
   b. sheet 9
   c. ninth sheet in the mechanical section
   d. none of the above
12. In the drawing key 4/A–4, what does the letter A stand for?
   a. architect's initial
   b. first edition of the drawings
   c. architectural
   d. first detail on the sheet
13. On which drawing would you expect to find the height of the foundation wall?
   a. site plan
   b. building elevation
   c. floor plan
   d. foundation plan
14. On which drawing would you expect to find the setback of the building?
   a. site plan
   b. building elevation
   c. floor plan
   d. foundation plan
15. On which drawing would you expect to find the height of the window heads?
   a. site plan
   b. building elevation
   c. floor plan
   d. window detail

F. Refer to the Two-Unit Apartment drawings in your textbook packet to answer these questions.

1. How far is the building from the west boundary?
2. What is the dimension from the finished floor to the top of the wall plate?
3. What is the overall length of the building at window height?
4. What is the overall length of the building at the eaves?
5. What is the north-to-south dimension inside the front bedrooms?
6. What is the slope of the roof?
7. What types of windows are used in the bedrooms?
8. How thick is the concrete footing?
9. What material is the foundation wall?
10. What is under the floor at its center?

READING DRAWINGS: LAKE HOUSE

In Part 2, you will examine all the information necessary to build a moderately complex single-family home. The sequence of the units in Part 2 follows the sequence of actual construction. In some cases, all the information necessary for a particular stage of construction can be found on one sheet of drawings. Other stages require cross-referencing among several drawings. The relationships among the various drawings are discussed as the need to cross-reference them arises.

The assignments in this part refer to the Lake House drawings provided in your drawing packet. This Lake House was designed as a vacation home on a lake. The design is moderately complex, involving several floor levels and some interesting construction techniques.

UNIT 9 Clearing and Rough Grading the Site

Objectives

After completing this unit, you will be able to perform the following tasks:

- Identify work to be included in clearing a building site according to site plans.
- Interpret contour lines and grading indications on a site plan.
- Interpolate unspecified site elevations.

Property Boundary Lines

The boundary lines of the building site are shown on the site plan. The direction of a property line is usually expressed as a bearing angle. The *bearing* of a line is the angle between the line and north or south. Bearing angles are measured from north or south depending on which keeps the bearing under 90° (see **Figure 9–1**). Angles are expressed in degrees (°), minutes (′), and seconds (″). There are 360 degrees in a complete circle, 60 minutes in a degree, and 60 seconds in a minute.

The *point of beginning* (POB) may or may not be shown on the site plan. If the point of beginning is not shown on the plan, start at a convenient corner. Corners are usually marked with an *iron pin* (IP) or some permanent feature. The approximate direction of the boundaries can be found with a handheld

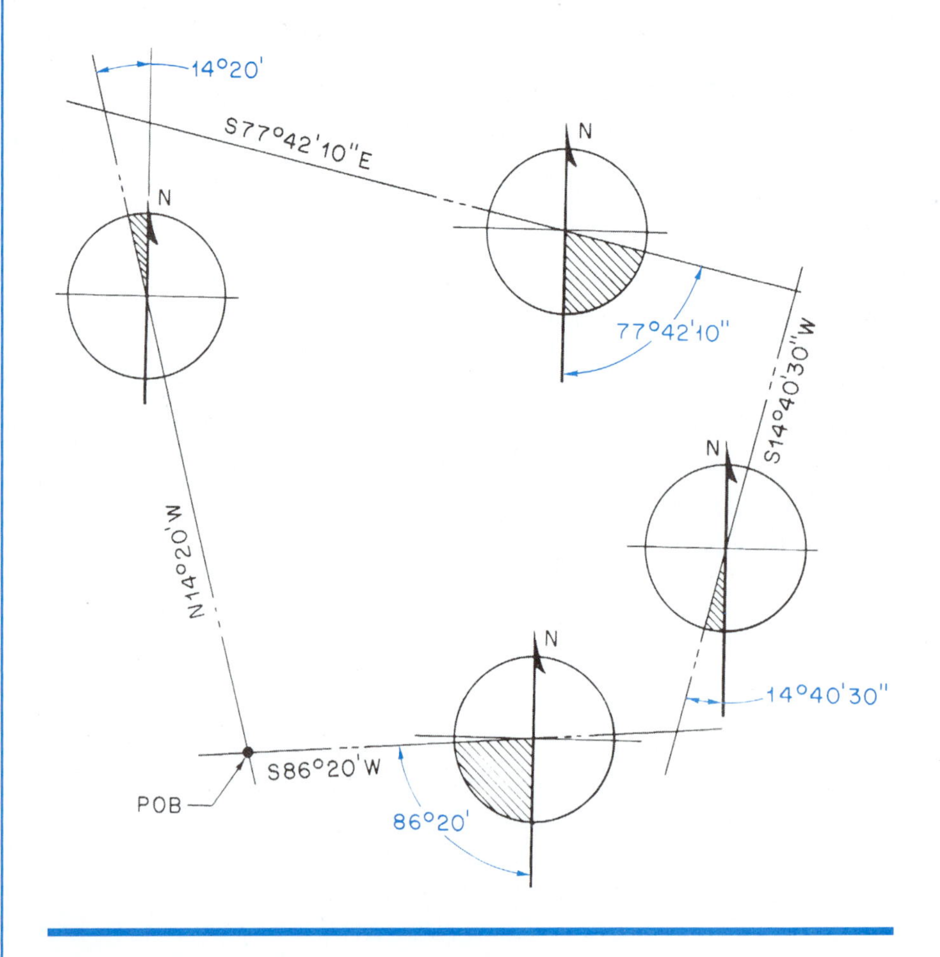

Figure 9–1. Bearing angles are always less than 90°.

compass. This approximation should be accurate enough to aid in finding the marker (iron pin, manhole cover, concrete marker, or similar item) at the next corner. Proceed around the perimeter in this manner to find all corners. All construction activity and debris should be kept within the property boundaries unless permission is first obtained from neighboring landowners.

Clearing the Site

The first step in actual construction is to prepare the site. This means clearing any brush or trees that are not to be part of the finished landscape. The architect's choice of trees to remain is based on consideration of many factors. Trees and other natural features can be an important part of architecture—not only for their natural beauty but for energy conservation. For example, deciduous trees, which lose their leaves in the winter, can be used to effectively control the solar energy striking a house during warmer months as shown in **Figure 9–2**, where the trees shade the south side of the house. In the winter, the sun shines through the deciduous trees on the south side of a house, thus taking advantage of this source of heat as shown in **Figure 9–3**. The Lake House in your drawing package offers a good example of the importance of the selection of trees to remain on a site. This house gets a large part of its heat from its passive-solar features, which are described more fully later.

Trees that are to be saved are shown on the plot plan by a symbol and a note indicating their butt diameter and species (see **Figure 9–4**). Areas that are too densely wooded to show individual trees are outlined and marked "woods" (see **Figure 9–5**). Removal of unwanted trees may require felling and stump removal, or may be accomplished with a bulldozer and dump truck. In either case, care must be exercised not to damage the trees that are to be saved.

Grading

Grading refers to moving earth away from high areas and into low areas. Site grading is necessary to ensure that water drains away from the building properly and does not puddle or run into the building. In some cases, grading may be necessary for access to the site. For example, if the site has a steep grade, it may be necessary to provide a more gradual slope for a driveway.

Figure 9–2. The summer sun is shaded by deciduous trees.

Figure 9–3. The winter sun passes through deciduous trees.

Figure 9–4. Typical note and symbol for individual tree.

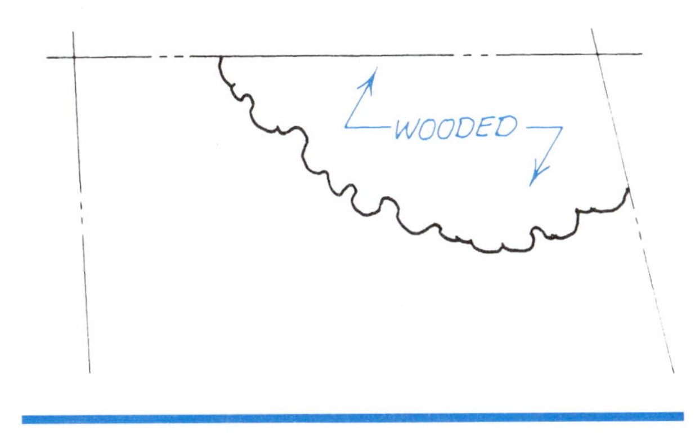

Figure 9–5. Typical note and symbol for wooded area.

GREEN NOTE

Surface water must be controlled in order to prevent erosion of the building site or neighboring property. Surface water should not be allowed to collect in puddles or overly damp areas of earth, as these can promote mosquito breeding and harm the growth of desirable vegetation. Proper grading of the site is the most effective way of controlling surface water. Some type of barrier should be placed prior to finished grading to prevent construction waste water from washing onto adjoining property. Hay bales are often used for this purpose.

Grade is measured in vertical feet from sea level or from a fixed object such as a manhole cover. This vertical distance is called ***elevation***. The term *elevation* to denote a vertical position should not be confused with elevation drawings that show the height of objects. The elevations of specific points are given as *spot elevations*. Spot elevations are used to establish points in a driveway, a walk, or the slope of a terrace, as shown in **Figure 9–6.** Spot elevations are often given for trees that are to be saved.

The grade of a site is shown by *topographic* ***contour lines***. These are lines following a particular elevation. The vertical difference between contour lines is the ***vertical contour interval***. For plot plans this is usually 1 or 2 feet. When the land slopes steeply, the contour lines are closely spaced. When the slope is gradual, the contour lines are more widely spaced.

The builder must be concerned with not only the grade or contour of the existing site but also that of the finished site. To show both contours, two sets of contour lines are included on the plot plan. Broken lines indicate natural grade (NG), and solid lines indicate finished grade (FG) (see **Figure 9–7**). **Figure 9–8** shows a cross section of the site in **Figure 9–7**.

When the natural-grade elevation is higher than the finished-grade elevation, earth must be removed. This is referred to as *cut*. When the natural grade is at a lower elevation than the finished grade, *fill* is required. To determine the amount of cut or fill required at a given point, find the difference between the natural grade and the finished grade (see **Figure 9–9**).

Interpolating Elevations

Sometimes it is necessary to find an elevation that falls between two contour lines. This can be done by interpolation. *Interpolation* is a method of finding an unknown value by comparing it with known values.

Example: To interpolate the elevation of the tree at point A in **Figure 9–10**, follow the listed steps of the procedure

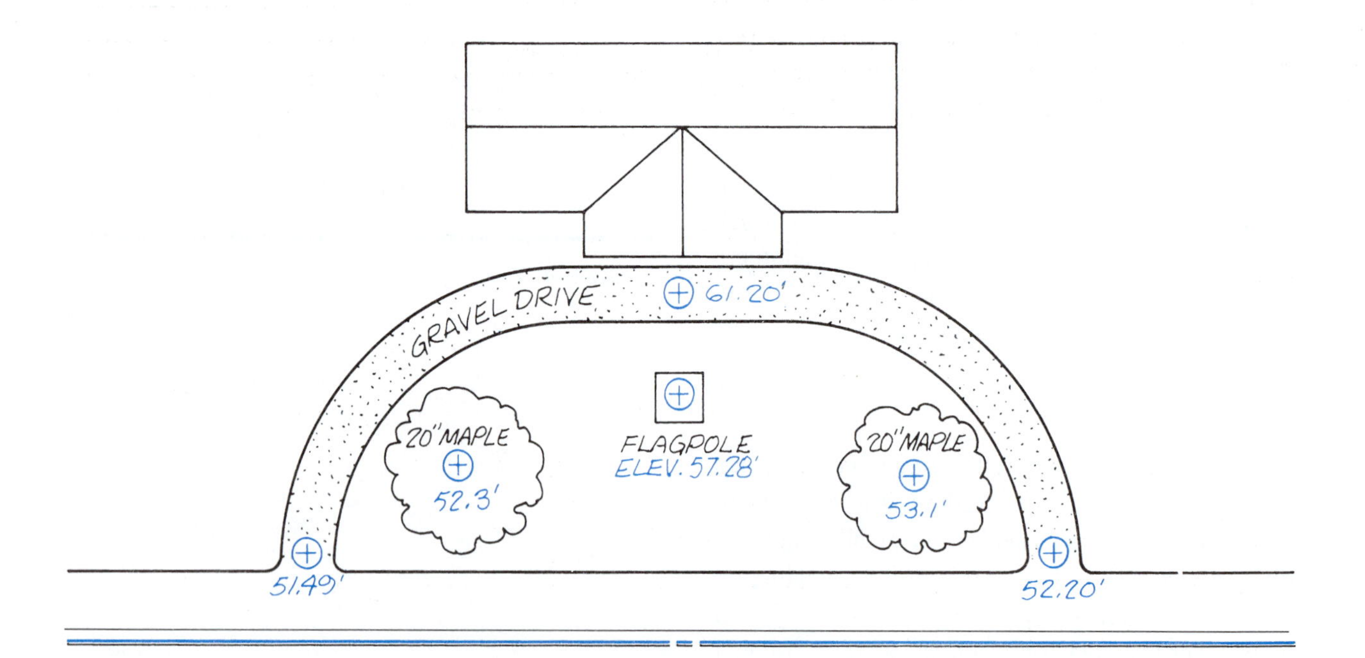

Figure 9–6. Spot elevations for specific locations.

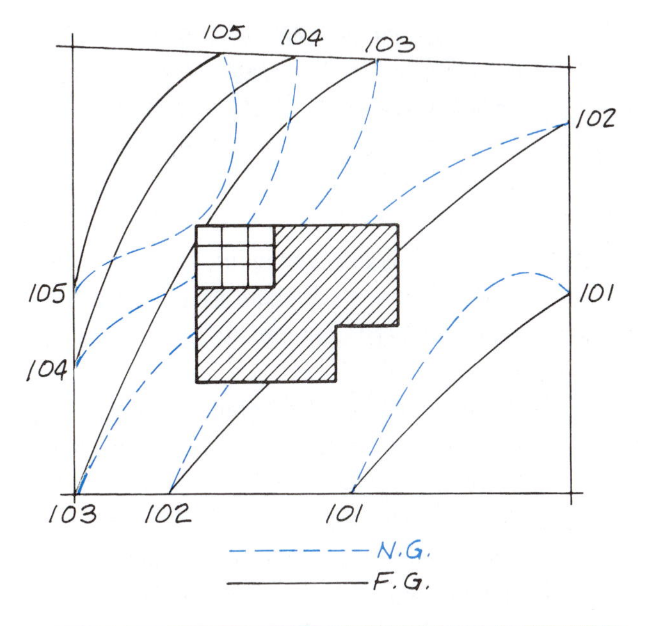

Figure 9–7. Two sets of contour lines show that this site will be graded to be more level in the area of the building.

using the information shown in the illustration and the numbers enclosed in parentheses.

Step 1. Scale the distance between the two adjacent contour lines (12 feet).

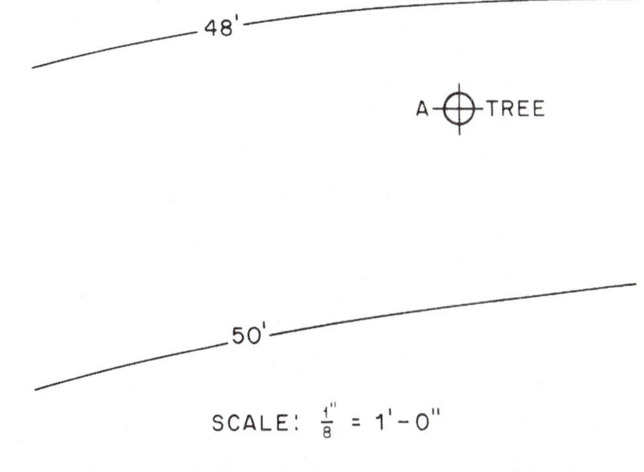

Figure 9–10. Interpolate the elevation of the tree.

Step 2. Scale the distance from the unknown point to the nearest contour line (4 feet).

Step 3. Multiply the contour interval by the fraction of the distance between the contour lines to the unknown point. (Contour interval = 2 feet; fraction of distance between contour lines = $^4/_{12} = {^1/_3}$. $2 \times {^1/_3} = {^2/_3}'$.)

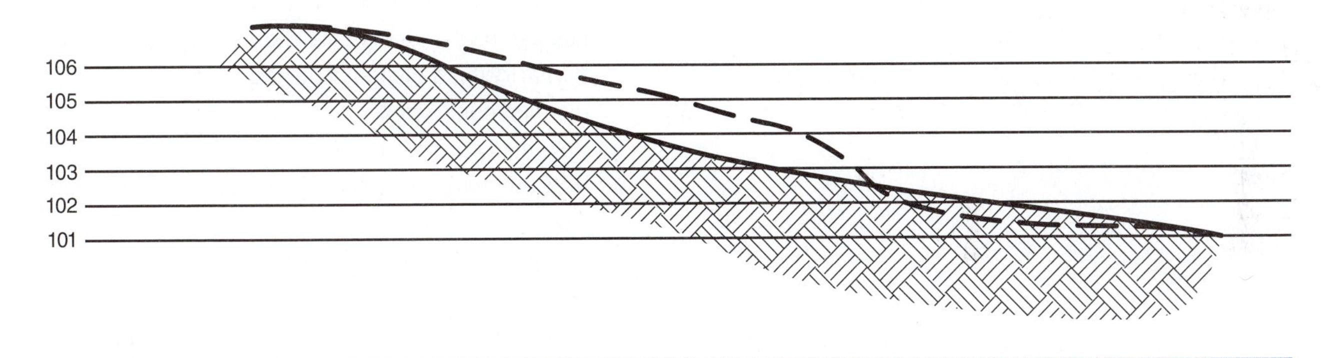

Figure 9–8. Cross section of the site showing natural grade and finished grade.

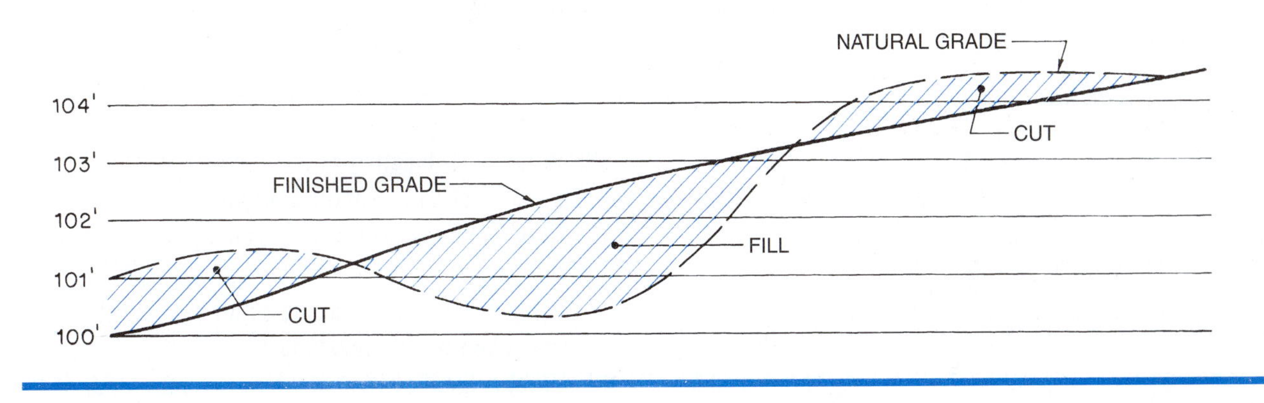

Figure 9–9. Cutting is required where NG is above FG. Fill is required where NG is below FG.

Step 4. If the nearest contour line is below the other one, add this to it. If the nearest contour line is above the other one, subtract this amount.

(Nearest contour = 48 feet. This is below the other contour line at 50 feet, so $^2/_3$ foot is added to 48 feet. 2/3′ + 48′ = 48.66′.)

USING WHAT YOU LEARNED

One of the first activities on a building site, after laying out the building lines, is excavating for the foundation and any concrete slabs. This requires that you know elevations of the foundations and slabs are compared to the elevation of the natural grade of the site. Let's determine how far the northwest corner of the garage floor is above or below the natural grade at that point.

The garage elevation is clearly marked as 343.0′. The corner falls between the natural-grade contour lines for 342′ and 344′. We will have to interpolate to find a close estimate of what that elevation is. Using the inch scale on an architectural scale, measure the distance between the two contour lines at the corner of the garage. The shortest distance between these two lines that crosses the corner of the garage is $1\,^9/_{16}$″. The corner is $^9/_{16}$″ from the 344′ contour line. These elevations are expressed in decimal parts of a foot, so we will find our elevation to the nearest 1/10th of a foot. Divide $^9/_{16}$ by $1\,^9/_{16}$ (or $^{25}/_{16}$) to find that the corner is 36 percent of the way from 344′ to 342′. Thirty-six percent of 2 feet (the contour interval) is 0.72′. Subtract 0.72′ from 344′ to find that the elevation of the natural grade at the corner is 343.28′. To the nearest 1/10th of a foot this is 343.3 feet, so the garage floor is 0.3 feet above the natural grade at that point.

Assignment

Refer to the site drawings of the Lake House in your textbook packet to complete the assignment.

1. What are the lengths of the north, east, and west boundaries of the Lake House site?

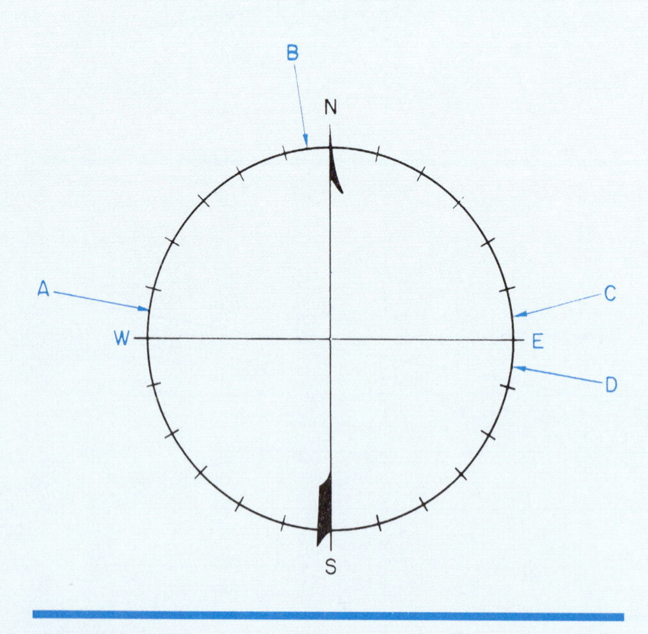

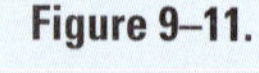

Figure 9–11.

2. Which of the compass points shown in **Figure 9–11** corresponds with the north boundary of the Lake House? Which compass point corresponds with the east boundary?
3. How many trees are indicated for removal?
4. How many trees are to remain on the site? (Do not include wooded areas.)
5. What is the finished-grade elevation at the tree nearest the Lake House?
6. What is the natural-grade elevation of the most easterly tree to be saved?
7. What is the elevation of the tree to be saved nearest the lake?
8. What is the natural-grade elevation at the southwest corner of the Lake House? Do not include the deck as part of the house.
9. What is the finished-grade elevation at the southwest corner of the Lake House?
10. How much cut or fill is required at the entrance of the garage?
11. Is cut or fill required at the southwest corner of the Lake House? How much?
12. What is the elevation at the northeast corner of the site?

Locating the Building

UNIT 10

Objectives

After completing this unit, you will be able to perform the following tasks:

- Lay out building lines according to a site plan.
- Use the 6-8-10 or equal-diagonals method to check the squareness of corners.
- Use a laser level to measure the depths of excavations.

Laying Out Building Lines

The position of the building is shown on the site plan. Dimensions show the distance from the street (or lake) to the building and from the side boundaries to the building. The location of one corner can easily be found by measuring with a long (100′ or 200′) steel tape. Refer to **Figure 10–1** as you read the following directions for finding the corners of a rectangular building. The lot boundaries are represented by A, B, C, and D. Along boundary line A-C measure the distance from the front edge of the property, usually a street. Mark this point with a stake (e). As you drive each stake, drive a nail in the top of the stake to accurately mark the location. Measure the same distance along the other side boundary (B-D) and mark that point with stake f. Stretch a string from stake e to stake f. The front of the building will fall on this string. The distance from the front boundary to the building line is called the front ***setback***. Both the front and side setbacks are nearly always regulated by zoning ordinances. Next measure the side setback, which is the distance from the side of the property to the side of the building at stake e to side of the building (line e-G). Drive a stake where the side setback and front setback meet. This stake represents one front corner of the building (corner G). Find the other front corner by measuring the width of the building from the first corner (line G-H). Check the distance from the corners to the front boundary of the lot to make sure it agrees with the setback shown on the site plan.

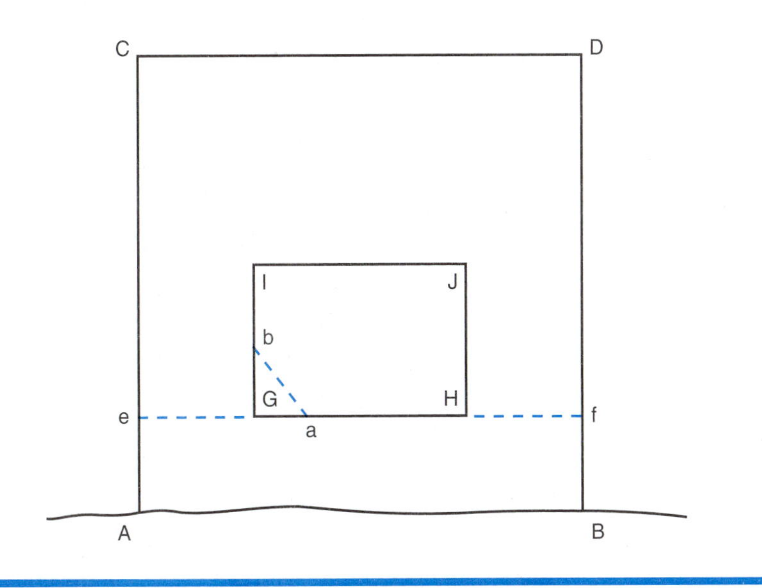

Figure 10–1. Laying out building lines.

Now measure the depth of the building from the front corners to the rear of the building (corner I). Use a framing square to make corner G as square as possible. Check the corner for square using the 6-8-10 method (see **Figure 10–2**). Measure 6 feet from the corner along line G-I to find point a. Measure 8 feet from the corner along line G-e to find point b. Line a-b should be exactly 10 feet. (See Math review 24.) Adjust the angle of the side building line as necessary to make a-b 10 feet. Drive a stake at corner I. Repeat the process to find the other rear corner J. Check to see that the rear building line (I-J) is the dimension shown on the plans. Make adjustments as necessary until all corners are square and all dimensions are correct.

Now check the final layout to ensure that rectangles have 90-degree corners by measuring the diagonals (see **Figure 10-3**). When the diagonals are equal, the corners of the rectangle are square.

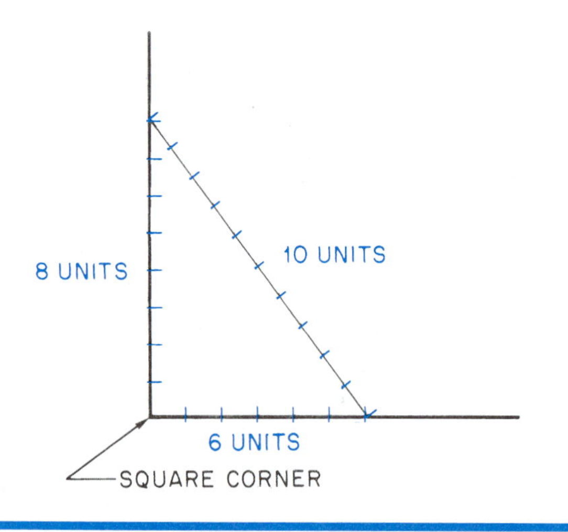

Figure 10–2. The 6-8-10 method of checking a square corner.

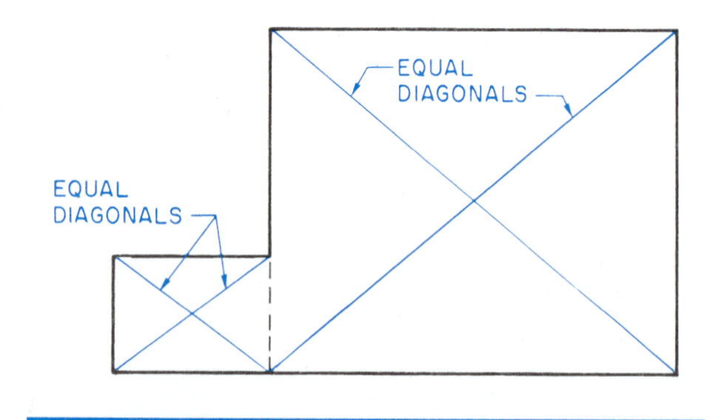

Figure 10–3. When the diagonals of a rectangle are the same length, the corners are square

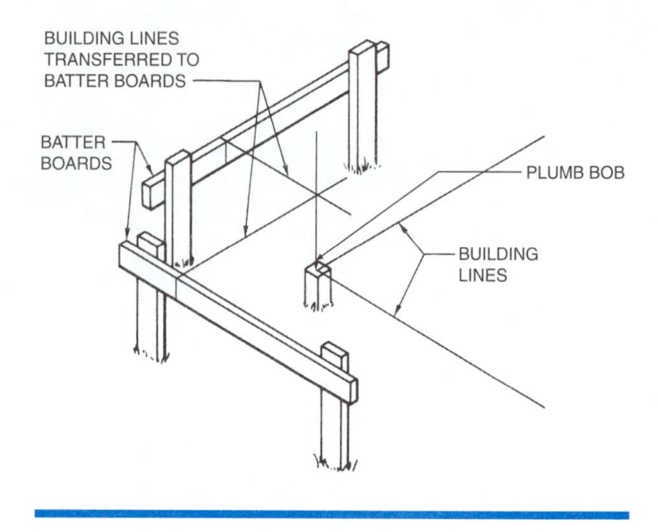

Figure 10–4. Batter boards.

The ***building lines*** can be saved, even after the corner stakes are removed for earthwork, by erecting batter boards (see **Figure 10–4**). ***Batter boards*** are sturdy horizontal boards fastened between 2 × 4 stakes, at least 4 feet outside the building lines. The building lines are extended and marked on the batter boards.

Excavating

Most buildings require some *excavation* (digging) to prepare the site for a foundation. The depth of the excavation is measured from a fixed benchmark. A benchmark can be any stationary object such as a surveyed point on a street or very large boulder that cannot be moved. All elevations (vertical distances) are measured from this benchmark. Only in the case of a real coincidence would the benchmark be at the same elevation as the surface of the ground where the excavation is to be done. The actual depth of the excavation is the difference between the elevation at the surface of the ground and the elevation at the bottom of the excavation (see **Figure 10–5**).

Concrete footings are placed in the bottom of the excavation to support the entire weight of the building (see **Figure 10–6**). These footings are placed on unexcavated earth to reduce the chance of the soil compacting under them. This means that the excavation contractor must measure the depth of the excavation accurately. The footings may be *stepped,* as in **Figure 10–7**, to accommodate a sloping site. This requires measuring the depth at each step of the footing. Information about

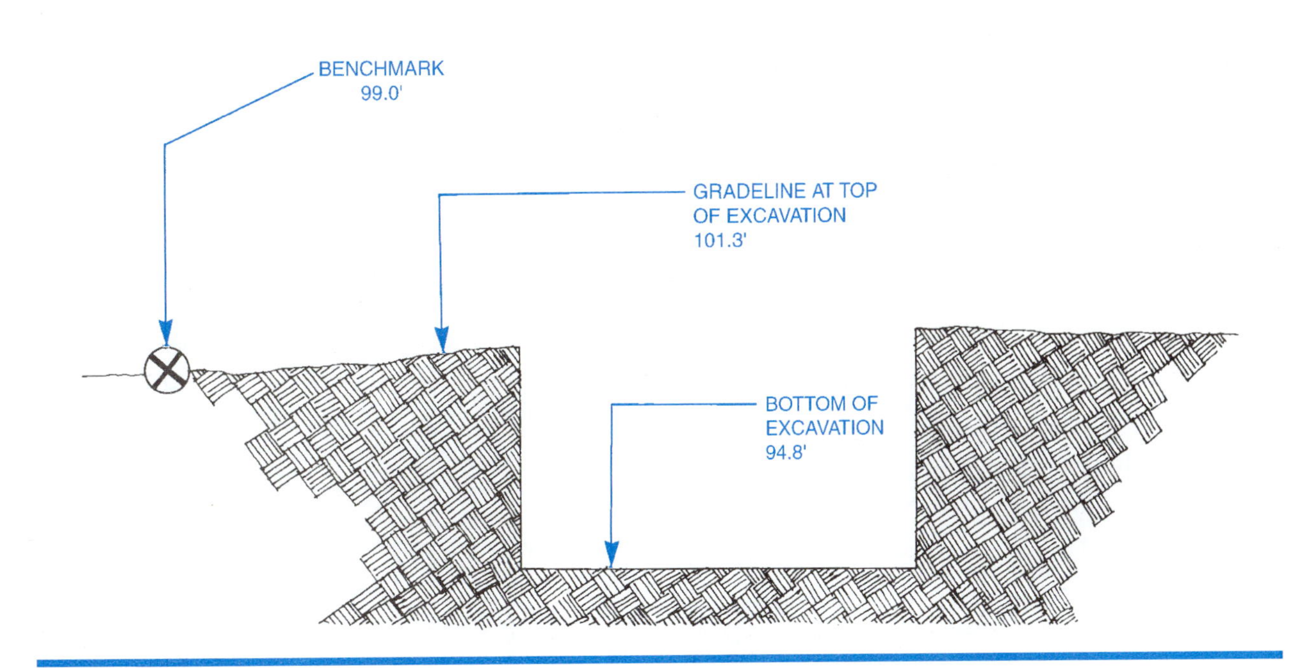

Figure 10–5. The top of this excavation is 2.3 feet above the benchmark. Its depth is 6.5 feet (101.3′ – 94.8′).

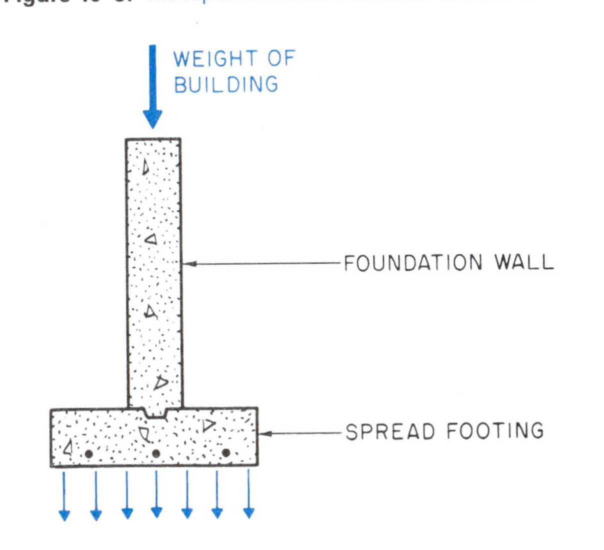

Figure 10–6. The footing spreads the weight of the building over a greater soil area.

GREEN NOTE

The different soil types found at varying depths during an excavation can be surprising. In some areas, there may only be a few inches of top soil to support healthy plant growth. Below that layer can be heavy clay, which contains few nutrients and does not drain well, or gravel that also has few nutrients and does not hold water at all. For this reason, the top soil should be stripped from the area to be excavated and stock piled in a safe place to be used later where landscaping and lawns are planted.

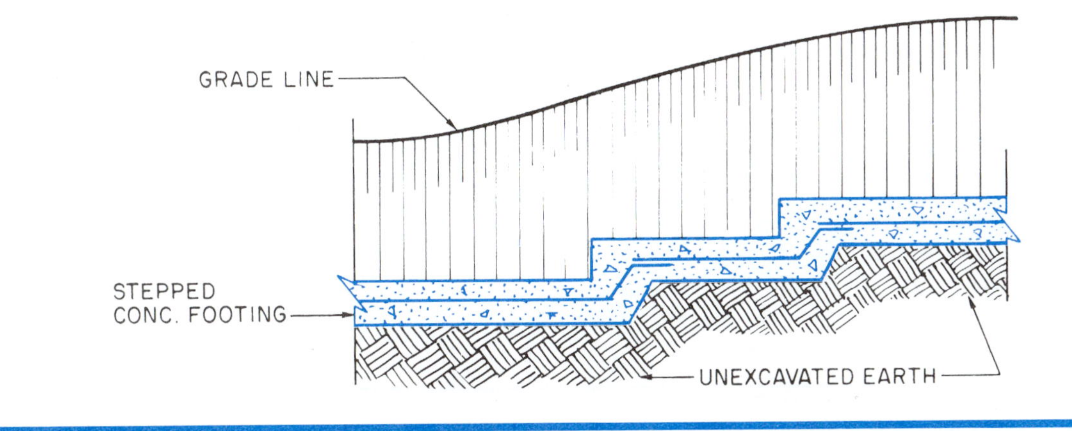

Figure 10–7. Footings can be stopped to accommodate a sloping site.

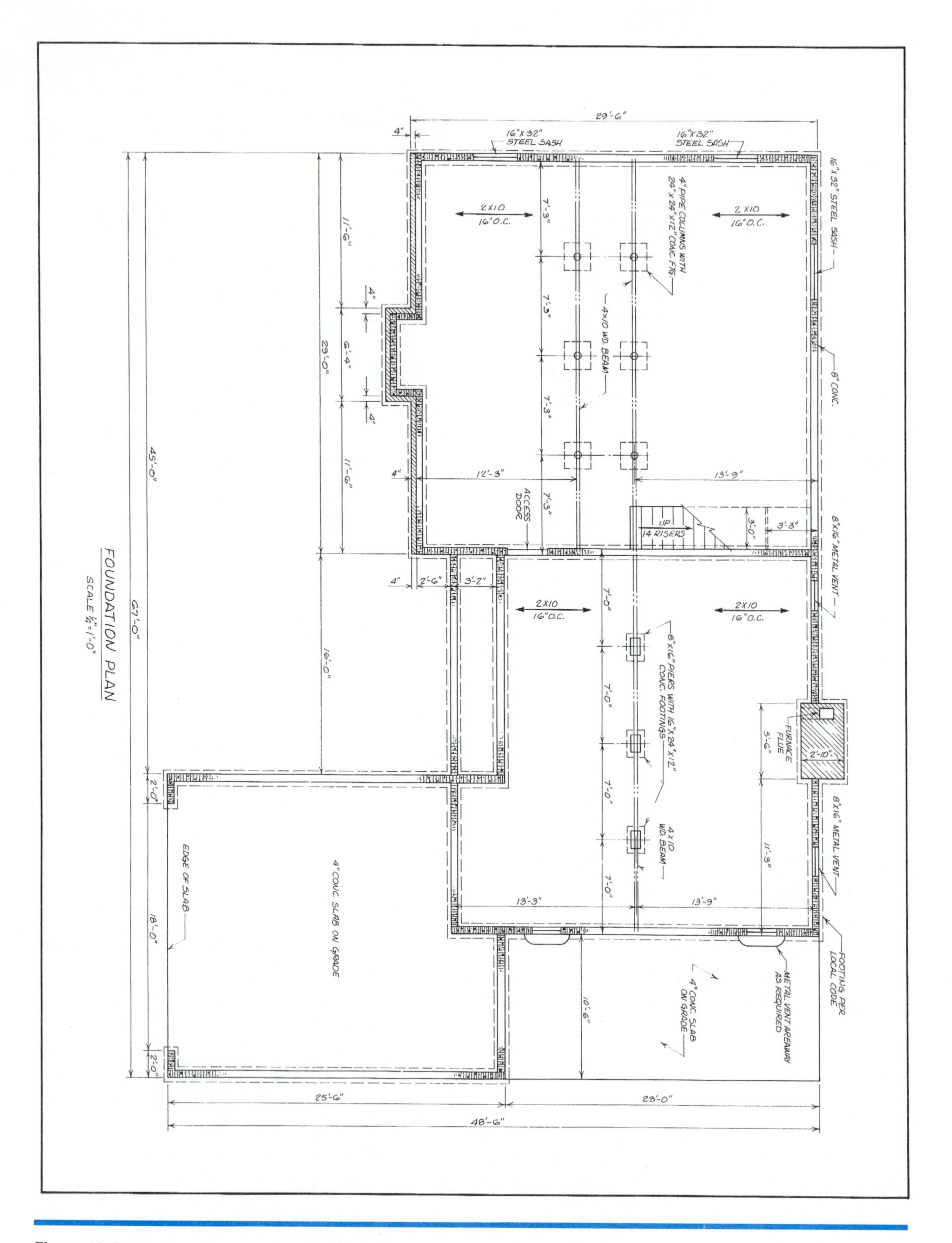

Figure 10–8. The foundation plan gives complete dimensions.

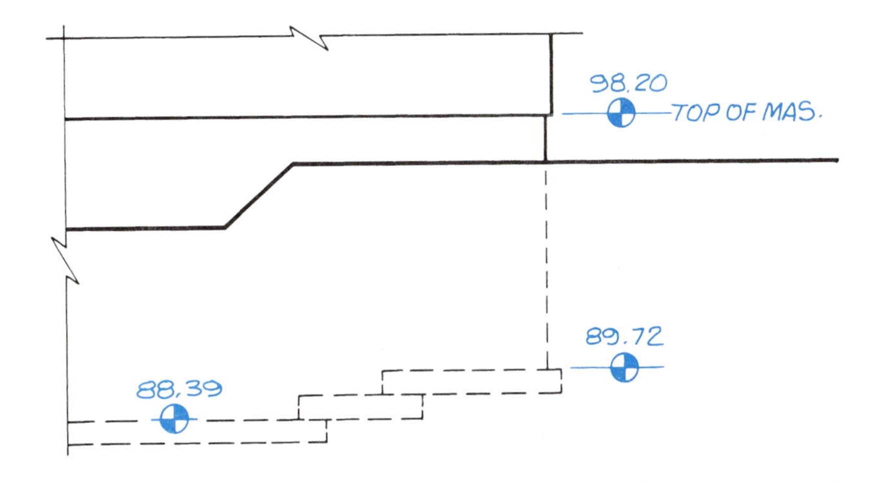

Figure 10–9. Spot elevations show key points on the footing and the foundation.

the footing design is found on the foundation plan and building elevations. The layout of the foundation walls and their footings is shown on the foundation plan. The foundation walls are shown by two solid lines with dimensions to indicate their sizes (see **Figure 10–8**). A dotted line on each side of the foundation indicates the concrete footing. The size of the footing may be omitted when the plan is developed for use in several locations. If the footing size is not indicated on the foundation plan, a consulting engineer will have to determine the size in accordance with the local building code and the conditions at the site.

The depth of the foundation, including its footing, is shown on the elevations. To simplify calculating excavation depths, many architects indicate the elevations as key points along the footings (see **Figure 10–9**). A section view through all or part of the building may show a typical depth, but it is wise to check all the elevations for steps in the footing. The footings may be shown on the elevation as a double or a single dotted line. In masonry foundations, steps in footings are usually in increments of 8 inches to conform to standard concrete block sizes.

Note: A laser level is often used to measure elevations and depths of excavations. Setting up a laser level is similar to setting up a builder's level. You should follow the manufacturer's instructions and obey the safety precautions provided for the use of the laser level.

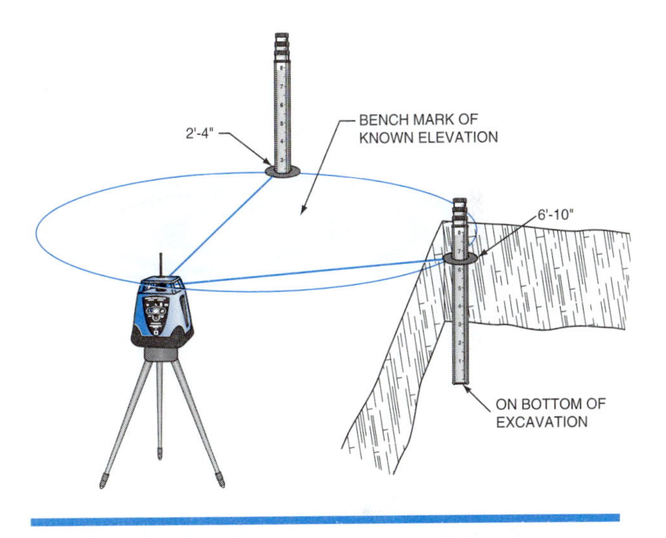

Figure 10–10. This excavation is 4'-6' deep—the difference between the two readings on the target rod.

Use the following procedure to measure differences in elevation, such as the depth of an excavation:

Step 1. Set the instrument up on a tripod, and level it in a convenient location.

Step 2. Hold the target rod on a known elevation, such as a benchmark or ground of known elevation. Note the reading on the target rod where the target registers the laser beam.

Step 3. Take a similar reading with the target rod at the bottom of the excavation. Subtract the target reading at the benchmark from the reading at the bottom of the excavation to find the difference in elevation at the benchmark and at the bottom of the excavation, **Figure 10–10.**

USING WHAT YOU LEARNED

In the old days, before people had much awareness of how site conditions affected the livability of a house, it was common for a builder to begin by removing all of the trees that were near the planned building. It was thought that those trees would just be in the way. Today we are much more conscious of how the site affects the performance of the building. What would be the result in how the house functions if all of the oak and maple trees near the front of the house were removed? Those trees, being on the south side of the house, provide cooling shade in the summer, when they have all of their leaves. They also help hold the soil in place on the steep slope down to the lake.

Assignment

Refer to the Lake House drawings (in the packet) to complete this assignment.

1. What is the distance from the Lake House to the nearest property boundary? (Do not treat the decks as part of the house for this question.)
2. What is the distance from the Lake House to the lake?
3. What is the distance from the north property line to the garage?
4. What is the area of the basement of the Lake House, including the foundation? Ignore slight irregularities in the shape of the foundation. For ease in calculating, divide the foundation into rectangles, **Figure 10–11.**
5. Find the highest and lowest natural grades meeting the house.
6. How much cut or fill is required for the basement at the northeast corner of the house?
7. Measuring from the natural grade, how deep is the excavation for the footing under the overhead garage door?
8. What is the elevation at the bottom of the deepest excavation for the Lake House? (Do not include the garage.)

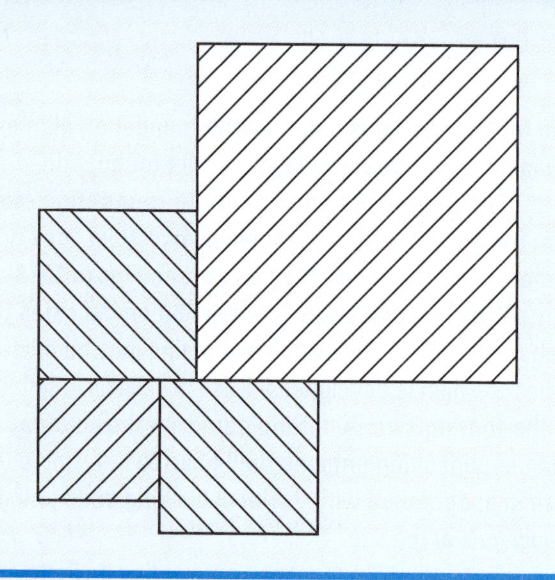

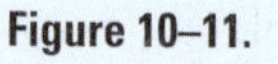

Figure 10–11.

9. Why would a row of large evergreen trees between the Lake House and the lake decrease the energy efficiency of the house?
10. What aspect of the location of a building is most often regulated by local ordinances?

Site Utilities

UNIT 11

Objectives

After completing this unit, you will be able to perform the following tasks:

- Interpret symbols and notes used to describe site utilities.
- Explain the septic system indicated on a site plan.
- Determine the pitch of drain lines.

Sewer Drains

The *building sewer* carries the waste to the municipal sewer or septic system (see **Figure 11–1**). Because sewer lines usually rely on gravity flow, they are large in diameter (4 inches, minimum) and are pitched to provide flow. Because water supply lines and gas lines are pressurized, pitch is not important in their installation. Therefore, the sewer is installed first, and other piping is routed around it as necessary. The size, material, and pitch of drains are usually given in a note on the site plan (see **Figure 11–2**). The pitch of a pipe is given in fractions of an inch per foot. A pitch of ¼ inch per foot means that for every horizontal foot, the pipe rises or falls ¼ inch.

In some cases, sewers may have to flow uphill. This is the case with the Lake House. Uphill flow is accomplished by a *grinder pump* (see **Figure 11–3**). A grinder pump grinds solids into small enough particles to be pumped and pumps the sewage at low pressure.

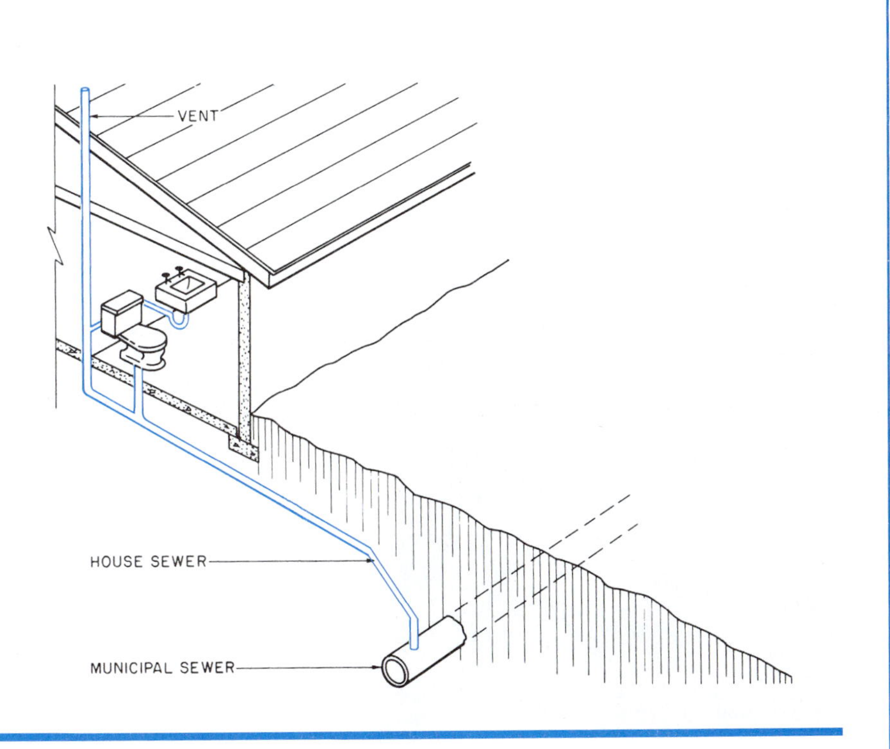

Figure 11–1. The house sewer carries waste from the house to the municipal sewer or septic systems.

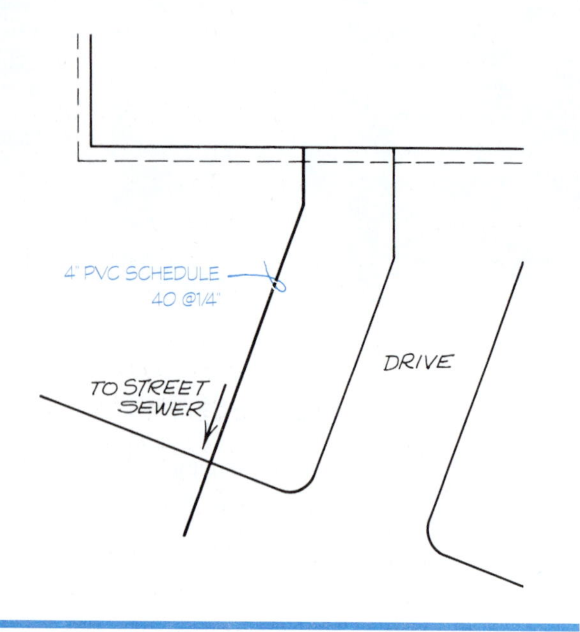

Figure 11–2. This note indicates cast iron pipe sloped ¼ inch for every foot of run.

Building Sewer

Any plumbing that is to be concealed by concrete work must be installed without fixtures or *roughed in* before the concrete is placed. Because the plumbing contractor installs all plumbing inside the building lines, this phase of construction is discussed later with mechanical systems. The sewer, however, is usually installed *after* the building is erected to prevent damage due to machinery on the site during construction. The sewer, too, is generally considered a site utility, which is why it is discussed here.

The workers who install the sewer must be able to determine the elevation at which it passes through the foundation and the pitch of the line outside the building. The sewer line may be shown on plans as a solid or broken line. Although it is usually labeled, this is not always true. When the sewer is not labeled as such, it can still be recognized by its material, pitch, and ending place. In light construction, the sewer is usually the

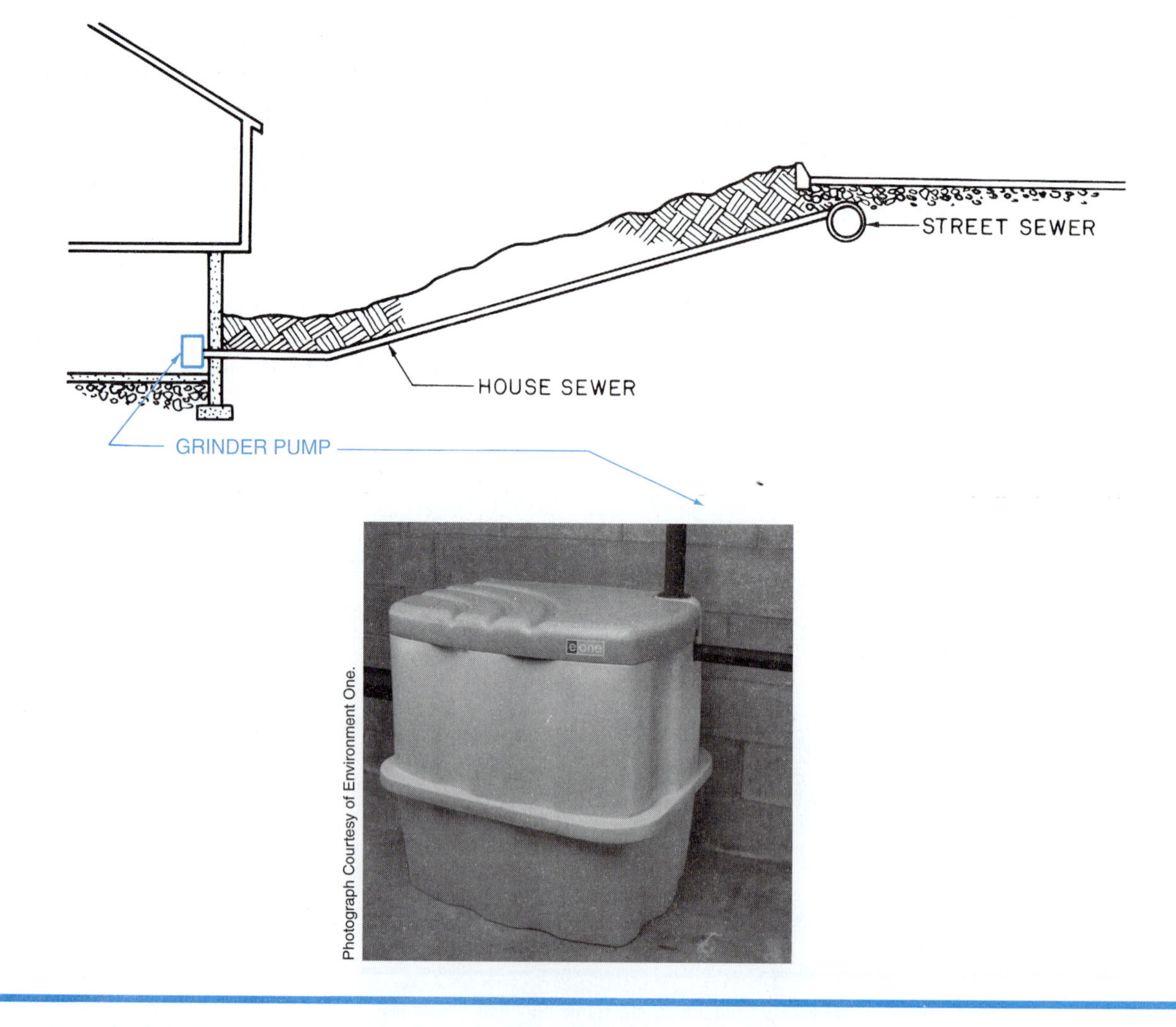

Figure 11–3. A grinder is a pump that can move sewage uphill.

only 4-inch pipe to the building. Also, the sewer is the only line with the pitch indicated.

Municipal Sewers and Septic Systems

In most developed areas, the building sewer empties into a municipal sewer line near the street. The contractor for the new building is responsible for everything from the municipal sewer to the house.

In less developed areas, the sewer carries the sewage to a septic system. The most common type of *septic system* includes a septic tank and drain field (see **Figure 11–4**). The septic tank holds the solid waste while it is decomposed by bacterial action. The liquids pass through the baffles and flow out of the tank to the distribution box. The distribution box (DB) diverts the liquid into *leach lines.* These perforated plastic or loose-fitting tile lines allow the liquid to be absorbed by the surrounding soil. The liquid gradually evaporates from or drains through the soil. The *drain field,* where the leach lines are laid, is usually made of a layer of crushed stone.

The design of septic systems is closely regulated by most health and plumbing codes. These local codes should be checked before designing or installing any septic system. The building code or health department code often requires a *percolation test* before the system can be approved. In a *percolation test,* holes are dug. Then a measured amount of water is poured into each hole. The amount of time required for the water to drain into the soil is an indication of how well the soil

GREEN NOTE

There are several materials used to make pipes and fittings. Traditionally sewer pipes were cast iron and supply pipes were galvanized iron or steel. As polyvinyl chloride (PVC) became available, it became the most common material for drainage, waste, and vent piping and for sewer pipes. It requires less energy to manufacture than cast iron or steel, it lasts almost indefinitely, and it requires less time to install. The downside is that it does not withstand heat, so it can't be used for hot water.

Copper, with soldered fittings, has been widely used for supply piping, because it is noncorrosive and will last much longer than iron or steel pipe. Well water is sometimes slightly acidic and the acid does corrode copper pipe, but that is not usually a concern because municipal water is treated to control the acidity.

In recent years, copper has become very expensive and two kinds of plastic pipe have been developed for use where copper has been the standard. Chlorinated polyvinyl chloride (CPVC) is inexpensive, noncorrosive, and easy to install. Cross-linked polyethylene (PEX) is more flexible, so it requires fewer fitting and is quick to install. Both CPVC and PEX are widely used in home construction today.

Manufacture of all plastic piping requires petrochemicals, but as a percentage of the total materials in a home the amount of piping material is very small, plastic pipe is not considered to have much of an environmental impact. There is no one best material for piping in green home construction.

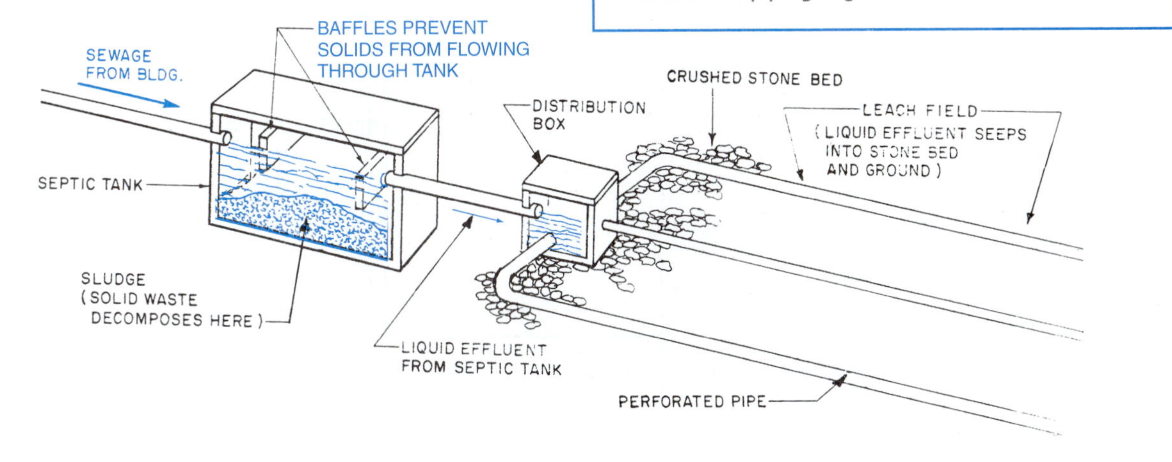

Figure 11–4. Septic system.

will accept water from the septic system. This ability to accept water is called *percolation.* The locations of key elements in the system are often shown on the site plan.

To ensure proper drainage for a septic system, the elevation of its drainage piping is critical. For this reason, it is necessary to know the *invert elevation* of a pipe, which is the elevation at the lowest point inside the pipe (see **Figure 11–5**).

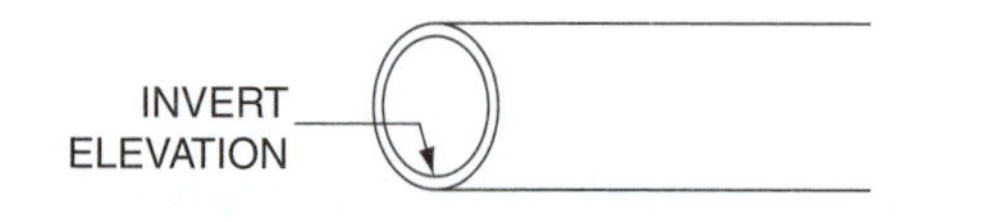

Figure 11–5. Invert elevation is the elevation at the lowest point inside the pipe.

Other Piping

Other utility piping, such as that for water supply or gas, is shown on the site plan. If these lines will pass beneath the concrete footings or be concealed under a concrete slab, they must be roughed in before the concrete is placed. Water supply pipes follow the most direct route from the municipal water main or well to the main shut-off valve or pump. Gas lines run from the main to the gas meter. All supply lines on the plot plan should be labeled according to type and size of piping (see **Figure 11–6**).

Electrical Service

The electrical service is the wiring that brings electricity to the house. There are two types of electrical service: overhead and underground (or buried).

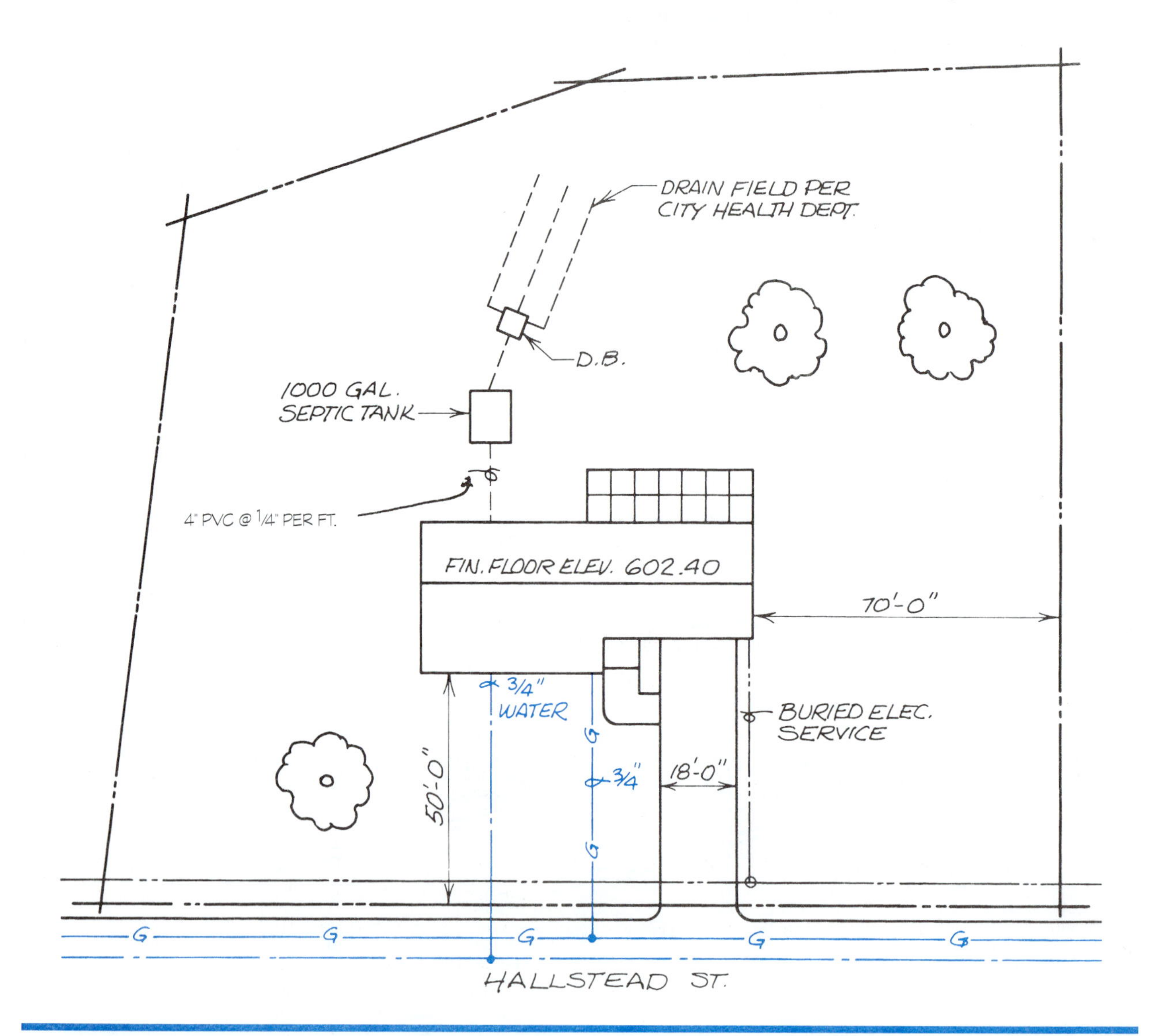

Figure 11–6. Partial site plan with utilities indicated.

Overhead service involves a cable from the utility company transformer or pole to a weather-tight fixture called a *service head* or *mast head* on the house (see **Figure 11–7**). The service head is mounted on the top of a pipe, which serves as a conduit to the meter receptacle. In an underground service, the cable is buried (see **Figure 11–8**). Although electrical service is a site utility, it is not usually installed until the building is enclosed.

The electrical service to residential and small commercial buildings is similar, as previously explained and as shown in **Figures 11–7** and **11–8.** This type of service has only one distribution point, the main electrical panel. Heavy commercial and industrial electrical services require multiconductor feeders, and typically these are underground services from a utility transformer to a service entrance section in the designated electrical or utility room. Additional utility coordination will be required within the building, which is discussed in Unit 14. The placement of these underground electrical utility lines must be coordinated with the locations of other site and building utilities prior to starting installation.

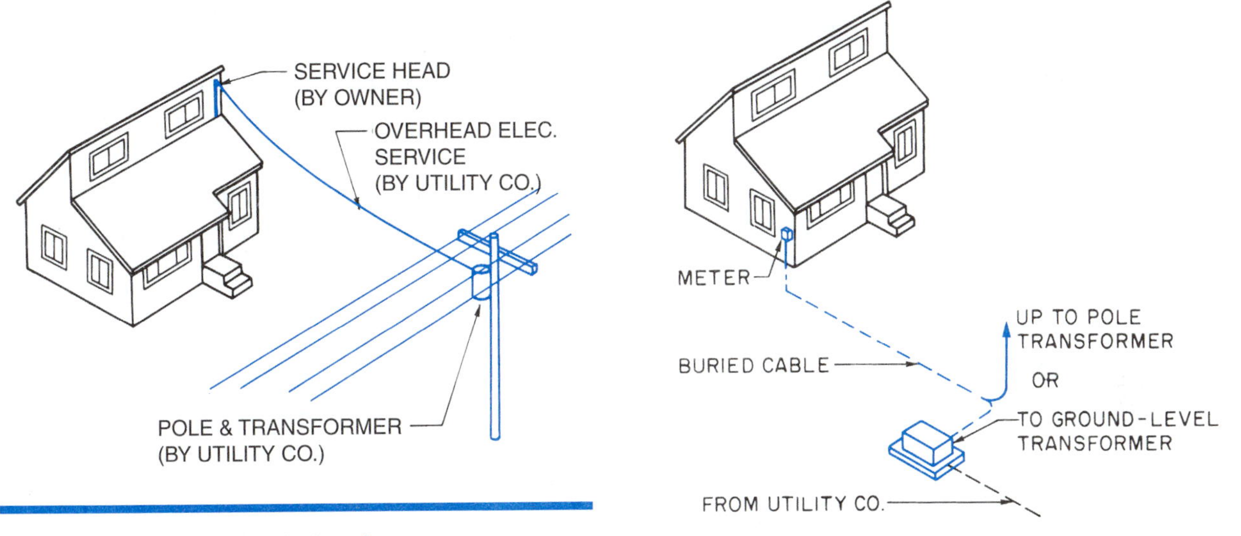

Figure 11–7. Overhead electrical service.

Figure 11–8. Underground electrical service.

USING WHAT YOU LEARNED

All building trades workers need to have an understanding of where the utilities will be located on the site. It is obvious that plumbers need to know where the water supply and sewer enter the building, but masons and carpenters also need this information. For example, if the piping is to run through a sleeve in the concrete, that sleeve must be in place before the concrete is placed. Where does the water supply enter the Lake House? The site plan shows it entering through the north wall of the play room. This is a case where the architect should be consulted to determine if it would be better to run the water supply closer to the sewer pipe, so it enters the utility room.

Assignment

Refer to the Lake House drawings (in the packet) to complete the assignment.

1. What size is the sewer for the Lake House?
2. How many lineal feet are required from the foundation wall to the septic tank?
3. What is the rise of the sewer from the house to the septic tank?
4. Where does the sewer pass through the foundation?
5. How many lineal feet of perforated pipe are needed for the drain field?
6. How many cubic yards of crushed stone are needed for the drain field?
7. What is the invert elevation of the sewer pipe at the point it enters the house?

UNIT 12 Footings

Objectives

After completing this unit, you will be able to perform the following tasks:

- Find all information on a set of drawings pertaining to footing design.
- Interpret drawings for stepped footings used to accommodate changes in elevation.
- Discuss applicable building codes pertaining to footing design.

All soil changes shape under force. When the tremendous weight of a building is placed on soil, the soil tends to compress under the foundation walls and allow the building to settle. To prevent settling, concrete footings are used to spread the weight of the building over more area. The footings distribute the weight of the building, so that there is less force per square foot of area.

The simplest type of footing used in residential construction is referred to as *slab-on-grade.* In this system, the main floor of the building is a single concrete slab, reinforced with steel to prevent cracking. The walls are erected on this slab so that it supports the weight of the building (see **Figure 12–1**). Slab-on-grade foundations are common in warm climates. This type of construction is indicated on the floor plan by a note (see **Figure 12–2**) and on section views of the construction (see **Figure 12–3**). If excavation is involved in the construction where a slab-on-grade is to be placed, it is very important to thoroughly compact all loose fill before placing the concrete. Tamping the fill prevents the soil from compacting under the concrete later, causing the concrete to settle or crack.

A thickened slab, sometimes called a haunch, is used to further strengthen the slab where concentrated weight, such as a bearing wall, will be located. A *haunch* is an extra thick portion of the slab that is made by ditching the earth before the concrete is placed (see **Figure 12–4**).

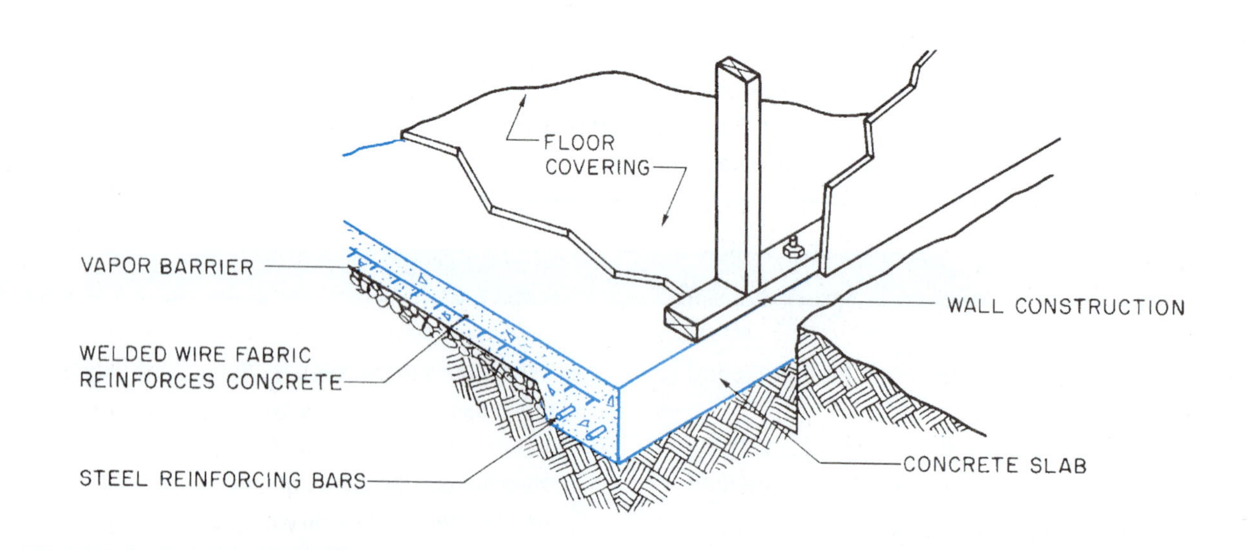

Figure 12–1. Slab-on-grade foundation.

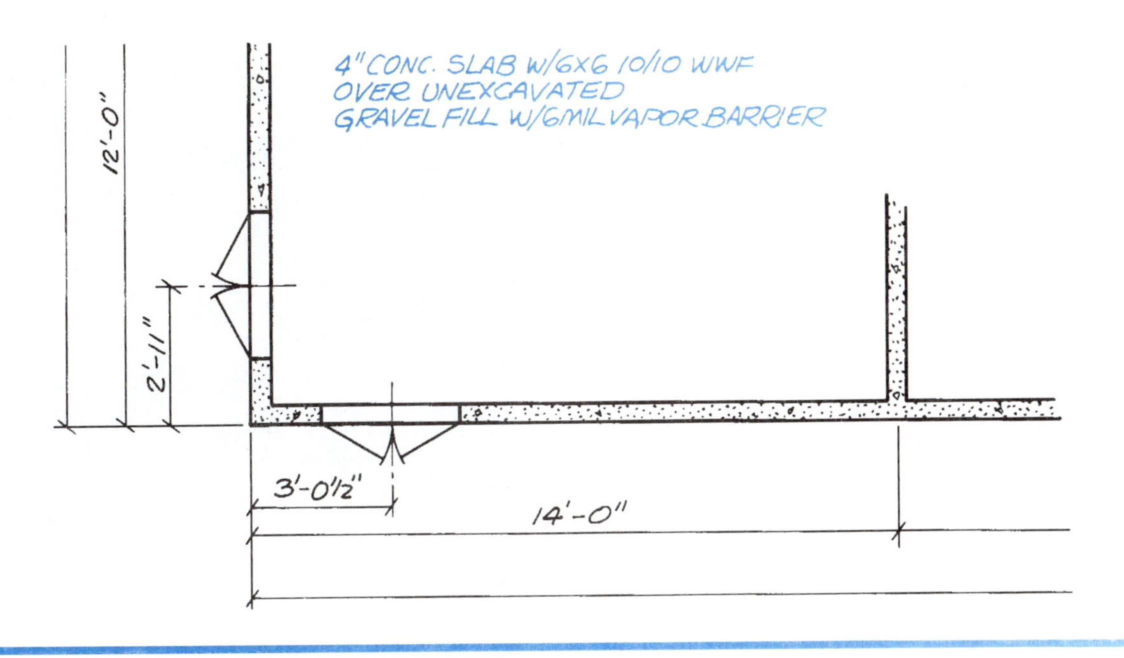

Figure 12–2. Note indicating slab-on-grade construction.

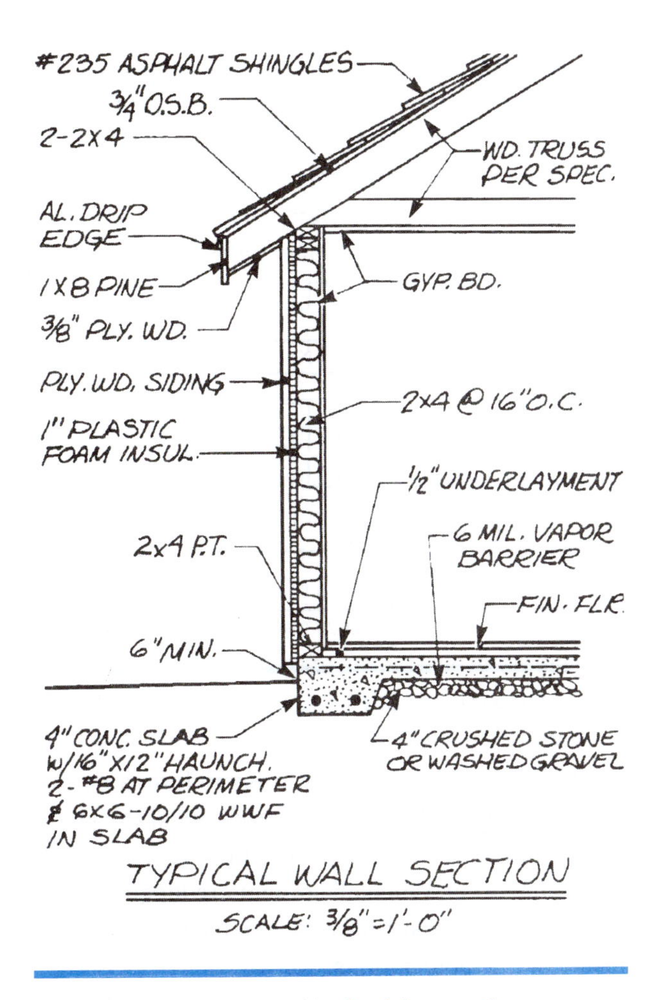

Figure 12–3. Typical wall section for slab-on-grade construction.

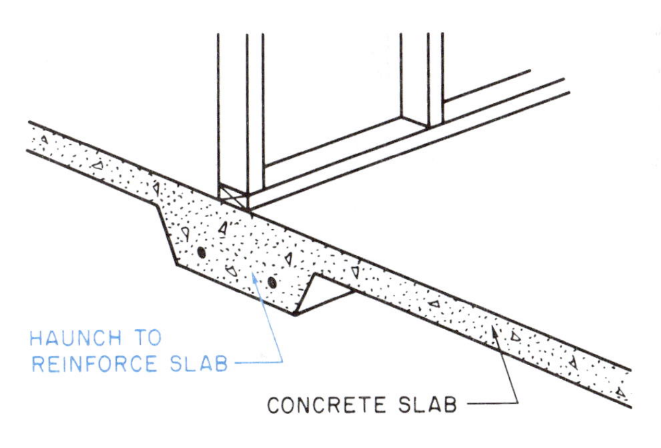

Figure 12–4. A haunch is a thickened part of a slab to reinforce it under a load-bearing wall.

Spread Footings

In most sections of the country, the foundation of the house rests on a footing separate from the concrete floor. This separate concrete footing is called a *spread footing* because it spreads the force of the foundation wall over a wider area (see **Figure 12–5**). Spread footings can be made by placing concrete inside wooden or metal forms (see **Figure 12–6**). The footing is shown on the foundation or basement plan by dotted lines outside the foundation wall lines (see **Figure 12–7**).

The dimensions of the footings can be determined from the dimensions shown for the foundation.

TABLE R403.1(3)
MINIMUM WIDTH AND THICKNESS FOR CONCRETE FOOTINGS
WITH CAST-IN-PLACE CONCRETE OR FULLY GROUTED MASONRY WALL CONSTRUCTION (inches)[a, b]

SNOW LOAD OR ROOF LIVE LOAD	STORY AND TYPE OF STRUCTURE WITH CMU	LOAD-BEARING VALUE OF SOIL (psf)					
		1500	2000	2500	3000	3500	4000
20 psf	1 story—slab-on-grade	14 × 6	12 × 6	12 × 6	12 × 6	12 × 6	12 × 6
	1 story—with crawl space	19 × 6	14 × 6	12 × 6	12 × 6	12 × 6	12 × 6
	1 story—plus basement	25 × 8	19 × 6	15 × 6	13 × 6	12 × 6	12 × 6
	2 story—slab-on-grade	23 × 7	18 × 6	14 × 6	12 × 6	12 × 6	12 × 6
	2 story—with crawl space	29 × 9	22 × 6	17 × 6	14 × 6	12 × 6	12 × 6
	2 story—plus basement	35 × 12	26 × 8	21 × 6	17 × 6	15 × 6	13 × 6
	3 story—slab-on-grade	32× 11	24 × 7	19 × 6	16 × 6	14 × 6	12 × 6
	3 story—with crawl space	38 × 14	28 × 9	23 × 6	19 × 6	16 × 6	14 × 6
	3 story—plus basement	43 × 17	33 × 11	26 × 8	22 × 6	19 × 6	16 × 6
30 psf	1 story—slab-on-grade	15 × 6	12 × 6	12 × 6	12 × 6	12 × 6	12 × 6
	1 story—with crawl space	20 × 6	15 × 6	12 × 6	12 × 6	12 × 6	12 × 6
	1 story—plus basement	26 × 8	20 × 6	16 × 6	13 × 6	12 × 6	12 × 6
	2 story—slab-on-grade	24 × 7	18 × 6	15 × 6	12 × 6	12 × 6	12 × 6
	2 story—with crawl space	30 × 10	22 × 6	18 × 6	15 × 6	13 × 6	12 × 6
	2 story—plus basement	36 × 13	27 × 8	21 × 6	18 × 6	15 × 6	13 × 6
	3 story—slab-on-grade	33 × 12	25 × 7	20 × 6	17 × 6	14 × 6	12 × 6
	3 story—with crawl space	39 × 14	29 × 9	23 × 7	19 × 6	17 × 6	14 × 6
	3 story—plus basement	44 × 17	33 × 12	27× 8	22 × 6	19 × 6	17 × 6
50 psf	1 story—slab-on-grade	17 × 6	13 × 6	12 × 6	12 × 6	12 × 6	12 × 6
	1 story—with crawl space	22 × 6	17 × 6	13 × 6	12 × 6	12 × 6	12 × 6
	1 story—plus basement	28× 9	21 × 6	17 × 6	14 × 6	12 × 6	12 × 6
	2 story—slab-on-grade	27 × 8	20 × 6	16 × 6	13 × 6	12 × 6	12 × 6
	2 story—with crawl space	32 × 11	24 × 7	19 × 6	16 × 6	14 × 6	12 × 6
	2 story—plus basement	38 × 14	28 × 9	23 × 6	19 × 6	16 × 6	14 × 6
	3 story—slab-on-grade	35 × 13	27 × 8	21 × 6	18 × 6	15 × 6	13 × 6
	3 story—with crawl space	41 × 15	31 × 10	24 × 7	20 × 6	17 × 6	15 × 6
	3 story—plus basement	47 × 18	35 × 12	28 × 9	23 × 7	20 × 6	17 × 6
70 psf	1 story—slab-on-grade	19 × 6	14 × 6	12 × 6	12 × 6	12 × 6	12 × 6
	1 story—with crawl space	25 × 7	18 × 6	15 × 6	12 × 6	12 × 6	12 × 6
	1 story—plus basement	30 × 10	23 × 6	18 × 6	15 × 6	13 × 6	12 × 6
	2 story—slab-on-grade	29 × 9	22 × 6	17 × 6	14 × 6	12 × 6	12 × 6
	2 story—with crawl space	34 × 12	26 × 8	21 × 6	17 × 6	15 × 6	13 × 6
	2 story—plus basement	40 × 15	30 × 10	24 × 7	20 × 6	17 × 6	15 × 6
	3 story—slab-on-grade	38 × 14	28 × 9	23 × 6	19 × 6	16 × 6	14 × 6
	3 story—with crawl space	43 × 16	32 × 11	26 × 8	21 × 6	18 × 6	16 × 6
	3 story—plus basement	49 × 19	37 × 13	29 × 10	24 × 7	21 × 6	18 × 6

For SI: 1 inch = 25.4 mm, 1 plf = 14.6 N/m, 1 pound per square foot = 47.9 N/m^2.

a. Interpolation allowed. Extrapolation is not allowed.

b. Based on 32-foot-wide house with load-bearing center wall that carries half of the tributary attic, and floor framing. For every 2 feet of adjustment to the width of the house add or subtract 2 inches of footing width and 1 inch of footing thickness (but not less than 6 inches thick).

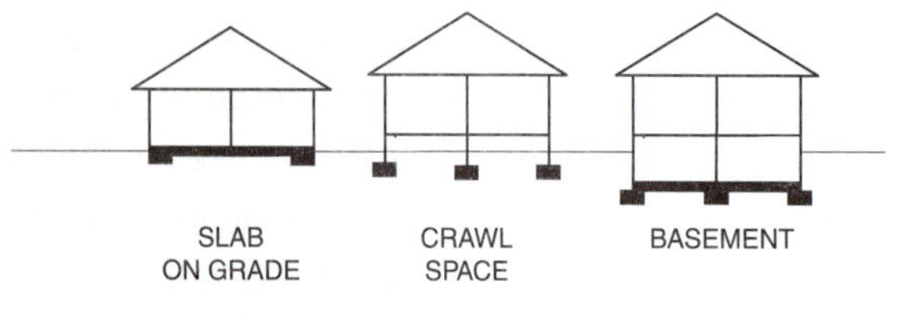

Figure 12–9. This is a table from the *International Residential Code*®. The Code also includes a table for light frame construction with brick veneer and those with hollow masonry foundations.

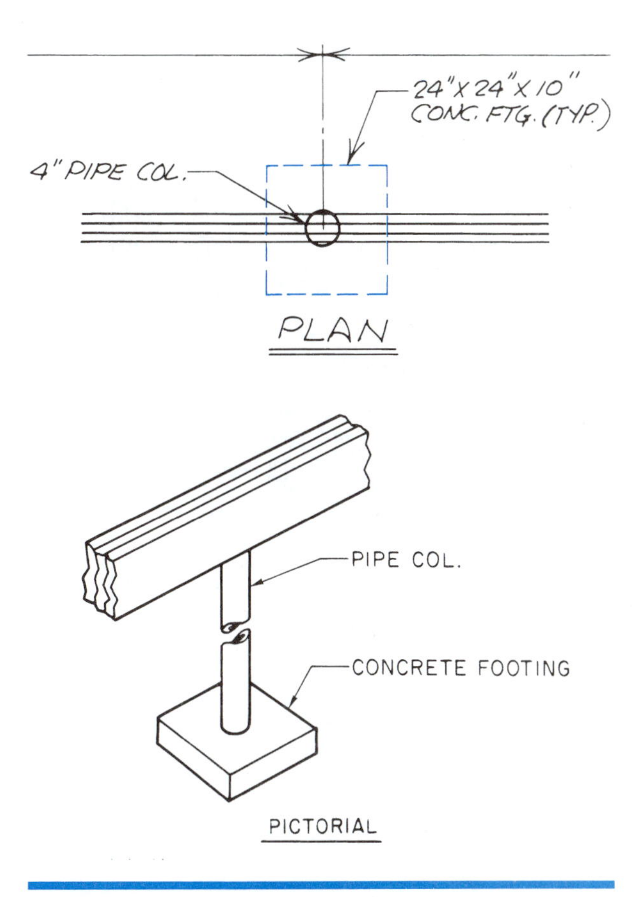

Figure 12–10. Column footings appear as a rectangle of broken lines on the plan.

Reinforcement

Footings, column pads, and other structural concrete frequently include steel reinforcement. Footing reinforcement is normally in the form of steel reinforcement bars, commonly called *rebars*. Reinforcement bars are designated by their diameters in eighths of an inch (see **Figure 12–11**).

Depth of Footings

In many sections of North America, the moisture in the surface of the earth freezes in the winter. As this frost forms, it causes the earth to expand. The force of this expansion is so great that if the earth under the footing of a building is allowed to freeze, it either cracks the footing or moves the building. To eliminate this problem, the footing is always placed below the depth of any possible freezing. This depth is called the ***frostline*** (see **Figure 12–12**).

The frost-depth map shown in **Figure 12–12** is only approximate and is not generally accurate enough for

REINFORCEMENT BARS

Size Designation	Diameter in Inches
3	.375
4	.500
5	.625
6	.750
7	.875
8	1.000
9	1.128
10	1.270
11	1.410
14	1.693
18	2.257

Figure 12–11. Standard sizes of rebars.

foundation design. The local building department will have specific footing depth requirements for their area.

Two methods are commonly used to indicate the elevation, or depth, of the bottom of the footings. The easiest to interpret is for spot elevations to be given at key points on the elevation drawings. Where elevations are given in this manner, the tops of the footing forms are leveled with a leveling instrument, using a benchmark for reference.

Another commonly used method is to dimension the bottom of the footing from a point of known elevation. This may be the finished floor or the top of the masonry foundation, for example. These dimensions are given on the building elevations. Footings and other features marked with a reference symbol ⯐, called a ***datum*** symbol, are to be used as reference points for other dimensions.

For example, see the South Elevation 3/3 of the Lake House (included in the packet). The left end of this view shows a footing with its bottom at an elevation of 334.83 feet. This is a variation of the usual practice of showing the elevation of the top of the footing. The top of this footing is 10 inches higher, or 335.66 feet. (See Math Review 16 in Appendix B.) What room of the Lake House is this footing under?

At the right end of the South Elevation 3/3, the footing is shown to be 5′–40″ below the finished floor and masonry. This is the basement floor, which the Site

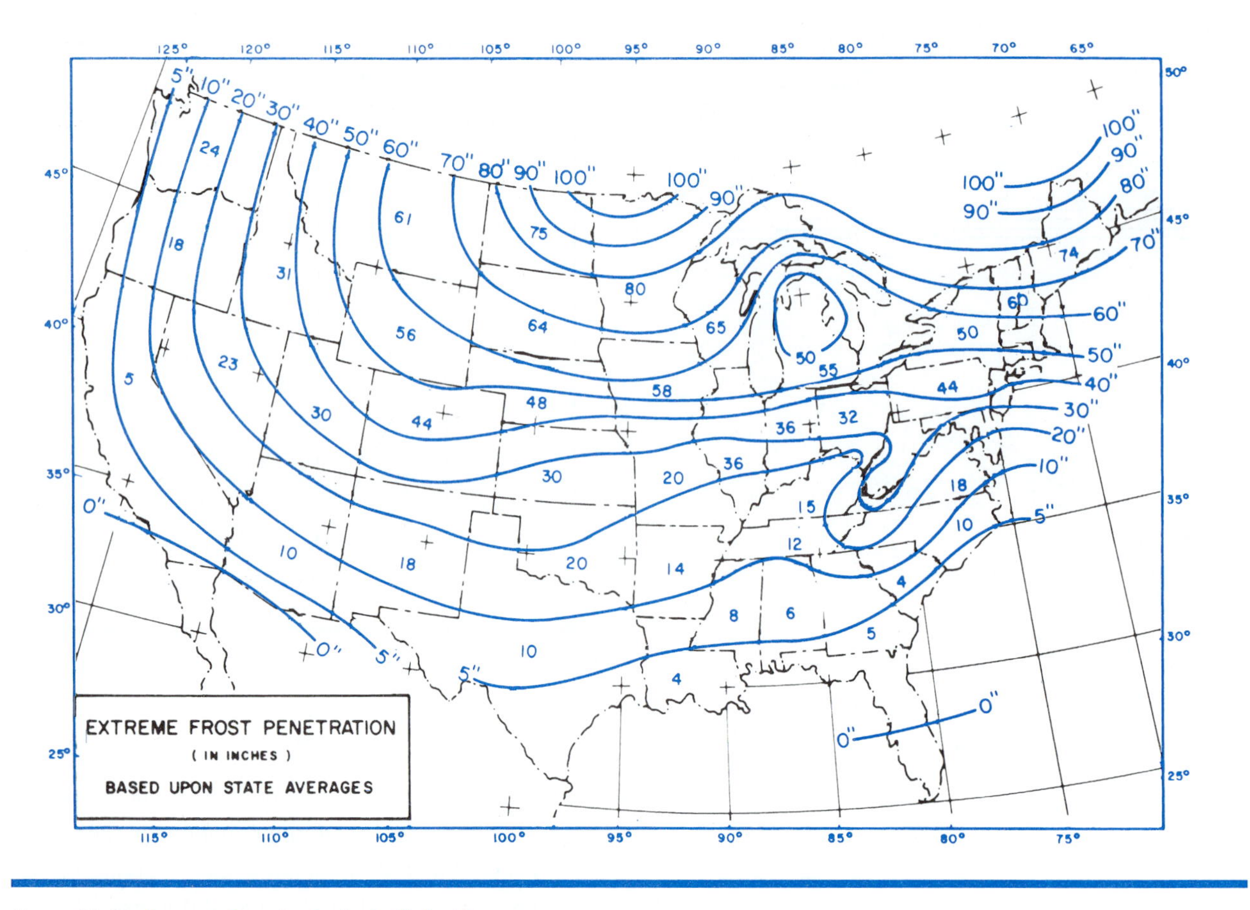

Figure 12–12. Average frost depths in the United States.

Plan shows to be at 337.0 feet. Therefore, the top of this footing is at 331′–8″ or 331.67′.

Stepped Footings

On sloping building sites, it is necessary to change the depth of the footings to accommodate the slope. This is done by stepping the footings. When concrete blocks are to be used for the foundation walls, these steps are normally in increments of 8 inches. This allows the concrete blocks to be laid so that the top of each footing step is even with a masonry course. Some buildings require several steps in the footing to accommodate steeply sloping sites. Steps in the footings are shown on the elevation drawings and on the foundation plan by a single line across the footing.

The Lake House has several steps in the footing. For example, see the east side of the garage. This is shown on the East Elevation 2/3. This step is also shown on the Foundation Plan. It is 8′–0″ from the north end of the garage. Normally the two levels of the footing should overlap each other (see **Figure 12–13**). The stepped footing shown on the East Elevation does not appear to overlap. This would be something to consult the engineer or architect about.

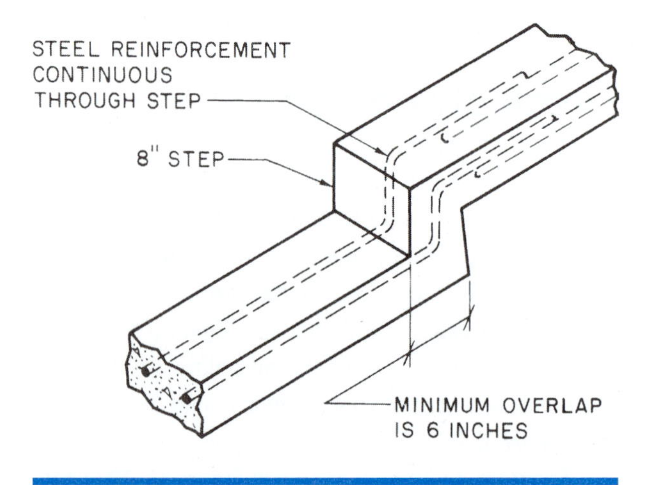

Figure 12–13. Stepped footing.

USING WHAT YOU LEARNED

Every component of a building must be installed or built in the exact location and with the exact dimensions intended by the designer. If a supporting column is only slightly too high or slightly too short, the structure that column supports will not align properly with the rest of the building. What should the elevation be at the top of the columns supporting the deck between the Lake House kitchen and the garage?

Refer to the upper level floor plan to find the elevation of the top surface of the deck. That elevation is 341.67′. Deck detail 3/6 on sheet 6 shows the sizes of materials used in the deck construction. Add all of the dimensions of the materials between the top of the column and the top of the deck: The first thing we encounter is a metal post base. Most post bases raise the wood 1 inch off the concrete, so let's assume that is the dimension in this case. In actual practice, the dimension should be confirmed by measuring the base or consulting the manufacturer's literature before the column is built. On top of that are three 2 × 10s, which would have a depth of 9¼ inches. The joists are 2 × 8s, which would have a depth of 7¼ inches. Finally, we have 1 × 6 composite decking, which is actually 1 inch thick. 1″ + 9¼″ + 7¼″ + 1″ = 18½″. Subtract 18.5″ from the elevation at the top of the deck, 341.67′, to find that the top of the column should be at 340.13′.

Assignment

Refer to the Lake House drawings in your textbook packet to complete this assignment.

1. What is the typical width and depth of the concrete footings for the Lake House?
2. What is the total length and width (outside dimensions) of the concrete footings for the garage of the Lake House? (Remember to allow for the footings to project beyond the foundation wall.)
3. How many concrete pads are shown for footings under columns or piers in the Lake House?
4. What are the dimensions of these pads?
5. What reinforcement is indicated for these pads?
6. What reinforcement is indicated for the spread footings under the Lake House?
7. What is indicated by the 2-inch dimension between the 12-inch round concrete footings?
8. What is the elevation of the top of the footing under the garage door?
9. What are the elevations of the tops of each section of concrete footing shown on the East Elevation 2/3?
10. How far outside the foundation walls are the typical footings?
11. Refer to the building code in your community (or the model code section shown in **Figure 12–9**), and list the specific differences between the Lake House footings and the minimum code requirements. Use the part of the Lake House footing that is under the greatest load. Assume a soil load-bearing capacity of 3,000 pounds per square foot and a snow load of 50 psf.

UNIT 13 Foundation Walls

Objectives

After completing this unit, you will be able to perform the following tasks:

- Determine the locations and dimensions of foundation walls indicated on a set of drawings.
- Describe special features indicated for the foundation on a set of drawings.

Laying Out the Foundation

When the ***concrete*** for the footings has hardened and the forms are removed, carpenters can begin erecting forms for concrete foundations, or masons can begin laying blocks or bricks for masonry foundations. Although the material differs, the drawings and their interpretation for each type of foundation are similar.

You referred to the dimensions on the foundation plan to lay out the footings in Unit 12. The same dimensions are used to lay out the foundation walls. The layout process is also similar. The outside surface of the foundation wall is laid out using previously constructed batter boards. Then the forms are erected or the masonry units are laid to these lines. The foundation plan includes overall dimensions, dimensions to interior corners and special constructions, and dimensions of special smaller features. It is customary to place the smallest dimensions closest to the drawing. The overall dimensions are placed around the outside of the drawing (see **Figure 13–1**).

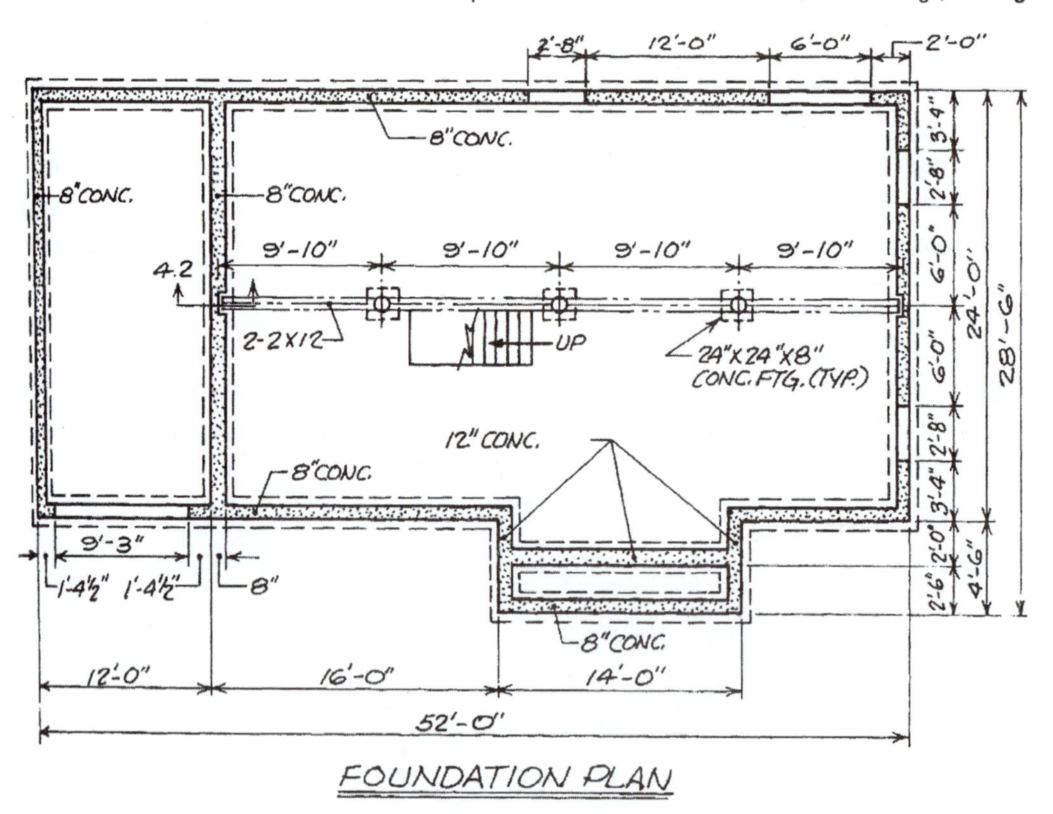

Figure 13–1. Dimensioning on a foundation plan.

All drawing sets include, at least, a wall section showing how the foundation is built, how it is secured to the footings, and any special construction at the top of the foundation wall (see **Figure 13–2**). Although a typical wall section may indicate the thickness of the foundation wall, you should carefully check around the entire wall on the foundation plan to find any notes that indicate varying thicknesses of the foundation wall. For example, the wall may be 12 inches thick where it has to support brick veneer above, while it is only 8 inches thick on the back of the building where there is no brick veneer. A careful check of the foundation plan for the Lake House shows that the house foundation is 10-inch thick concrete, except the portion under the

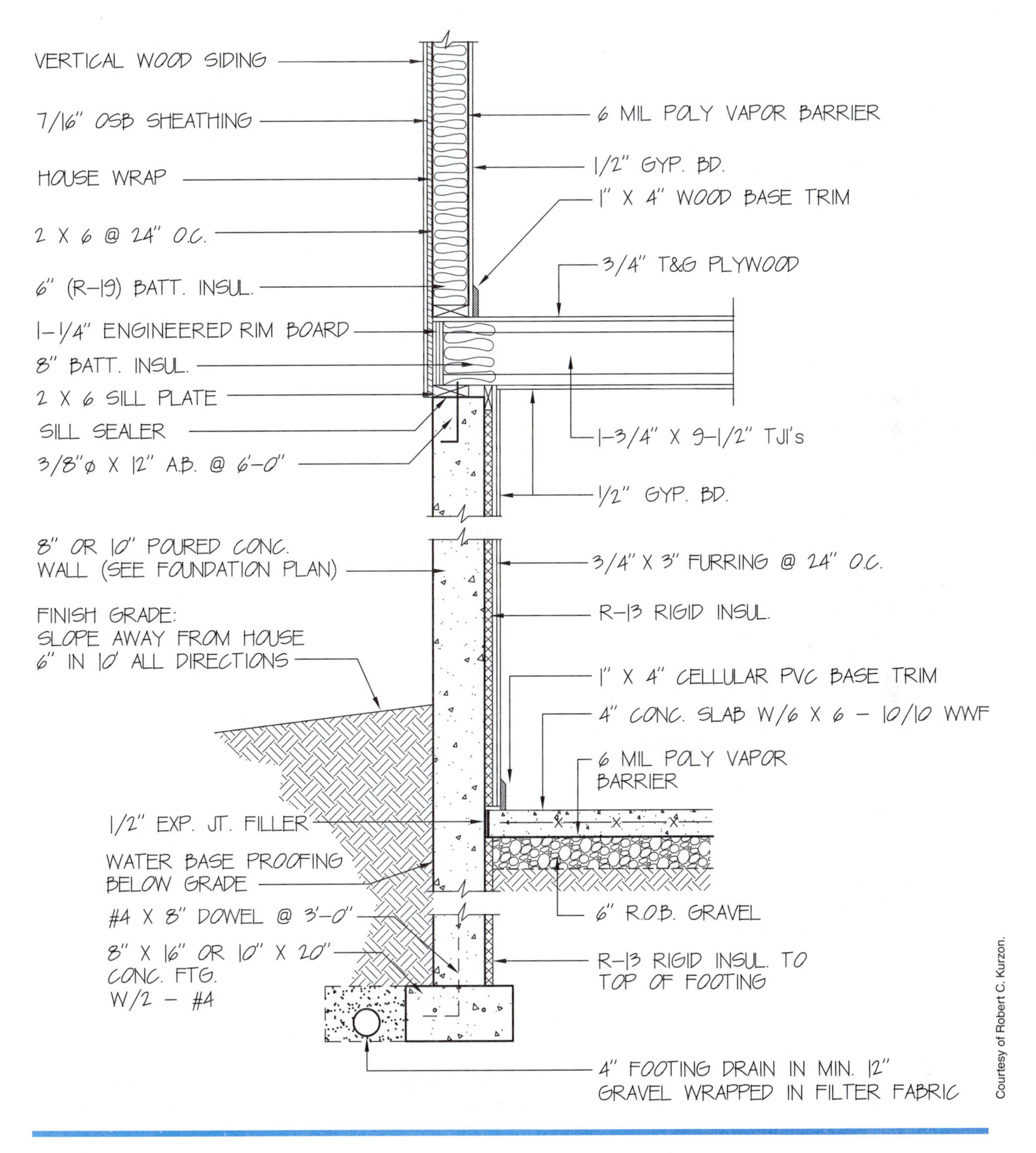

Figure 13–2. Section through foundation.

GREEN NOTE

Concrete and concrete masonry units are the traditional materials for building foundations. These materials do not provide particularly good thermal insulation, and the processes for the manufacture of concrete and concrete products are not environmentally friendly. In recent years, many new developments in the construction of foundations are taking place. One such innovation that is gaining popularity is the use of insulated concrete forms (ICFs). ICFs are typically made of rigid foam, such as Styrofoam®, but they can also be made of other materials, such as mineralized wood chips.

ICFs are made in a variety of sizes that are small enough to be easy to handle. The individual ICF units lock together, so that the result is a wall form into which concrete can be placed, resulting in a wall that is insulated with several inches of foam yet has the structural properties of concrete.

fireplace, which is 12 inches thick. Some portions of the foundation wall have 6-inch concrete masonry units at the top to create a ledge for special construction and the garage foundation calls for 8-inch concrete.

Structural concrete is the most common foundation material because it is strong and concrete foundations can be built quickly. Some buildings have masonry (usually concrete block) foundations. There are some additional topics to be discussed for masonry (concrete block) foundations.

The details for a masonry foundation may call for horizontal reinforcement in every second or third course. This is usually prefabricated wire reinforcement to be embedded in the mortar joints. Prefabricated wire reinforcement is available in varying sizes for different sizes of concrete blocks.

The height of the foundation wall is dimensioned on the building elevations. These are the same dimensions as of those used to determine the depth of the footings in the preceding unit. Just as the footing was stepped to accommodate a sloping building site, the top of the foundation wall may be stepped to accommodate varying floor levels in the *superstructure* (construction above the foundation) as shown in **Figure 13–3.**

The top of a foundation may be built with smaller concrete blocks to form a ledge upon which later brickwork will be built. It is also common practice to use one course of 4-inch solid block as the top course of a masonry foundation wall.

In concrete foundations, ***anchor bolts*** are placed in the top of the foundation (see **Figure 13–4**). These bolts are left protruding out of the top of the foundation so that the wood superstructure can be fastened in place later. Anchor bolts are not normally shown on the foundation plan, but a note on the wall section indicates their spacing. On masonry walls, anchor bolts can be placed in the hollow cores of the concrete blocks. They are held in place by filling the core with mortar grout. ***Grout*** is a ***Portland cement*** mixture that has high strength.

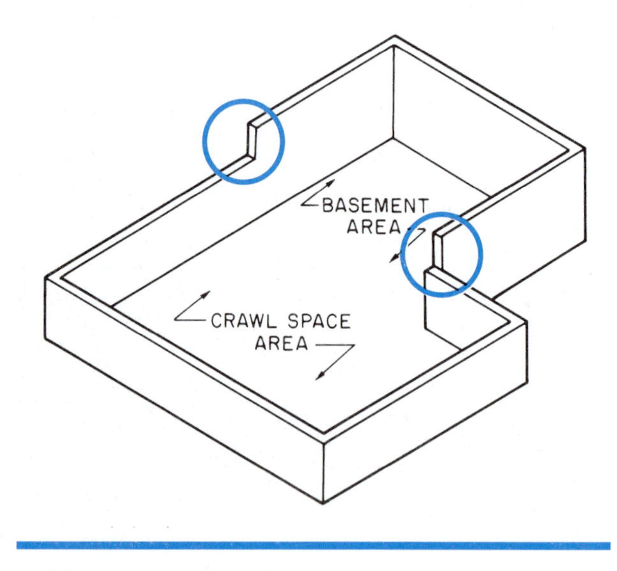

Figure 13–3. The top of the foundation may be stepped to allow for a partial basement or varying floor levels.

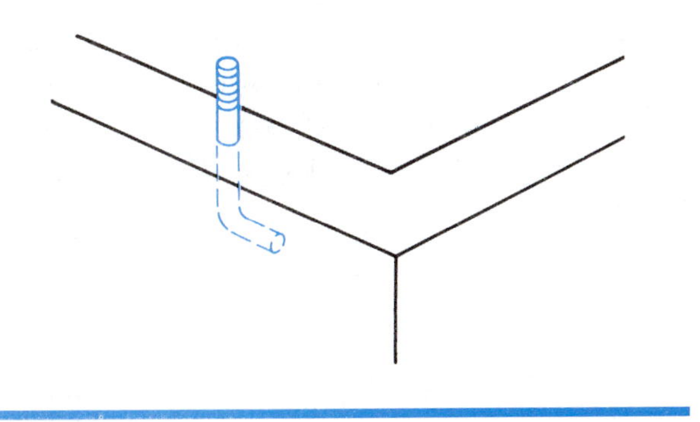

Figure 13–4. Anchor bolt.

In areas where there is a threat of extremely high winds or earthquakes, additional hold-down straps may be called for (see **Figure 13–5**). These hold-downs are normally used only with concrete foundations. Hold-downs are discussed in more detail in Unit 36.

Figure 13–5. Hold-down straps are used in addition to anchor bolts in an earthquake zone.

Special Features

Many foundations include steel or wooden beams, which act as girders to support the floor framing over long spans (see **Figure 13–6**). When the girder is steel, it is indicated by a single line with a note specifying the size and type of structural steel. A wood girder is usually indicated by two or more lines and a note specifying the number of pieces of wood and their sizes in a built-up girder (see **Figure 13–7**).

If the top of the girder is to be flush with the top of the foundation beam, pockets must be provided in the foundation (see **Figure 13–8**). ***Beam pockets*** are shown on the details and sections of the construction drawings. The locations of these beam pockets are dimensioned on the foundation plans.

If windows are to be included in the foundation, the form carpenter or mason must provide openings of the proper size. The locations of windows should be dimensioned on the foundation plan. The sizes of the windows may be shown by a note or given on the window schedule. Window sizes and window schedules are discussed in Unit 25. It is important, however, to get the masonry opening size from the window manufacturer before forming the opening in the foundation wall. The *masonry opening* is the size of the opening required in the foundation wall to accommodate the window. This size may be different from the ***nominal size*** given in a note on the foundation plan.

The foundation may include pilasters for extra support. A *pilaster* is a thickened section of the foundation, which helps it resist the pressure exerted by the earth

Figure 13–6. The girder supports the floor framing.

COMMON SIZES OF STRUCTURAL STEEL GIRDERS

SHAPE:
S = STANDARD BEAM
W = WIDE-FLANGE BEAM
NOMINAL DEPTH IN INCHES
WEIGHT PER FOOT IN POUNDS

Shape	Nominal depth		Weight per foot
S	10	X	35
S	8	X	23
S	8	X	18.4
W	10	X	33
W	10	X	21
W	10	X	11.5
W	8	X	31
W	8	X	28
W	8	X	17
W	8	X	13

COMMON SIZES OF WOOD BUILT-UP GIRDERS

NUMBER OF PIECES
THICKNESS EACH (NOMINAL) IN INCHES
WIDTH EACH (NOMINAL) IN INCHES

Number of pieces		Thickness each		Width each
2	–	2	X	10
2	–	2	X	12
3	–	2	X	10
3	–	2	X	12

Figure 13–7. Typical specifications for structural steel and wood built-up girders—other types are described in the project specifications.

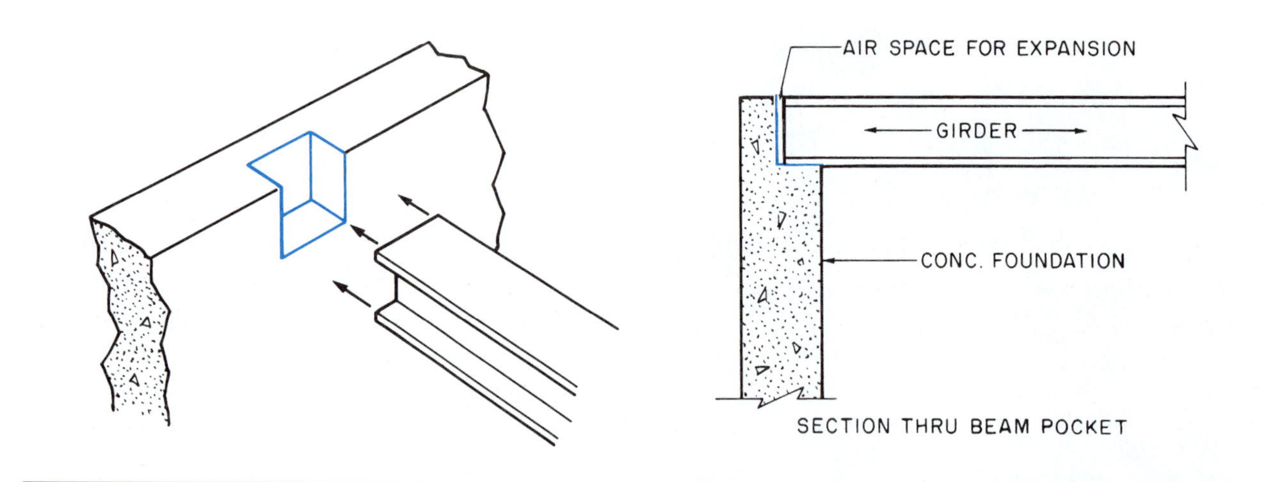

Figure 13–8. A beam pocket is a recess in the wall to hold the girder.

on the outside. The location and size of pilasters are shown on the foundation plan.

The Lake House drawing includes a special feature not commonly found on foundation plans for houses. There are four notes that read 3½″ STD. WT. STL. COL. W/8 × 8 × ½ B.PL. These notes indicate a 3½-inch square, standard weight, steel column with 8-inch-by-8-inch-by-½-inch-thick base plates. This structural steelwork is explained in more detail later, but to completely understand the foundation plan, it is necessary to know that the steel will be erected.

Permanent Wood Foundation

Foundations are usually constructed of concrete or concrete block. However, a type of specially treated wood foundation is sometimes used (see **Figure 13–9**). These *permanent wood foundations* do not use concrete footings.

Instead, they are built on 2 × 8s or laid on gravel fill below the frostline. The foundation walls are framed with lumber that has been pressure treated to make it rot resistant and insect resistant. The framing is covered with a plywood skin, and the plywood is covered with

Courtesy of Southern Forest Products Association.

Figure 13–9. Permanent wood foundation.

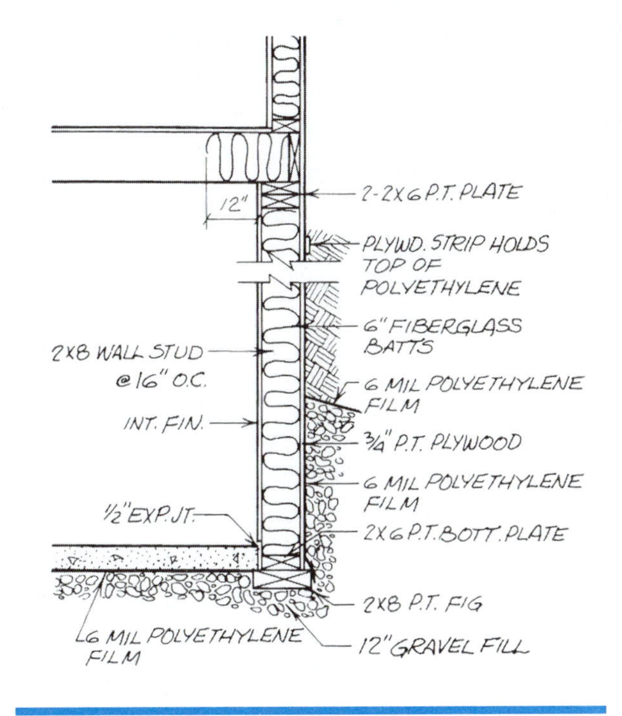

Figure 13–10. Section of wood foundation.

polyethylene film for complete moisture proofing (see **Figure 13–10**). All the nails in a wood foundation are stainless steel to prevent rusting. The IRC includes separate provisions to cover wood foundations.

USING WHAT YOU LEARNED

Before any concrete form construction begins, it is necessary to have a good picture in your mind of what the finished product will be. The Lake House foundation has many variations from the typical wall shown in the Typical Wall Section on sheet 4. For example, near the southeast 3½″ ⧄ steel column, the Foundation Plan has a note that says "6″ CONC. BLK. – TOP OF WALL TO BELOW STAIR." What does this note mean and why is that detail necessary? The floor plans show stairs from the living room down to the playroom at this point. The stairs are also shown by the hidden lines on the South Elevation 3/3 and that elevation includes a note that says "6″ CMU FDN WALL (BELOW STAIR)." The typical foundation wall is 10 inches thick, so the 6-inch concrete masonry units create a 4-inch ledge to support the stairs.

Assignment

Refer to the Lake House drawings in your textbook packet to complete the assignment.

1. What is the thickness of the concrete foundation in the kitchen area of the Lake House?
2. Approximately how many lineal feet of concrete wall are included in the foundations of the Lake House?
3. How thick is the south foundation of the fireplace?
4. What is the elevation at the top of the north end of the east foundation wall of the garage?
5. What is the highest elevation in the entire foundation?
6. How high is the foundation wall at the highest elevation of the foundation?
7. What size anchor bolts are indicated at the top of the Lake House foundation?
8. What spacing is indicated for the anchor bolts?
9. What secures the foundation wall to the footing?
10. What is the elevation of the top of the concrete wall at the southwest 3½″ steel column?

UNIT 14 Drainage, Insulation, and Concrete Slabs

Objectives

After completing this unit, you will be able to perform the following tasks:

- Locate and explain information for control of groundwater as shown on a set of drawings.
- Locate and describe subsurface insulation.
- Determine the dimensions of concrete slabs and the reinforcement to be used in concrete slabs.

Drainage

After the foundation walls are erected and before the excavation outside the walls is ***backfilled*** (filled with earth to the finished grade line), footing drains, if indicated, must be installed. ***Footing drains*** are usually perforated plastic pipe placed around the footings in a bed of crushed stone (see **Figure 14–1**). If the site has a natural slope, the footing drains can be run around the foundation wall to the lowest point, then away from the building to drain by gravity. In areas where there is no natural drainage, the drain is run to a dry well or municipal storm drain.

At one time, clay drain tile was the most common type of pipe for this purpose. However, perforated plastic pipe is used in most new construction. Plastic drain pipe is manufactured in 10-foot lengths of rigid pipe and in 250-foot rolls of flexible pipe (see **Figure 14–2**). An assortment of plastic fittings is available for joining rigid plastic pipe. When footing drains are to be included, they are shown on a wall section or footing detail (see **Figure 14–3**). A note on the drawing indicates the size and material of the pipe.

If the floor drains are to be included in concrete-slab floors, they are indicated by a symbol on the appropriate floor plan (see **Figure 14–4**). If these floor

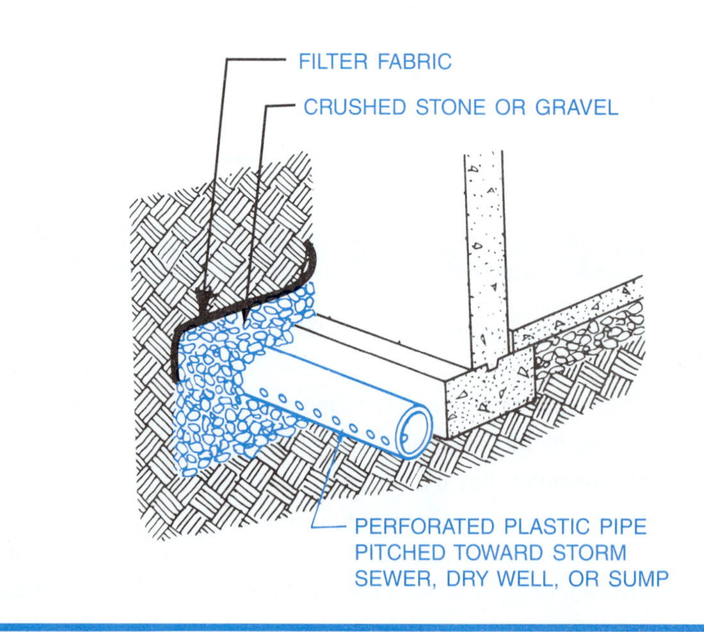

Figure 14–1. Footing drain.

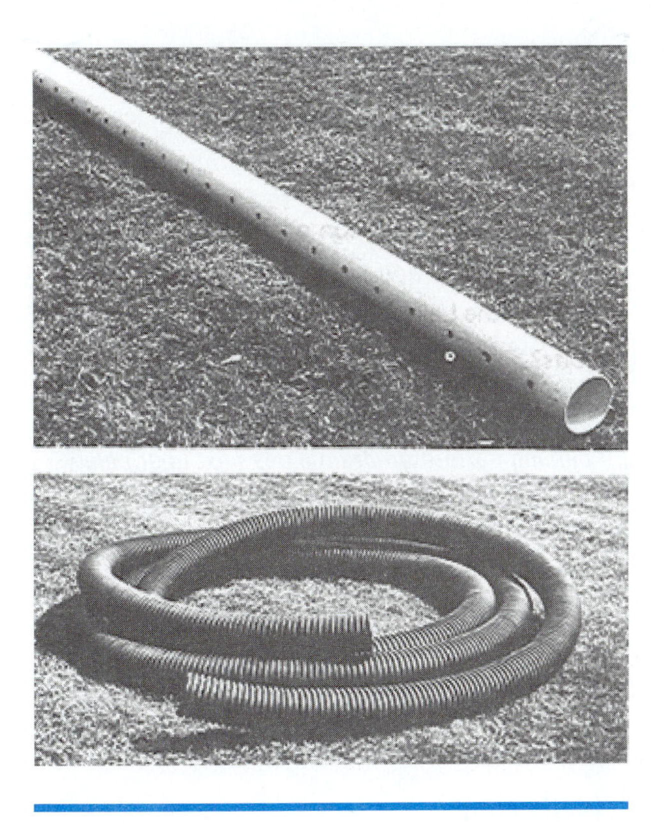

Figure 14–2. Plastic drain pipe.

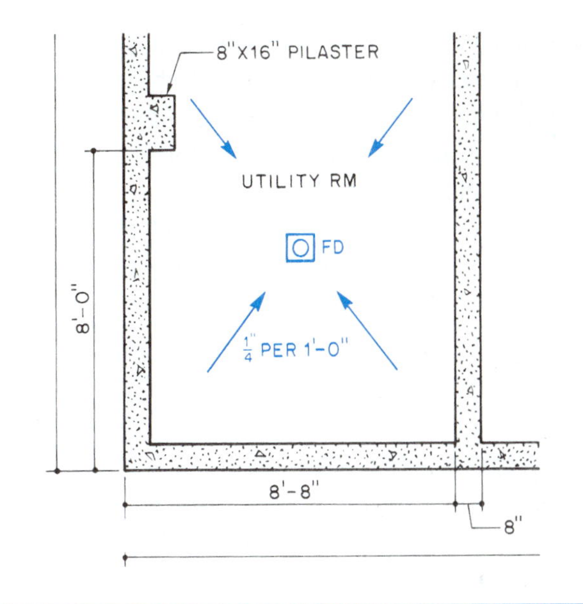

Figure 14–4. The floor drain is shown by a symbol. (Notice that the floor is pitched toward the drain.)

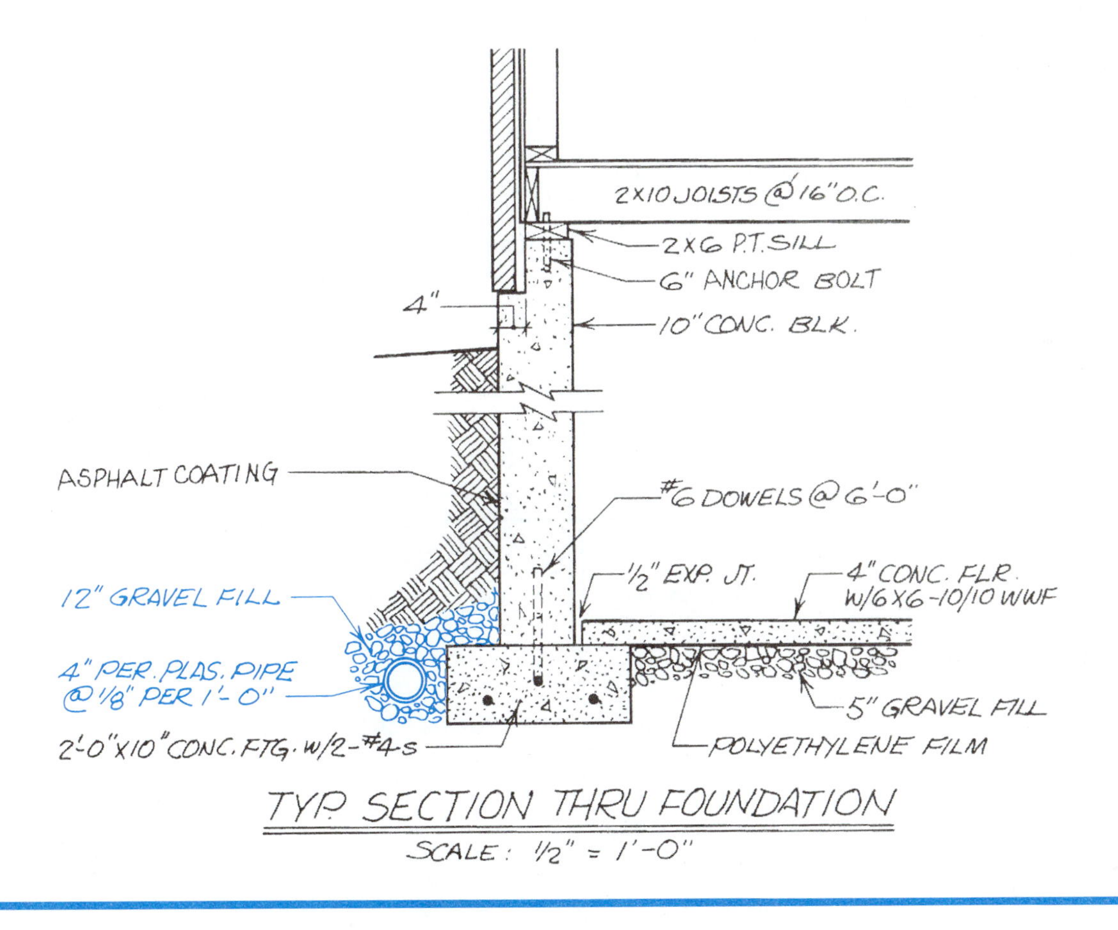

Figure 14–3. Footing drains are shown outside the footing.

drains run under a concrete footing, the piping had to be installed before the footing was placed (see Unit 11). The ***riser*** (vertical part through the floor) and drain basin are usually set at the proper elevation just prior to placing the concrete floor.

The floor plan may include a spot elevation for the finished drain, or it may be necessary to calculate it from information given for the pitch of the concrete slab. Pitch of concrete slabs is discussed later in this unit.

Vapor Barriers

Another technique often used to prevent groundwater from seeping through the foundation is coating the foundation wall with asphalt foundation coating.

At this point, subsurface work outside the foundation wall is completed, but backfilling should not be done until the superstructure is framed. The weight and rigidity of the floor on the foundation wall help the wall to resist the pressure of the backfill. If the backfilling must be done before the framing, the foundation walls should be braced. To retard the flow of moisture from the earth through the concrete-slab floor, the drawings may call for a layer of gravel over the entire area before the concrete is placed. A polyethylene ***vapor barrier*** is laid over the gravel underfill. The thickness of polyethylene sheeting is measured in ***mils.*** One mil equals 1/1,000 of an inch. For vapor barriers, 6-mil polyethylene is generally used.

GREEN NOTE

One of the most basic principles of green home construction is the control of groundwater to keep it from penetrating the foundation and causing structural damage as well as interior finish damage and mold. Groundwater is kept away from the foundation by proper grading of the site, careful damp-proofing, and providing a path for groundwater to escape without penetrating the foundation. This path involves placing perforated drain pipes next to or below the footings and pitching it, so all water flows downhill to a daylight exit point. Poor workmanship on any of these aspects of groundwater control can result in devastating water damage and serious implications for the health of the occupants.

Insulation

In cold climates, it is desirable to insulate the foundation and concrete slab. This insulation is usually rigid plastic foam board placed against the foundation wall or laid over the gravel underfill (see **Figure 14–5**). Rigid insulation under the floor may only be specified for the outer 2 feet of the slab. The heat of the building keeps the frost from penetrating more than 2 feet into the perimeter of the floor slab.

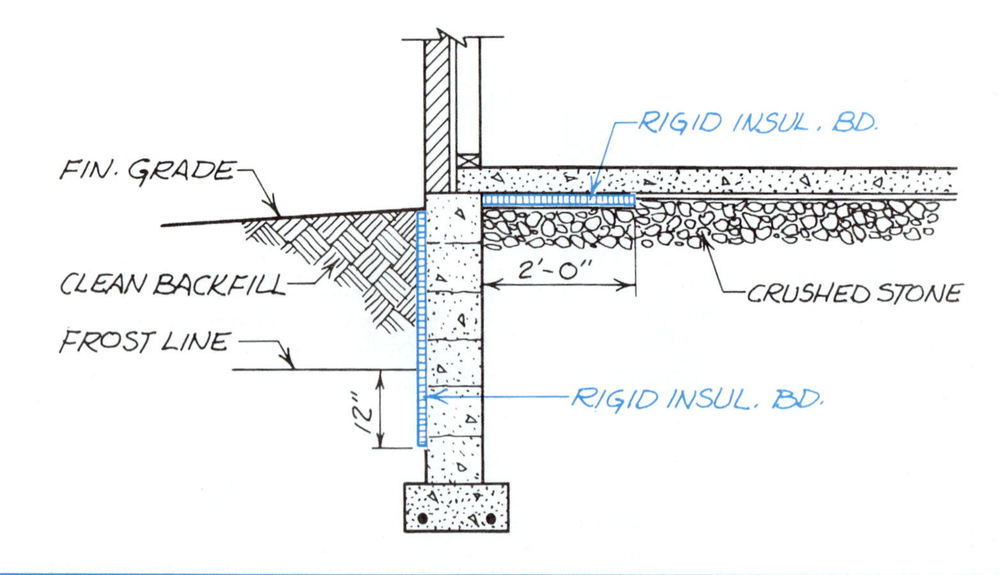

Figure 14–5. Rigid plastic foam insulation may be laid under the perimeter of the floor or against the foundation wall.

Like all materials, concrete and masonry expand and contract slightly with changes in temperature. To allow for this slight expansion and contraction, the joint between the concrete slab and foundation wall should include a compressible expansion joint material. Expansion joints can be made from any compressible material such as neoprene or composition sheathing material. This expansion joint filler is as wide as the slab is thick and is simply placed against the foundation wall before the concrete is placed.

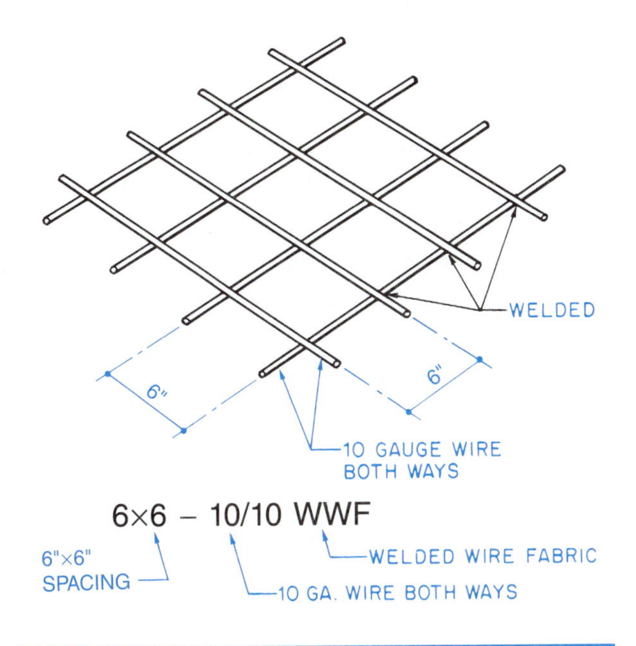

Figure 14–6. The callout for welded wire fabric explains the size and spacing of the wires.

Concrete Slabs

When the house has a basement, the floor is a concrete slab-on-grade. The areas to be covered with concrete are indicated on the foundation plan or basement floor plan. This is usually done by an area not giving the thickness of the concrete slab and any reinforcing steel to be used. To help the concrete resist minor stresses, it is usually reinforced with welded wire mesh. Welded wire mesh is sometimes abbreviated WWM and sometimes WWF for welded wire fabric. The specifications for welded wire mesh are explained in **Figure 14–6.** Where the slab must support bearing walls or masonry partitions, it may be haunched, as discussed in Unit 12.

When floor drains are included or where water must be allowed to run off, the slab is *pitched* (sloped slightly). A note on the drawings indicates the amount of pitch. One-quarter inch per foot is common. When there is any possibility of confusion about which way the slab is to be pitched, bold arrows are drawn to show the direction in which the water will run (see **Figure 14–7**).

When floor drains or forms are set for pitched floors, it is necessary to find the total pitch of the slab. This is done by multiplying the pitch per foot by the number of feet over which the slab is pitched. (See Math Review 8 in Appendix B.) For example, if the note on a concrete ***apron*** in front of a garage door indicates a pitch of ½ inch per foot and the apron is 4 feet wide, the total pitch is 2 inches. The proper elevation for the form at the outer edge of the apron is 2 inches less than the finished floor elevation.

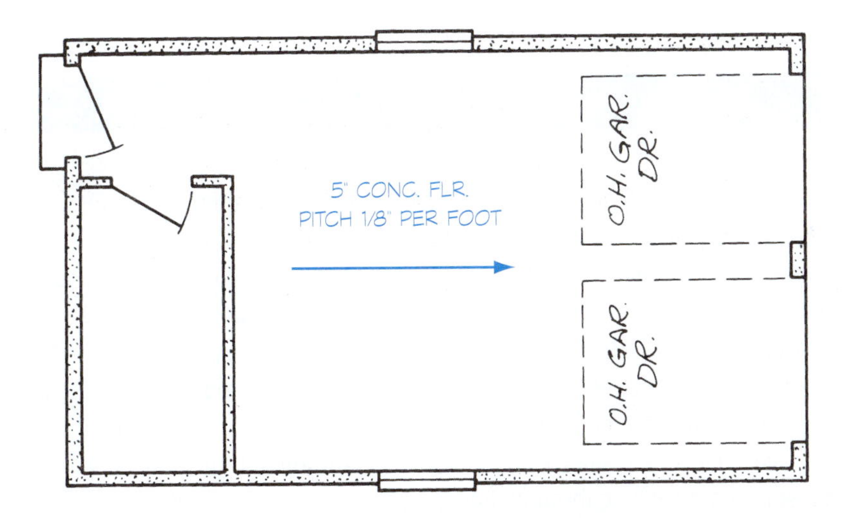

Figure 14–7. A bold arrow indicates the way that water will run off a pitched surface.

Slab-on-grade installations may require electrical raceways to be installed in or just below the concrete slab in the crushed or gravel fill. These installations must be made so that they do not reduce the structural integrity of the concrete slab. An oversized raceway in the concrete slab may cause the slab to crack and settle unevenly. An electrician should be present during concrete placement to observe and correct any damage to these electrical raceways.

All underground utility systems must be coordinated prior to starting installation, especially in commercial and industrial buildings, as discussed in Unit 11. The service entrance section may have up to six subdistribution sections or panels located throughout the building. Electrical feeders are required from the electrical service entrance section to these subdistribution sections or panels. This requires detailed coordination between the electrical installer and the other utility installers prior to starting the installation of the utility systems. The electrical installation must be laid out around the other utilities, with the other utility layouts normally having priority. The other utilities to be coordinated may include plumbing, fire sprinkler, heating and air conditioning, and specialty systems. This coordination must be done for both the underground and above ground systems.

Some of the concrete work in the Lake House is of particular interest, because it is a part of the passive solar heating system the Lake House uses. Section 1/4 and the lower-level floor plan 1/2 indicate that the area under the living room and dining room floors is a heat sink. A *heat sink* is a mass of dense material that absorbs the energy of the sun during the day and radiates it at night. The living room and dining room are on the south side of the Lake House. In the winter, when the leaves are off the deciduous trees, the sun shines in the large areas of glass in these rooms and warms the heat sink. At night, this heat is radiated into the house to provide additional heat when it is needed most. The floor over the heat sink is a concrete slab similar to that used in the playroom. Detail 2/5 helps to explain this concrete slab.

USING WHAT YOU LEARNED

All of the reinforcement for the concrete slabs must be in place before the concrete is delivered to the site. In the Lake House how many rebars of what size are required for the haunch in the concrete slab between the two north 3½ ⊠ steel columns? The haunch is shown on Foundation Plan 4/1 and Detail 3/1 shows the reinforcement. The first callout on the detail is "2 -- #4 CONTINUOUS" with a leader pointing to the symbols for reinforcing bars.

Assignment

Refer to the Lake House drawings in your textbook packet to complete this assignment.

1. What is the thickness of the concrete slab over the heat sink in the Lake House?
2. Describe the reinforcement used in the concrete slab in the playroom of the Lake House.
3. How many square feet of 2-inch rigid insulation are needed for the Lake House heat sink?
4. What prevents moisture from seeping through the concrete slab floor in the Lake House?
5. What is the finished floor elevation of the Lake House garage?
6. What is the elevation of the floor drain in the utility room of the Lake House?
7. What is the purpose of the 8-inch-thick concrete haunch in the middle of the Lake House slab?
8. How many cubic yards of concrete are required for the garage floor? The basement floor including the crawl space?

UNIT 15 Framing Systems

Objectives

After completing this unit, you will be able to identify each of the following types of framing on construction drawings:

- Platform
- Balloon
- Post-and-beam
- Energy-saving

Platform Framing

Platform framing, also called ***western framing***, is the type of framing used in most houses built in the last 70 years (see **Figure 15–1**). It is called platform framing because as the floor is built at each level, it forms a platform on which to work while erecting the next level (see **Figure 15–2**).

A characteristic of platform framing is that all wall *studs*, the main framing members in walls, extend only the height of one story. Interior walls, called *partitions*, are the same as exterior walls. The bottoms of the studs are held in position by a *bottom* (or *sole*) plate. The tops of the studs are held in position by a *top plate*. Usually, a *double top plate* is overlapped at the corners to tie intersecting walls and partitions together (see **Figure 15–3**). In some construction, the second top plate is not used. Instead, metal framing clips are used to tie

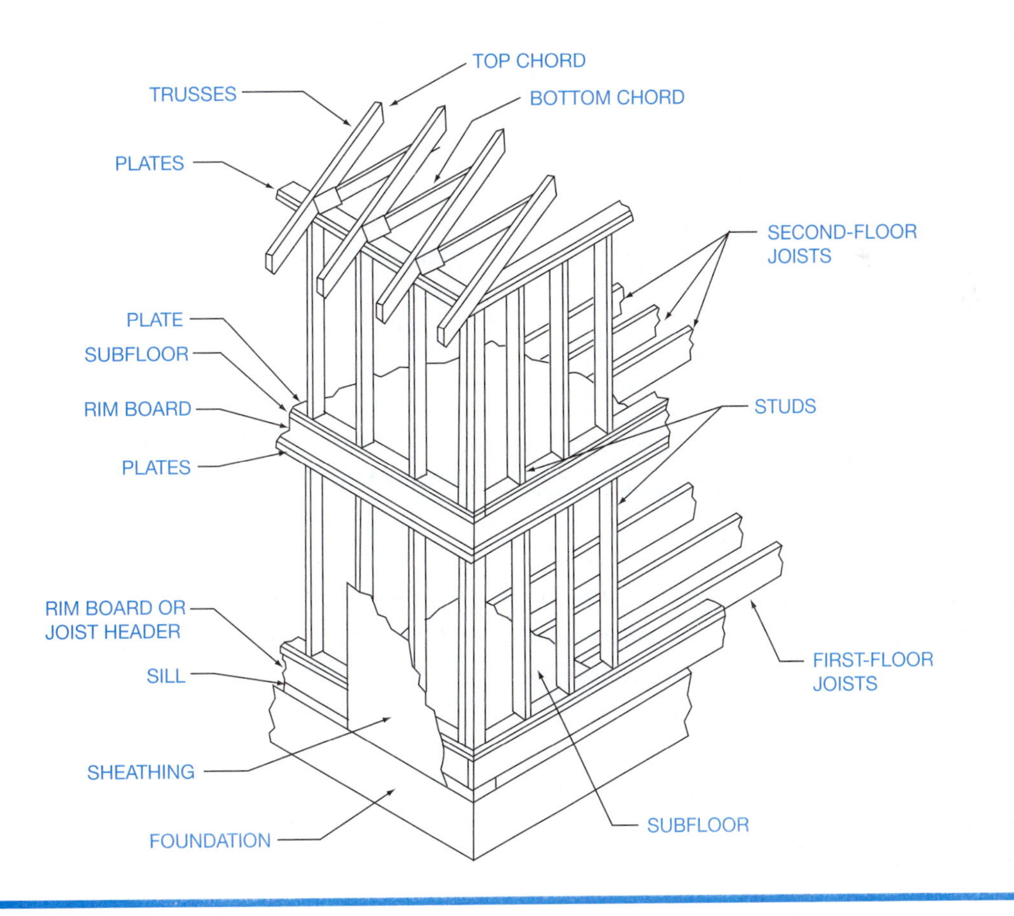

Figure 15–1. Platform, or western, framing.

Figure 15–2. Platform, or western, framing provides a convenient work surface during construction.

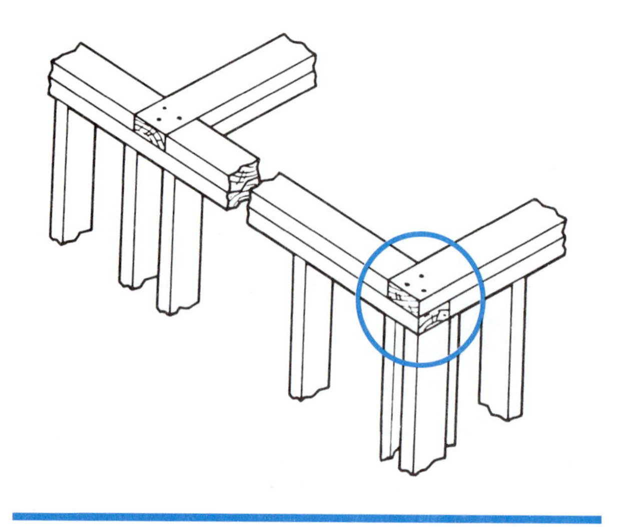

Figure 15–3. The double top plate overlaps at the corners.

intersecting walls together. Upper floors rest on the top plate of the walls beneath. The framing members of the upper floors or roof are positioned over the studs of the wall that supports them.

Platform construction can be recognized on wall sections (see **Figure 15–4**). Notice that the studs extend only from one floor to the next.

Balloon Framing

In ***balloon framing***, the exterior wall studs are continuous from the foundation to the top of the wall (see **Figure 15–5**). Floor framing at intermediate levels is supported by *let-in ribbon boards.* This is a board that fits into a notch in each joist and forms a support for the joists. Although balloon framing is not as widely used as it once was, some balloon-framing techniques are still used for special framing situations.

In both platform-frame and balloon-frame construction, the structural frame of the walls is covered with sheathing. ***Sheathing*** encloses the structure and, if a structural grade is used, prevents wracking of the wall. *Wracking* is the tendency of all the studs to move, as in a parallelogram, allowing the wall to collapse to the side (see **Figure 15–6**). There are two ways to prevent wracking. Plywood or other structural sheathing at the corners of the building prevents this movement. Also, diagonal braces can be attached to the wall framing at the corners to prevent wracking (see **Figure 15–7**).

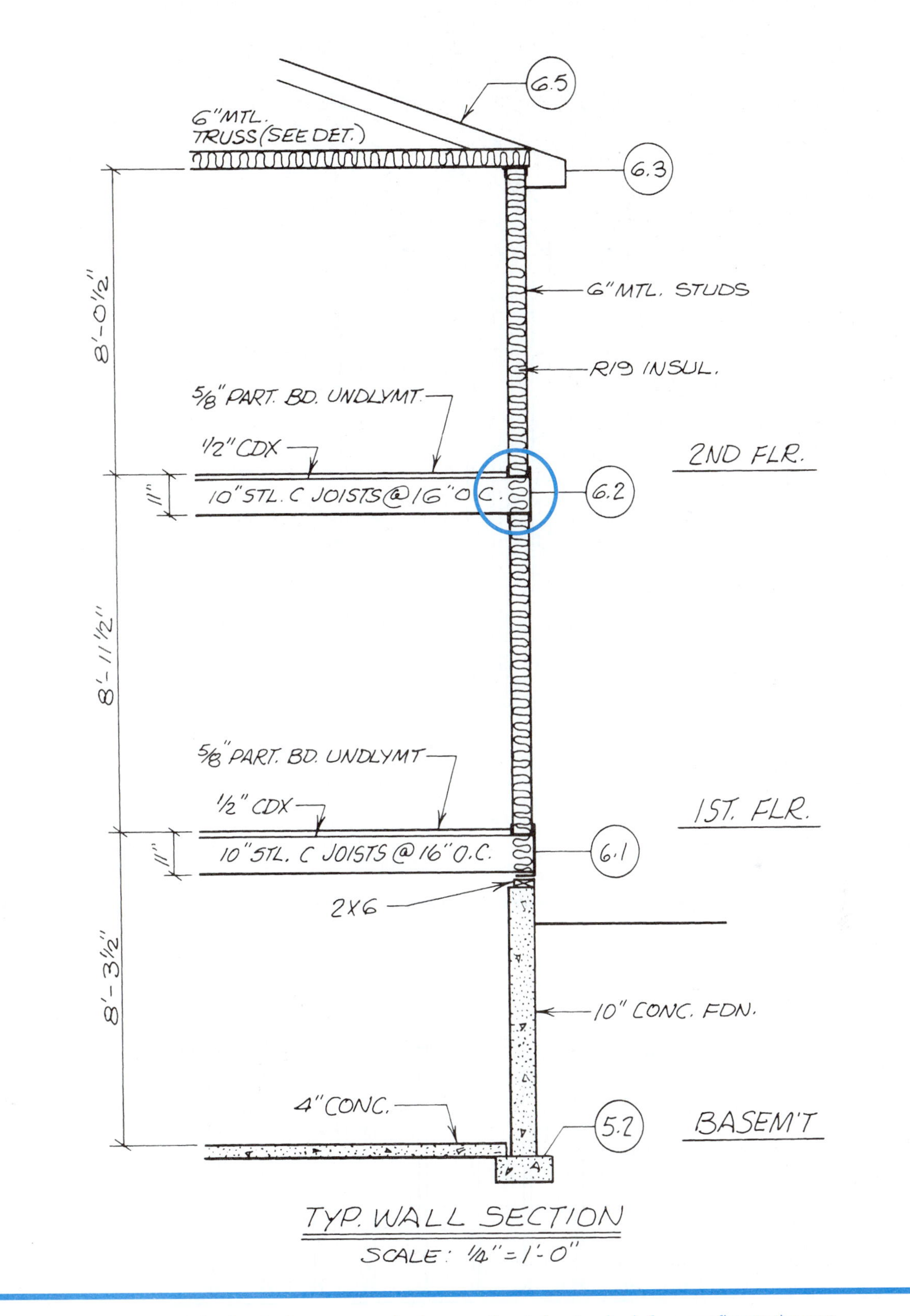

Figure 15–4. This can be recognized as platform construction because the studs extend only from one floor to the next.

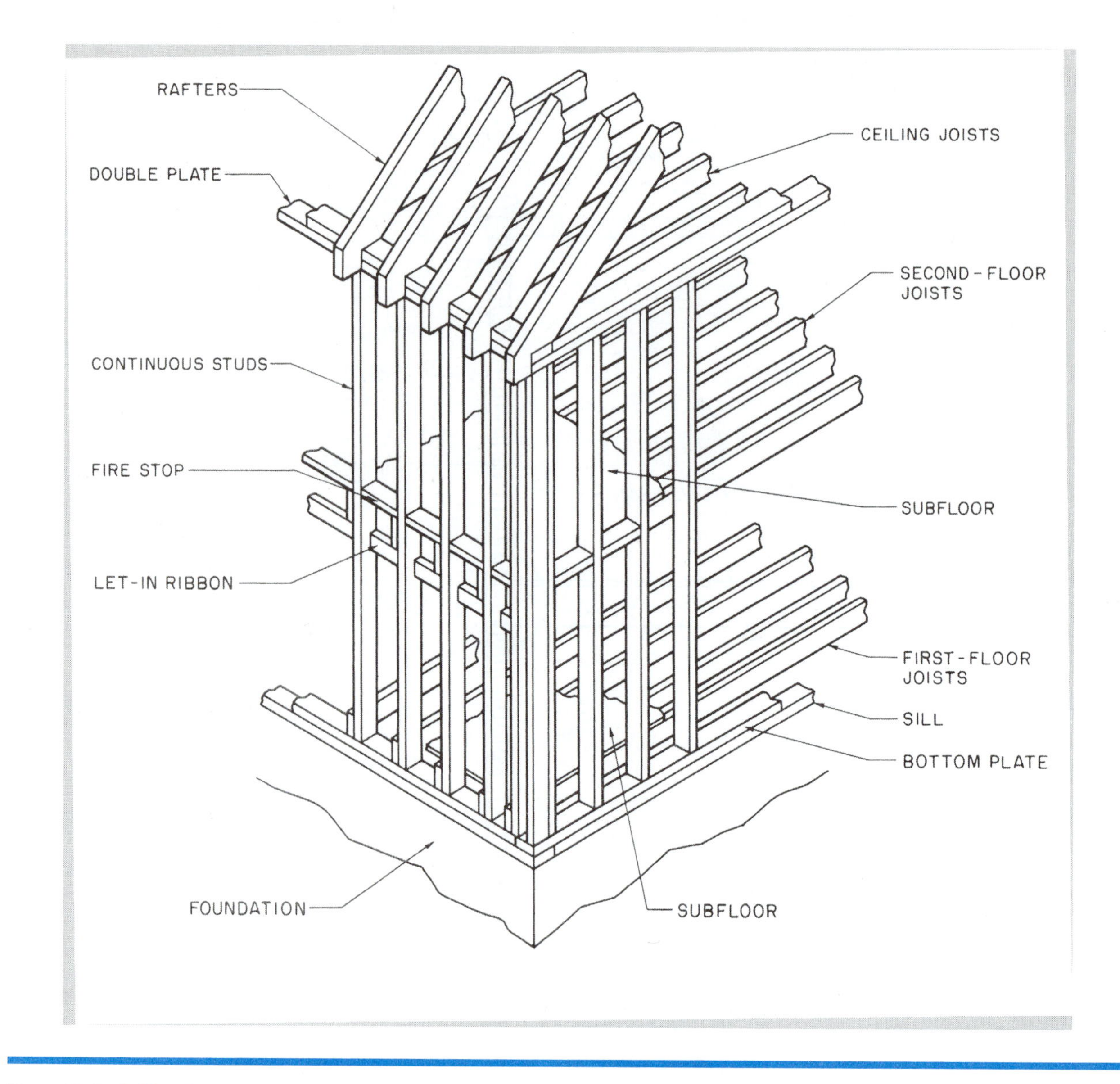

Figure 15–5. Balloon framing.

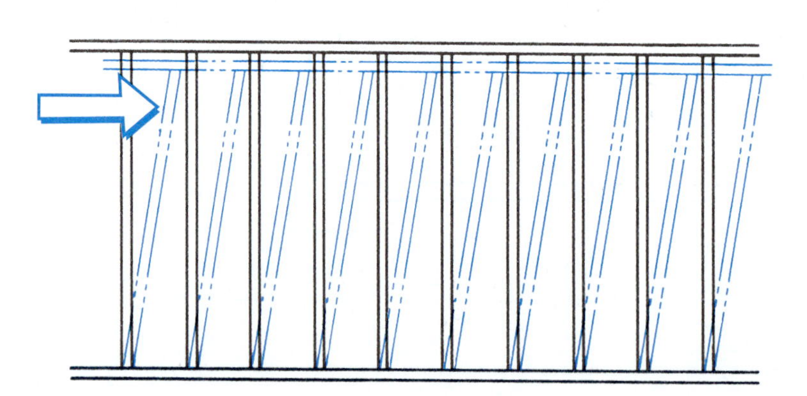

Figure 15–6. Wracking.

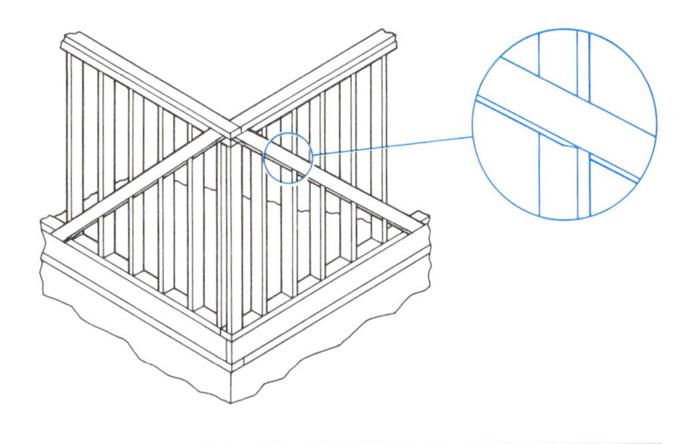

Figure 15–7. Let-in bracing prevents wracking.

Post-and-Beam Framing

Platform framing and *balloon framing* are characterized by closely spaced, lightweight framing members (see **Figure 15–8**). *Post-and-beam* framing uses heavier framing members spaced farther apart (see **Figure 15–9**). These heavy timbers are joined or fastened with special hardware (see **Figure 15–10**). Because post-and-beam framing uses fewer pieces of material, it can be erected more quickly. Also, although the framing members are large—they can range from 3 inches by 6 inches to 5 inches by 8 inches—their wider spacing results in a savings of material. However, to span this wider spacing, floor and roof decking must be heavier. Post-and-beam

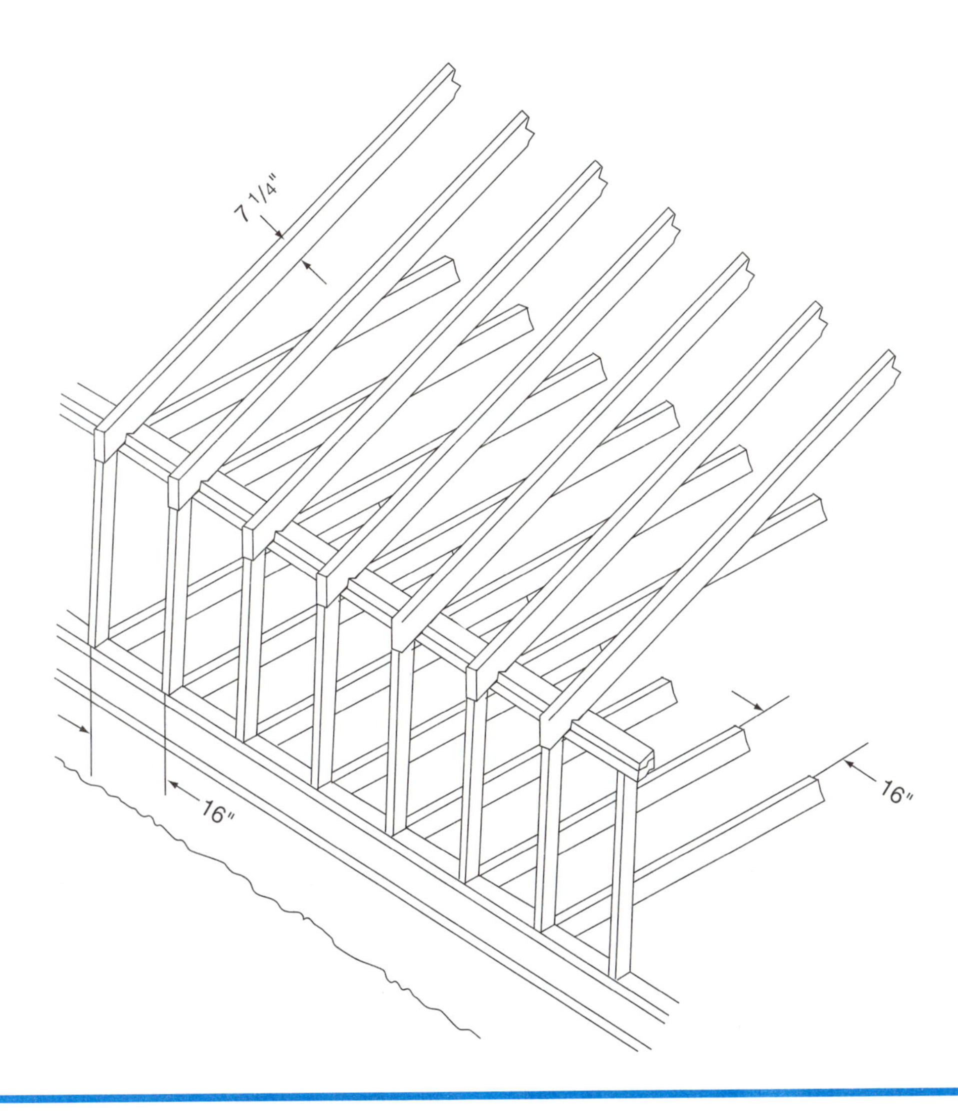

Figure 15–8. Conventional framing. 16" OC.

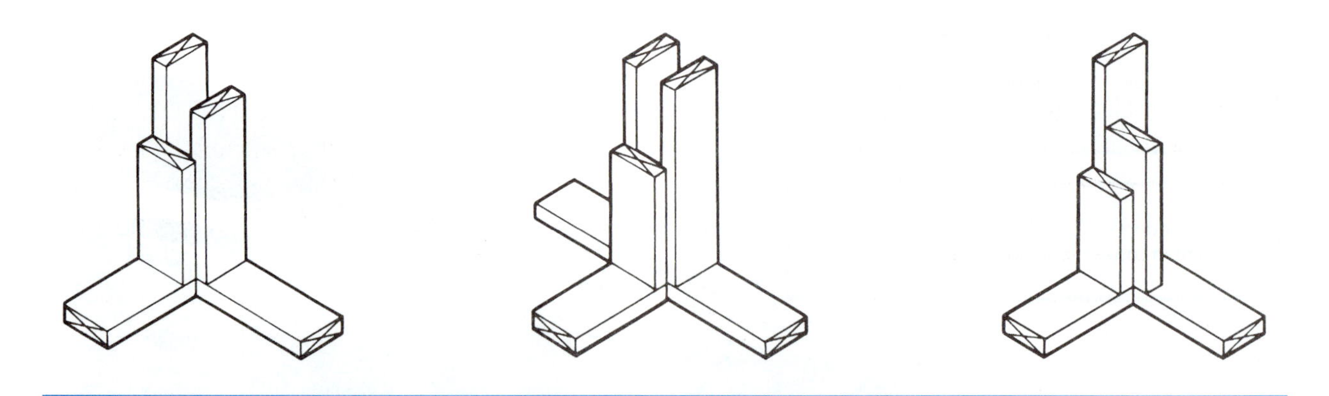

Figure 15–18. Conventional corner posts for 2 × 4 framing.

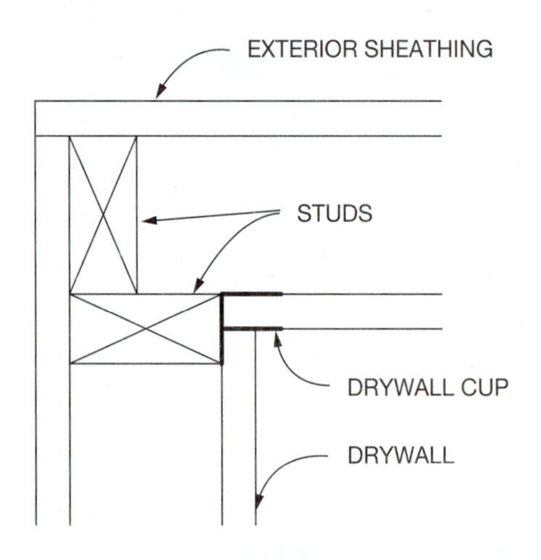

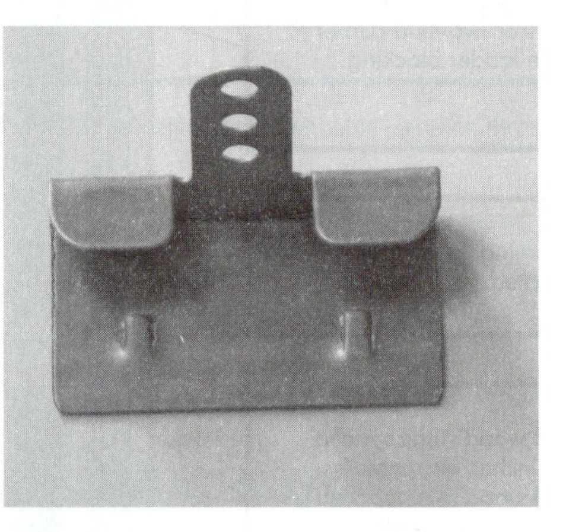

Figure 15–19. Energy-efficient corner construction.

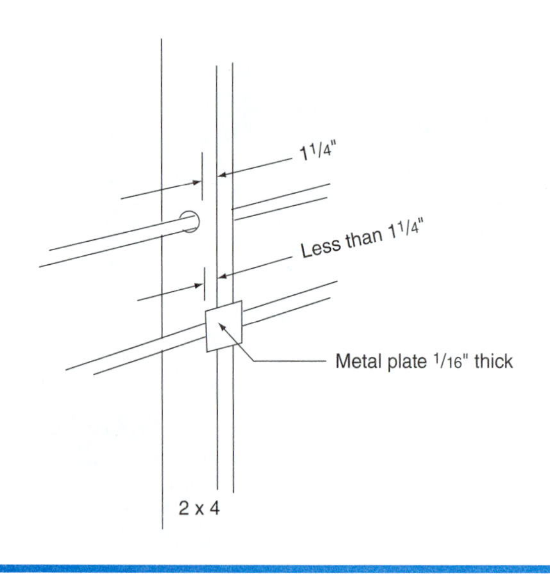

Figure 15–20. Nonmetallic-sheathed cable installed in wood framing.

Many frame building are constructed using some of the elements of advanced framing, but not all of them. Notes on plans and details will describe these elements where they are to be used.

Examination of the first floor plan and the detail drawings of the Lake House shows that the walls are framed for maximum efficiency. Several details indicate that the studs are 2 × 6s @ 24 OC. This allows room for more insulation in the wall. Also notice that the house is sheathed with 3/4-inch insulation sheathing. Interior partitions do not need insulation, and so they are framed with 2 × 4s.

When installing nonmetallic-sheathed cable through wood wall studs and floor joists, the cable must be installed at least 1¼ inches from the nearest edge. Where this 1¼ inches cannot be met, a 1/16-inch-thick metal plate must be installed to protect the cable from nails and screws (see **Figure 15–20**).

USING WHAT YOU LEARNED

A good familiarity of the framing system for a house will be a big help in understanding all of the details necessary to frame the building. One aspect of the framing system is the method for framing corners. In a typical outside corner of the Lake House, what size and how many studs are used? This information is clearly shown in Detail 4/4 on Sheet 4 in your textbook packet, the first sheet showing framing details. There are three 2 × 6 studs in a typical corner.

Assignment

1. Identify *a* through *h* in **Figure 15–21.**
2. What kind of framing is shown in **Figure 15–21**?
3. Identify *a* through *c* in **Figure 15–22.**
4. Identify *a* through *c* in **Figure 15–23.**
5. Sketch a plan view of a conventional corner detail. Include drywall and sheathing.
6. Sketch a plan view of an energy-efficient corner detail. Include drywall and sheathing.
7. What two materials are most often used for framing homes?
8. What framing material requires bushings or grommets for nonmetallic-sheathed cable installations?

Note: Refer to the Lake House drawings (in the packet) to complete the rest of the assignment.

9. Which of the types of framing discussed in this topic is used for the exterior walls of the Lake House?
10. What supports the west ends of the floor joists in bedroom #1?
11. What supports the east ends of the kitchen rafters?
12. How are the LVL beams fastened to the 3½-inch square posts?
13. How are the steel posts anchored?
14. What does the northwest square steel post rest on?
15. What supports the north edge of the living room floor?
16. List the dimensions (thickness × depth × length) of all LVL beams.

Note: The length can be found by subtracting the outside dimension of the posts from the centerline spacing shown on the plan views.

17. List the length of each piece of 3½² ▨ steel.

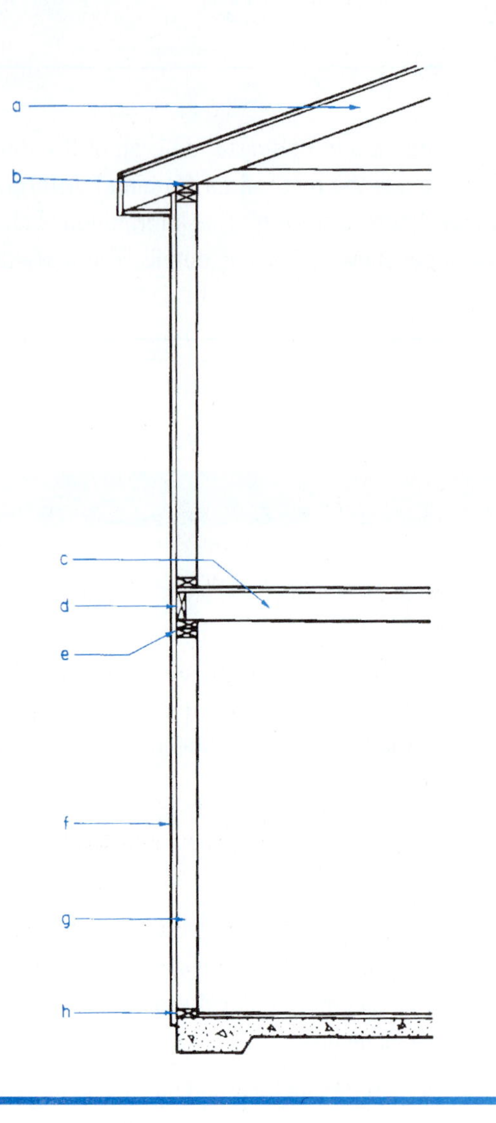

Figure 15–21.

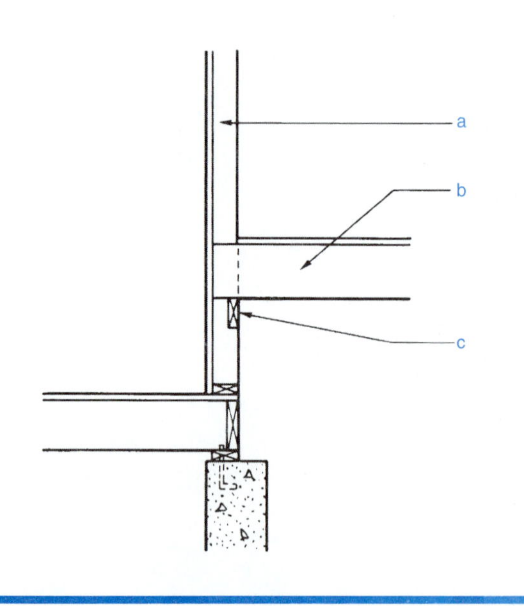

Figure 15–22.

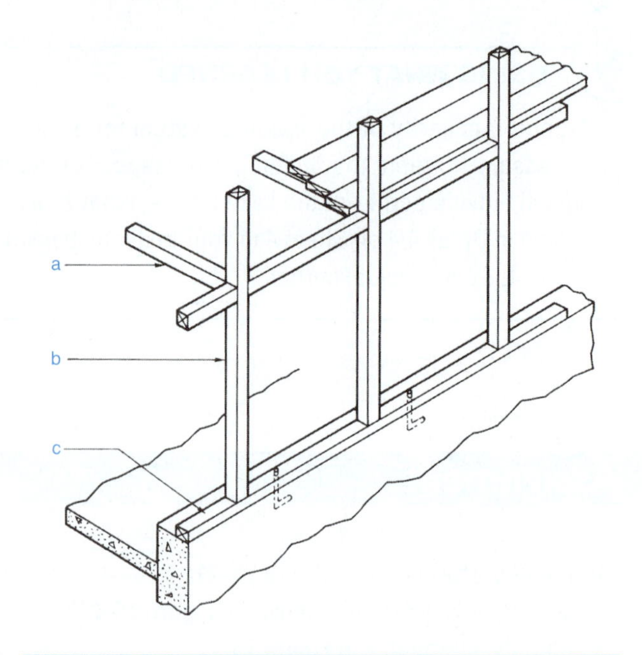

Figure 15–23.

Columns, Piers, and Girders

UNIT 16

The most common system of floor framing in light construction involves the use of joists and girders. ***Joists*** are parallel beams used in the floor framing (see **Figure 16–1).** Usually buildings are too wide for continuous joists to span the full width. In this case, the joists are supported by one or more ***girders*** (beams) running the length of the building. The girder is supported at regular intervals by wood or metal posts or by masonry or concrete piers.

An exception to the above system of girders and columns might be when the floor is framed with floor trusses (**Figure 16-2**). Floor trusses can often span a much greater distance without a supporting girder than can solid wood joists or I-joists. Joists and trusses are covered in Unit 17.

Columns and Piers

Metal posts called *pipe columns* are the most common supports for girders. However, masonry or concrete piers may be specified. The locations of columns, posts, or piers are given by dimensions to their centerlines. When metal or wooden posts are indicated, the only description may be a note on the foundation plan (see **Figure 16–3).** This note may give the size and material of the posts only, or it may also specify the kind of bearing plates to be used at the top and bottom of the post (see **Figure 16–4).** A *bearing plate* is a steel plate

Figure 16–1. Joist-and-girder floor framing.

Objectives

After completing this unit, you will be able to perform the following tasks:

- Locate columns and piers, and describe each from drawings.
- Locate and describe the girders that support floor framing.
- Determine the lengths of columns and the heights of piers.

found on a foundation plan indicating the type and size of girder to be used are shown in **Figure 16–3**. The girder in **Figure 16–3** is to be built up of three 2 × 10s.

Determining the Heights of Columns and Piers

The length of the columns or height of the piers depends on how the joists will be attached to the girder. The floor joists may rest directly on top of the girder or may be butted against the girder so that the top surface of the floor joist is flush with the top surface of the girder (see **Figure 16–6**).

To find the height of the columns or piers, first determine the dimension from the basement floor to the finished first floor. Then subtract from this dimension the

> **GREEN NOTE**
>
> *Designers, especially those designing with an eye toward green homes, have to weigh the strengths and weaknesses of every material choice they make. In choosing material for a girder, the green designer may see some benefits in using laminated veneer lumber (LVL) instead of built-up lumber girders or steel girders. LVL can be produced in any length and a wide range of thicknesses and depths, yet even the longest LVL girder is made up of smaller pieces of material. This means that high strength can be achieved while using material that might otherwise have been destined for the scrap pile.*

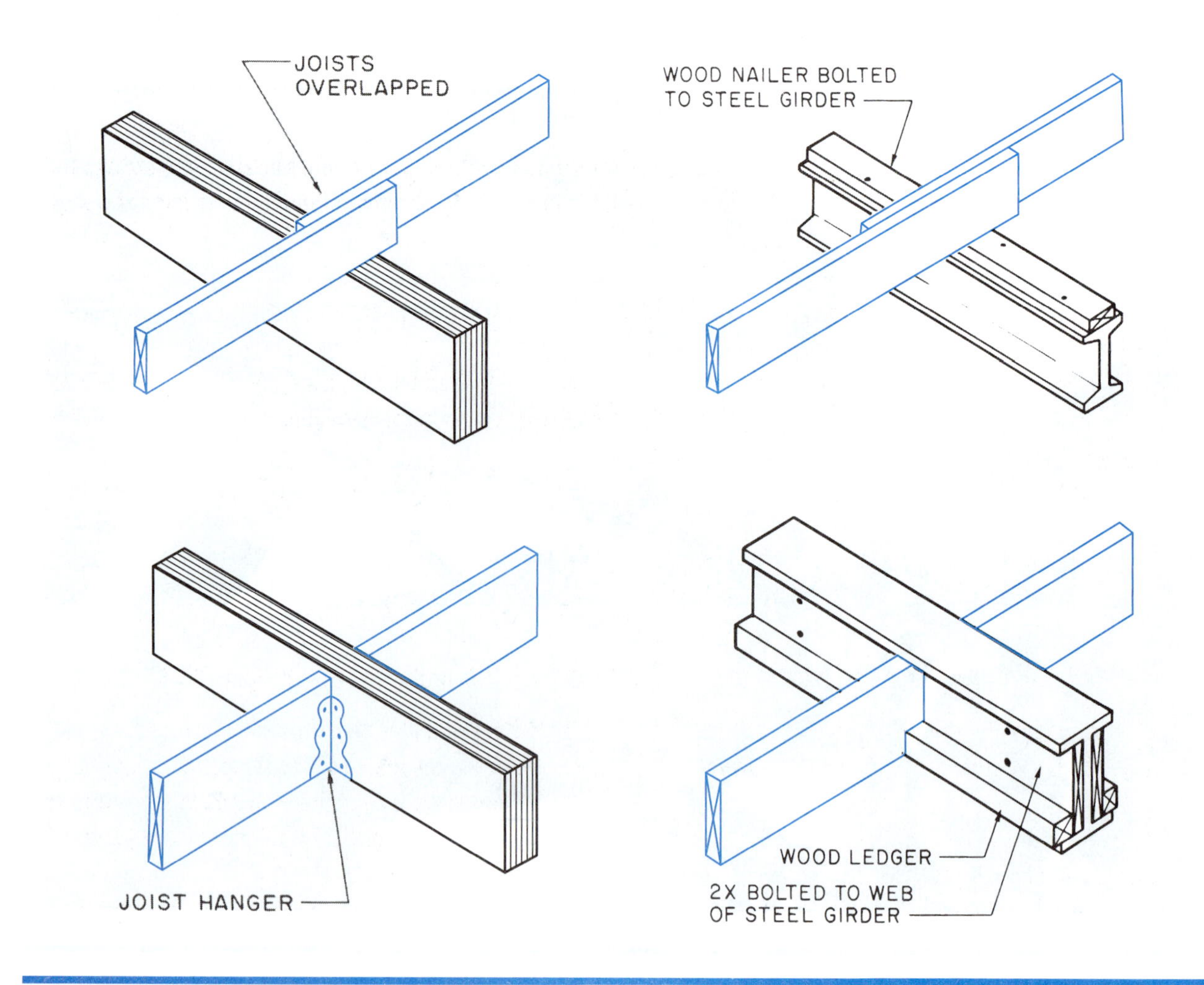

Figure 16–6. Several methods of attaching joists to girders.

depth of the first floor, including all of the framing and the girder. Then add the distance from the top of the basement floor to the bottom of the column. The result equals the height of the column or pier. (See Math Reviews 5 and 6 in Appendix B.) For example, the following shows the calculation of the height of the steel column in **Figure 16–7**:

- Dimension from finished basement floor to finished first floor = 8′-10½″
- Allowance for finished floor = 1″
- Nominal 2 × 8 joists = 7¼″
  2 × 4 bearing surface on girder = 1½″
  W8 × 31 = 8″
- Total floor framing = 17¾″ or 1′-5¾″
- Subtract total floor framing
  8′-10½″ minus 1′-5¾″ = 7′-4¾″
- Add thickness of concrete slab
  7′-4¾″ plus 4″ = 7′-8¾″

Figure 16–7 Calculate the height of the steel column.

USING WHAT YOU LEARNED

Laminated veneer lumber (LVL) is a popular material for girders in light construction, because it is strong and available in any length desired. Describe three ways floor joists made of dimensional lumber might be supported by an LVL girder.

1. The joists can rest on top of the girder.
2. A ledger piece can be nailed along the bottom of the LVL to support the joists.
3. Metal joist hangers can be used.

If *I-joists* (often called TJIs, which are discussed in Unit 17) are used for the joists, they would most likely span the full width of the floor and rest on top of the girder, because I-joists are available in greater lengths than dimensional lumber.

Assignment

Questions 1 through 5 refer to **Figure 16–8.**

1. What is the length of the girder?
2. Describe the material used to build the girder, including the size of the material.
3. What supports the span of the girder? (Include material and cross-sectional size.)
4. How many posts, columns, or piers support the girder?
5. What is the height of the columns or piers supporting the girder, including bearing plates?

Questions 6 through 9 refer to **Figure 16–9.**

6. What is the length of the girder?
7. Describe the material used to build the girder, including the size of the material. What supports the span of the girder? (Include material and cross-sectional size.)
8. How many posts, columns, or piers support the girder?
9. What is the dimension from the top of the footing under the pier to the top of the steel beam?

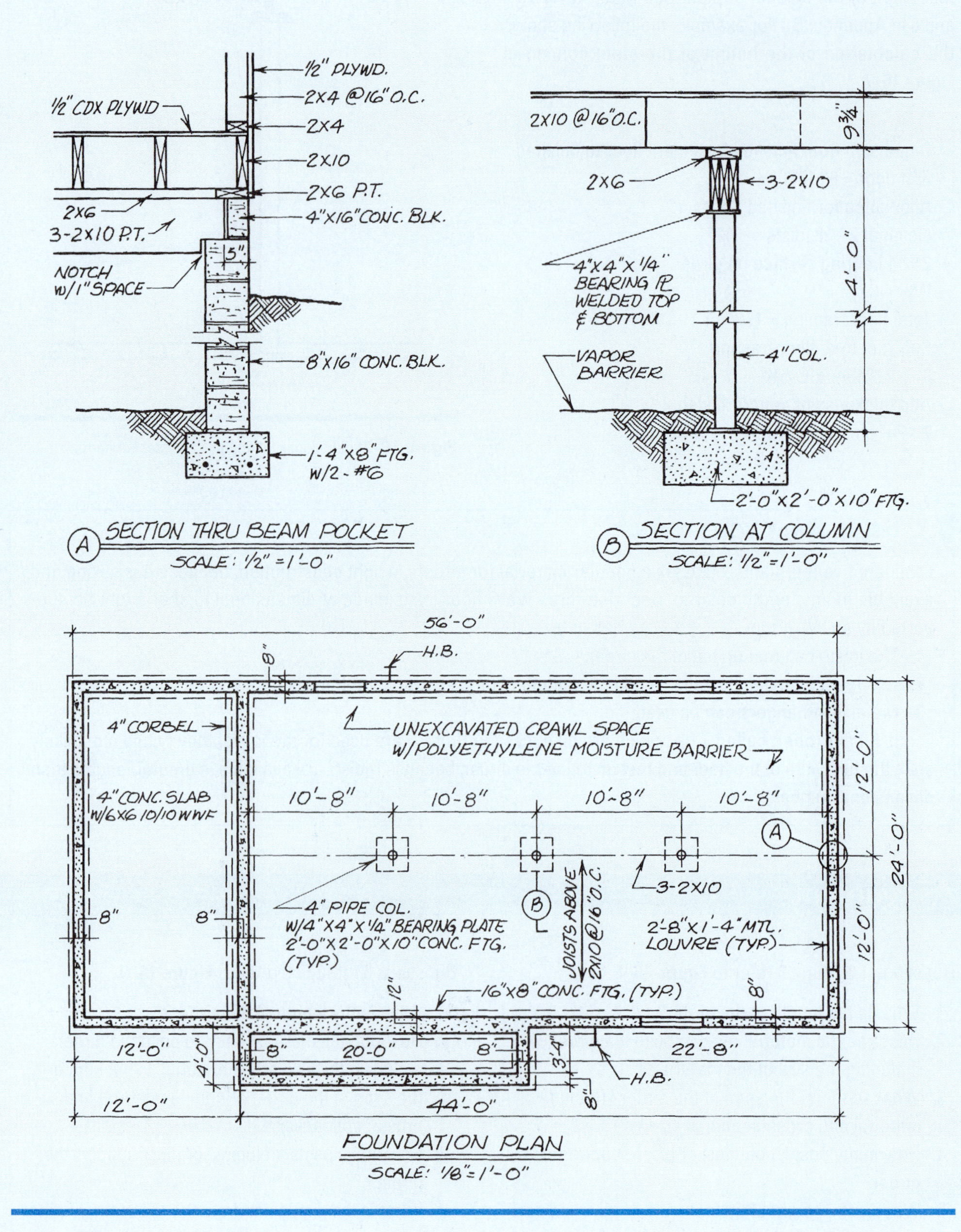

1/2" PLYWD.
2X4 @16" O.C.
1/2" CDX PLYWD
2X4
2X10
2X6 P.T.
2X6
3-2X10 P.T.
4"X16" CONC. BLK.
5"
NOTCH W/1" SPACE
8"X16" CONC. BLK.
1'-4"X8" FTG. W/2-#6
A SECTION THRU BEAM POCKET
SCALE: 1/2" = 1'-0"
2X10 @16" O.C.
9 3/4
2X6
3-2X10
4'-0"
4"X4"X1/4" BEARING PL WELDED TOP & BOTTOM
VAPOR BARRIER
4" COL.
2'-0"X2'-0"X10" FTG.
B SECTION AT COLUMN
SCALE: 1/2"=1'-0"
56'-0"
H.B.
8"
4" CORBEL
UNEXCAVATED CRAWL SPACE W/POLYETHYLENE MOISTURE BARRIER
4" CONC. SLAB W/6X6 10/10 WWF
10'-8"
10'-8"
10'-8"
10'-8"
12'-0"
24'-0"
12'-0"
A
3-2X10
B
8"
8"
4" PIPE COL. W/4"X4"X1/4" BEARING PLATE 2'-0"X2'-0"X10" CONC. FTG. (TYP.)
JOISTS ABOVE 2X10 @16" O.C.
2'-8"X1'-4" MTL. LOUVRE (TYP.)
12"
16"X8" CONC. FTG. (TYP.)
8"
12'-0"
4'-0"
8"
20'-0"
8"
3'-4"
22'-8"
H.B.
8"
12'-0"
44'-0"
FOUNDATION PLAN
SCALE: 1/8"=1'-0"

Figure 16–8

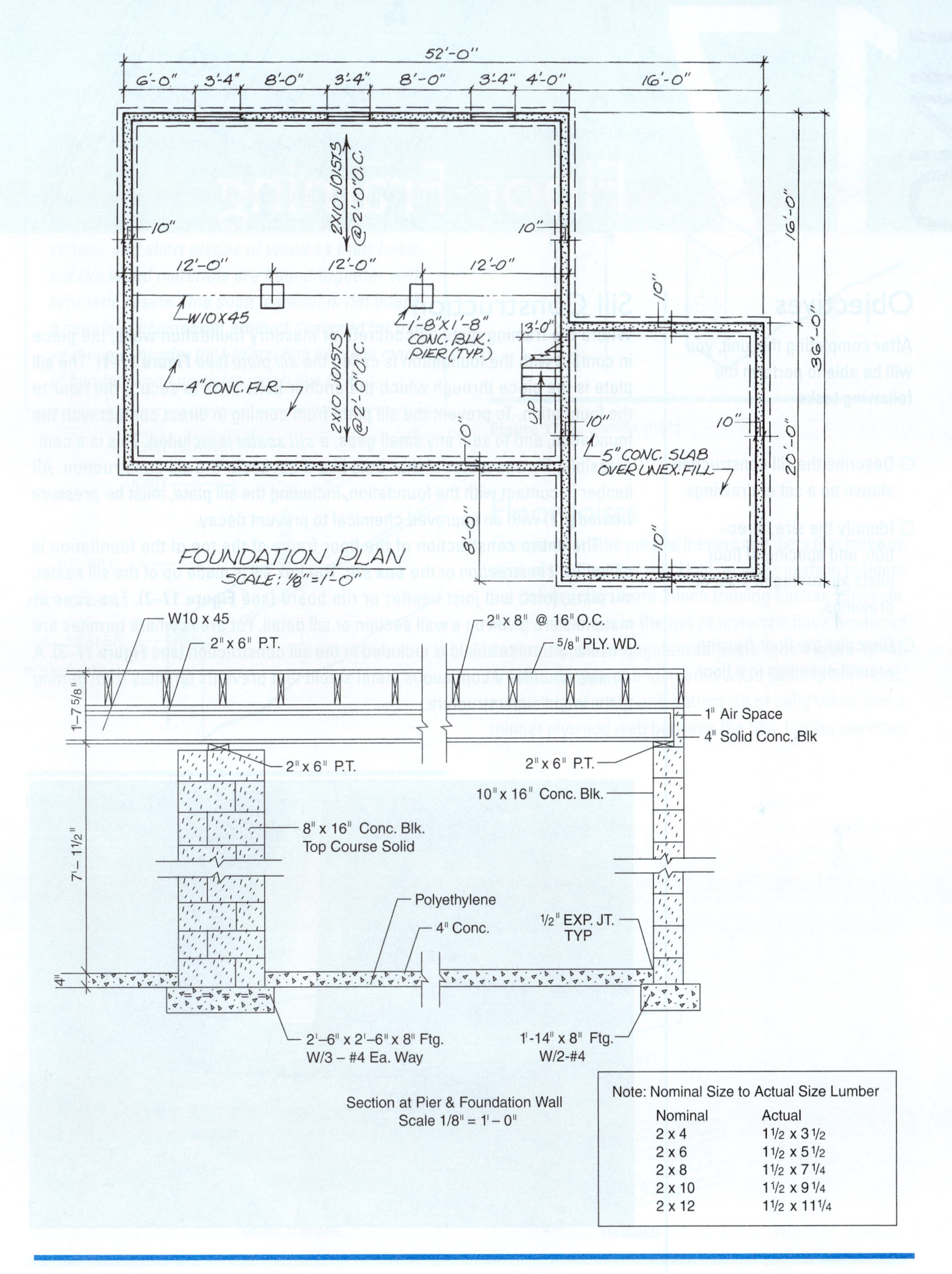

Note: Nominal Size to Actual Size Lumber

Nominal	Actual
2 x 4	1 1/2 x 3 1/2
2 x 6	1 1/2 x 5 1/2
2 x 8	1 1/2 x 7 1/4
2 x 10	1 1/2 x 9 1/4
2 x 12	1 1/2 x 11 1/4

Figure 16–9

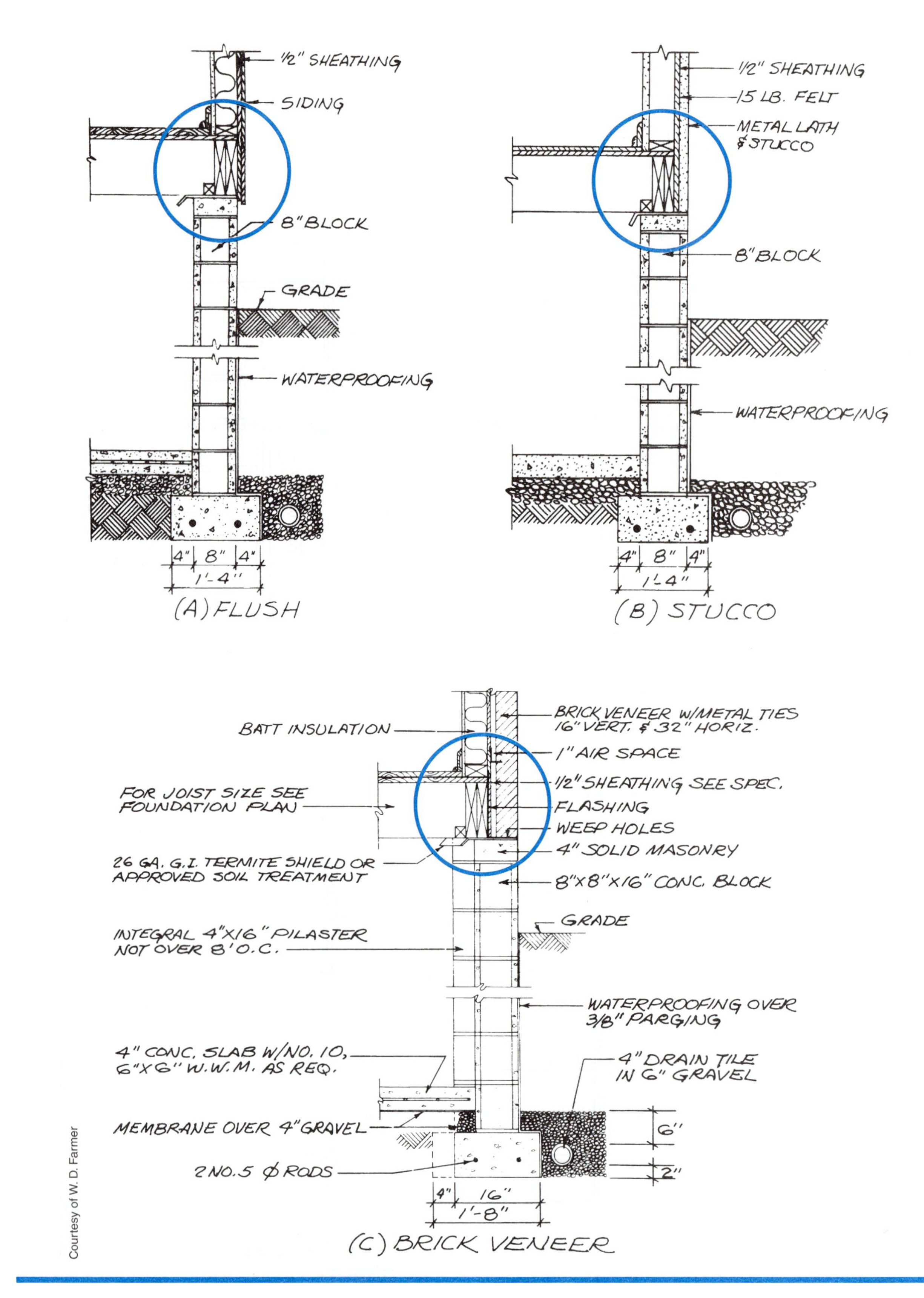

Figure 17–10. The joist headers are positioned according to the exterior finish to be used.

LVL beams rest in beam pockets that are welded to the posts. The structural steel channel (C8 × 11.5) is bolted directly to the posts.

The inner ends of the remaining floor joists in the Lake House are supported by bearing walls in the lower level. For example, some of the kitchen floor joists are supported by the west wall of the playroom, near the fireplace, and some by the LVL beam that spans the distance from the playroom wall to the foundation in the crawl space.

The size and spacing of material to be used are given on the framing plan by a note. They can also be found on the wall section. Lengths of framing members are usually not included on framing plans. However, these lengths can be found easily by referring to the floor plan that shows the location of the walls or beams on which the joists rest. For example, refer to **Figures 17–11** and **17–12** and find the length of the floor joists in bedroom #2 as follows:

1. The dimension from the outside of the north wall to the centerline of the 3½″ ⊠ steel post is 14′-1″.
2. The dimension from the centerline of the 3½′ ⊠ post is 2″, so the overall dimension of the bedroom floor is 14′-3″.
3. According to Wall Section 3/4, the rim board is flush with the north foundation wall, so subtract 1¼″(the thickness of the rim board) from each end.
4. 14′-3″ minus 2½″ (1¼″ at each end) equals 14′-1/2″ (the length of the joists).

Some floor framing is cantilevered to create a seemingly unsupported deck.

Cantilevered framing consists of joists that project beyond the bearing surface to create a wide overhang. This technique is used extensively for balconies, under bay windows, and for garrison-style houses (see **Figure 17–13**).

Framing at Openings

Where stairs and chimneys pass through the floor frame, some of the joists must be cut out to form an opening. The ends of these joists are supported by headers made of two or more members. The full joists at the sides of the opening have to carry the extra **load** of the shortened joists and headers, so they are also doubled or tripled (see **Figure 17–14**). The number of joists and headers required around openings of various sizes is spelled out in building codes and shown on framing plans.

framing (see **Figure 19–9**). These systems use top and bottom wall plates, studs, and floor joists. To make light-gauge framing compatible with wood, the metal members are made in common sizes for wood framing. The greatest difference is that metal framing is joined with screws instead of nails.

Headers over openings in light-gauge metal framing are usually very similar to those in wood framing. However, the system designed by the metal framing manufacturer should always be followed.

GREEN NOTE

The APA—The Engineered Wood Association publishes several Design and Construction Guides on a variety of topics suggesting ways to use wood products more efficiently and to make structures more durable. Most of these guides are fairly recent, so the techniques described in them have not been adopted by all designers and builders. Anyone in a position to make decisions about construction techniques would do well to become familiar with these and other modern construction developments.

Figure 19–9. The parts of this steel frame are the same as those of a wood frame. Notice the window header and cripple studs.

USING WHAT YOU LEARNED

Carpenters must calculate the lengths of all of the headers over windows and doors. Assuming the rough opening is to be 2 inches greater than the width of the door, how long must the header be over the door into the bath on the lower level of the Lake House?

This is door #002. The Schedule of Doors on Sheet 6 tells us that it is a 2′-0″ × 6′-8″ door. (That is a narrow doorway, but there is not enough space for a larger door.) The header rests on a 2 × 4 at each end, so the header length will be 3 inches longer than the rough opening (1½ inches at each side of the opening). Add the dimensions to find the length of the header: 2′-0″ + 2″ + 3″ = 2′-5″.

Assignment

Refer to the Lake House drawings in your textbook packet to complete this assignment.

1. What type of header should be used over the door from the hall to bedroom #1?
2. What type of header should be used over the door from the deck to the kitchen?
3. Why should these two headers be made differently?
4. What is the length of the header over the garage overhead door? Allow for two trimmers at each side and 1 inch at each side for jambs.
5. What are the R.O. dimensions for the door from the deck into the garage?
6. Name the location and give the R.O. dimensions for each interior door on the Upper Level Floor Plan.
7. According to **Figure 19–3**, what are the R.O. dimensions for the windows in bedroom #2?
8. How long is the header over the window in bedroom #2?
9. How many cripple studs are needed beneath the windows in the south wall of the living room?
10. What is the length of the cripple studs beneath the window in bedroom #2? (Assume the bottom of the header is 6′–8½″ from the top of the subfloor.)

The packet of engineering drawings usually includes a ***truss layout plan*** like the one shown in **Figure 21–4.** This plan shows the position of each truss on the building, with a label to indicate the truss ID. If you know from where in the building the section view in **Figure 21–2** is taken, you can find that truss on the truss layout. The section view is labeled as A/5, so look at the second floor plan in **Figure 21–5** to see where the cut was made to take section A/5. The section was cut through the master bedroom and master bath, which are on the left side of the house.

Looking at the front building elevations in **Figure 21–6** you can see that this house has a small gable on the roof above the main entrance. Refer to the truss layout, **Figure 21–4**, to see how this gable is to be framed. There is a triangle with three

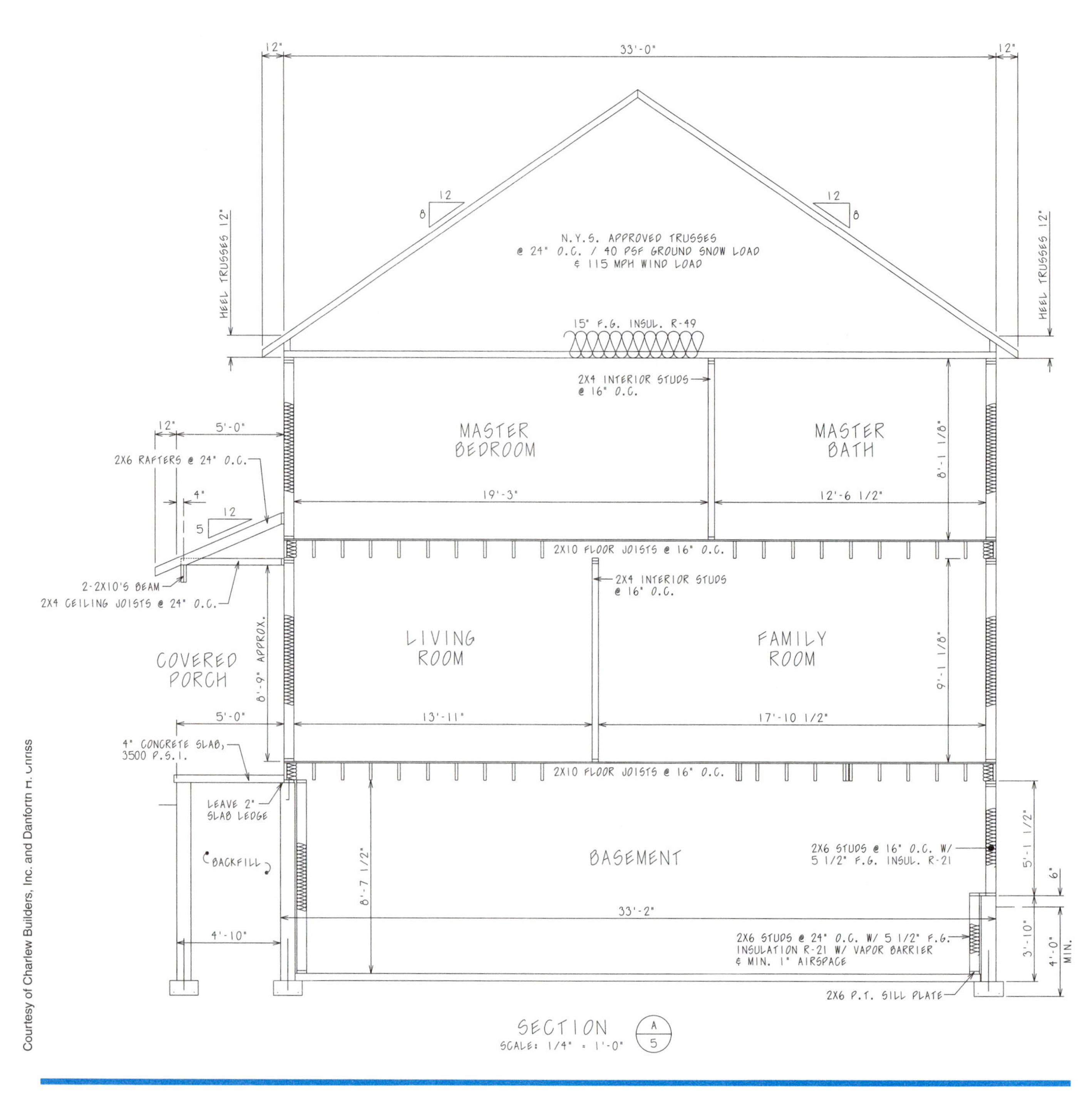

Figure 21–2. Building section.

trusses: V2, V1, and T2GE. The main roof trusses, T1A run beneath the trusses for the small gable, so the small trusses are applied on the main trusses after the main trusses are secured in place. Refer to the truss delivery sheet to see what the trusses look like and how big they are. V1 and V2 are the last two trusses on the delivery sheet and T2GE is the fourth truss from the bottom.

Bellevue Builders Supply **500 Duanesburg Road** **Schenectady, NY 12306** **(518) 355-7190 Fax: (518) 355-1371**	To: Charlew Constr. 130 Princetown Plaza Schenectady, NY 12306	Delivery Job Number: **J16-02632** Page: 1 Date: 01/17/17 09:53:28
Project: Charlew/83 Tamarack Block No: Model: Lot No: Contact: Site: Office: Name: Phone: Fax:	Deliver To: 83 Tamarack Lane Niskayuna, NY	Account No: 000000003 Designer: VB / VB Salesperson: Mike Scheidel Quote Number: B16-02632 P.O. Number: 9010847

Profile:	Qty:	Truss Id:	Span:	Truss Type:	Slope	LOH	ROH		Load By:
	7	**T3**	**38-10-00**	PIGGYBACK	**8.00**	00-10-08	00-10-08	**Bundle: 02632-A**	
	1	**T3SGE**	**38-10-00**	GABLE	**8.00**	00-10-08	00-10-08	**Bundle: 02632-B**	
	6	**T1A**	**32-10-08**	COMMON	**8.00**	00-10-08		**Bundle: 02632-C**	
	10	**T1**	**33-00-00**	COMMON	**8.00**	00-10-08	00-10-08	**Bundle: 02632-D**	
	2	**T1GE**	**33-00-00**	GABLE	**8.00**	00-10-08	00-10-08	**Bundle: 02632-E**	
	3	**T4**	**26-06-00**	COMMON	**8.00**	00-10-08	00-10-08	**Bundle: 02632-F**	
	1	**T4GE**	**26-06-00**	GABLE	**8.00**	00-10-08	00-10-08	**Bundle: 02632-F**	
	7	**PB1**	**08-05-12**	GABLE	**8.00**			**Bundle: 02632-G**	
	1	**PB1GE**	**08-05-12**	GABLE	**8.00**			**Bundle: 02632-G**	
	1	**T2GE**	**10-00-00**	COMMON	**10.00**			**Bundle: 02632-H**	
	1	**M1**	**05-09-08**	MONOPITCH	**8.00**	00-10-08		**Bundle: 02632-H**	
	1	**V1**	**06-09-09**	VALLEY	**10.00**			**Bundle: 02632-I**	
	1	**V2**	**03-07-03**	VALLEY	**10.00**			**Bundle: 02632-I**	

Above listed items have been received in good condition. (Exceptions listed to right)

Received by: ______________________________

Date: ______________________________

Thank you for your business.

Courtesy of Bellevue Builders Supply

Figure 21–3. Truss delivery list.

Assignment

A. Refer to Figure 23–14 to complete questions 1–11.

1. What is the run of the common rafters at A?

   Note: Do not include the overhang.

2. How much overhang does the roof have?
3. What is the actual length of the common rafters at A?
4. What is the actual length of the hip rafter at B?
5. What is the run of the common rafters at C?
6. What is the actual length of the common rafters at C?
7. What is the length of the short valley rafter?
8. What is the actual length of the shortest hip jack rafter?
9. What is the actual length of the second shortest hip jack rafter?
10. What is the actual length of the shortest valley jack rafter?
11. What is the actual length of the second shortest valley jack rafter?

B. Refer to the Lake House drawings (in the packet) to complete this part of the assignment.

12. What is the length of the structural steel hip rafter over the dining room?

    Note:

    - This hip rafter is a steel channel shown on Roof Framing Plan 2/6 and Details 4/6 and 5/6, and marked as MC8 × 8.7
    - Remember to allow for the distance from the column centerline to the end of the rafter as dimensioned on the detail drawing.
    - This roof has an unusual pitch of 2.96 in 12. This is close enough to use 3 in 12 for calculating rafter lengths.

Cornices

UNIT 24

Types of Cornices

The *cornice* is the construction at the place where the edge of the roof joins the sidewall of the building. On hip roofs, the cornice is similar on all four sides of the building. On gable and shed roofs, the cornice follows the pitch of the end (***rake***) rafters. The cornice on the ends of a gable or shed roof is sometimes called simply the *rake* (see **Figure 24–1**). The three main types of cornice are the *box cornice,* the *open cornice,* and the *close cornice.*

Box Cornice

The box cornice boxes the rafter tails. This type of cornice includes a fascia and soffit (see **Figure 24–2**). The ***fascia*** covers the ends of the rafter tails. The ***soffit*** covers the underside of the rafter tails. There are three types of box cornices. These types vary in the way the soffit is applied.

Sloping Box Cornice. In the sloping box cornice, the soffit is nailed directly to the bottom edge of the rafter tails. This causes the soffit to have the same slope or pitch as the rafter, (see **Figure 24–3**).

Narrow Box Cornice. In the narrow box cornice, the rafter tails are cut level. The soffit is nailed to this level-cut surface (see **Figure 24–4**).

Wide Box Cornice. In a wide box cornice, the overhang is too wide for a level cut on the rafter tails to hold the full width of the soffit. In conventional wood framing, *lookouts* are installed between the rafter ends and the sidewall. The lookouts provide a nailing surface for the soffit (see **Figure 24–5**). For a metal soffit, special metal channels fastened to the sidewall and the back of the fascia hold the soffit (see **Figure 24–6**).

Open Cornice

In an open cornice, the underside of the rafters is left exposed (see **Figure 24–7**). Blocking is installed between the rafters and above the wall plate to seal the cornice from the weather. An open cornice may or may not include a fascia.

Objectives

After completing this unit, you will be able to perform the following tasks:

- Describe the cornice construction shown on a set of drawings.
- List the sizes of the individual parts of the cornice shown on a set of drawings.
- Describe the provisions for attic or roof ventilation as shown on a set of drawings.

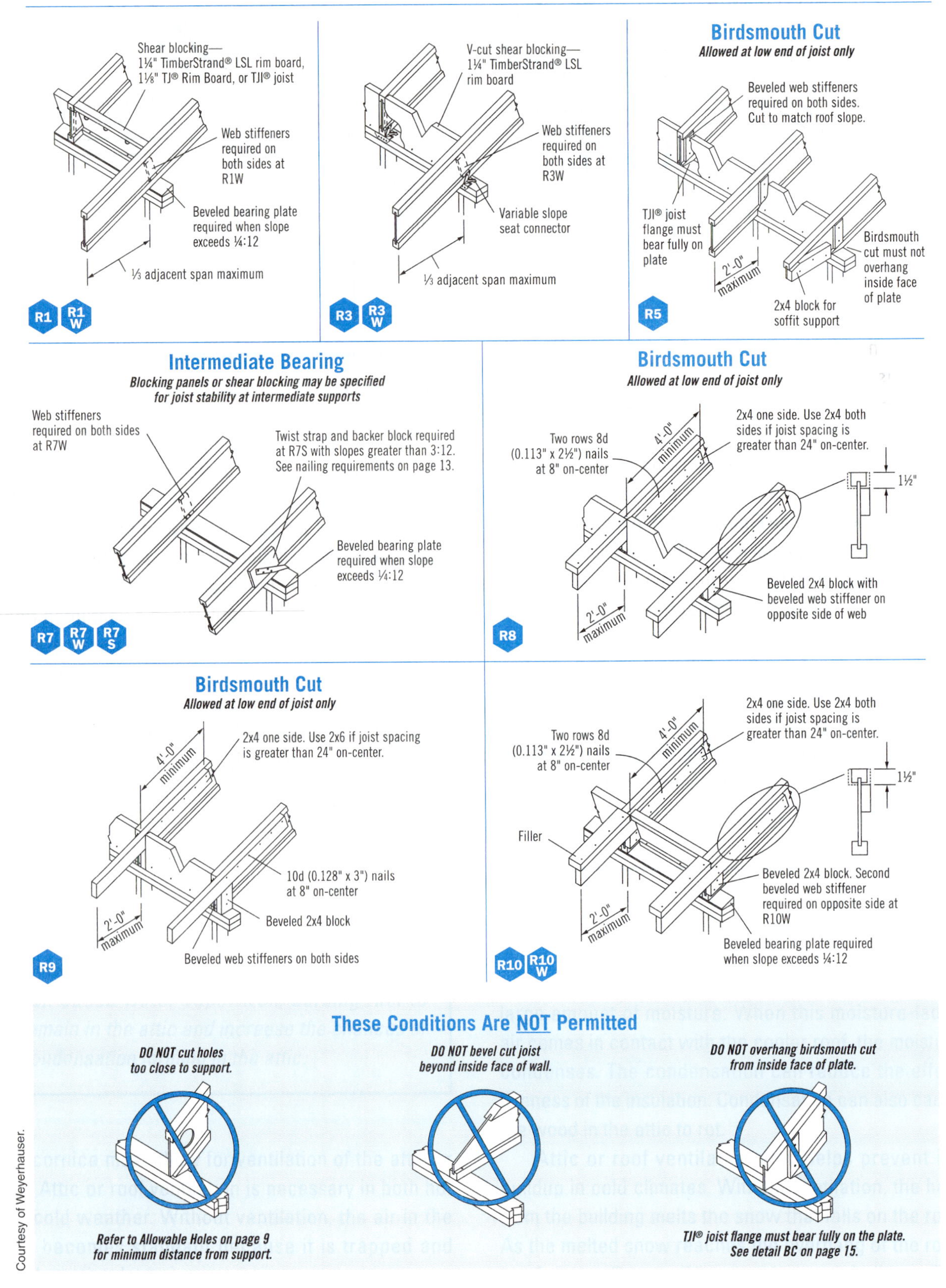

Figure 24–15. Roof details for framing with TJIs.

USING WHAT YOU LEARNED

It is important to allow ventilating air to flow along the bottom of the roof from the cornice to the upper ends of the rafters. In the Lake House, what prevents birds and rodents from entering with the air?

The typical cornice construction is shown in Roof Detail 1/7. That drawing shows a continuous vent above the OSB sheathing and outside the web stiffeners. This continuous vent has either screening or small enough openings so that only air can enter.

Assignment

Refer to the Two-unit Apartment and the Lake House drawings in your textbook packet to complete the assignment.

1. Which type of cornice does the Apartment have?
2. What material is used for the Apartment cornice?
3. How wide is the Apartment soffit?
4. The Apartment fascia is made of two parts. What are they?
5. What provision does the Apartment cornice have for ventilation?
6. How does attic air exit from the Apartment?
7. Sketch the Lake House cornice and show where air enters for ventilation.
8. There are two ways that air can escape from the Lake House roof. Describe one.

UNIT 25 Windows and Doors

Objectives

After completing this unit, you will be able to perform the following tasks:

- Interpret information shown on window and door details.
- Find information in window and door manufacturers' catalogs and online listings.

Window Construction

Most windows are supplied by manufacturers as a completely assembled unit. However, the carpenters who install windows often have to refer to window details for information. Some special installations require knowledge of the construction of the window unit. Also, when a special window is required, the carpenter may build parts of it on the construction site.

Wood Windows

The major types of windows are briefly discussed in Unit 19. All these windows include a frame and sash. The ***sash*** is the glass and the wood (or metal) that holds the glass. The sash is made of ***rails*** (horizontal parts) and ***stiles*** (vertical parts) (see **Figure 25–1**). The sash may also include muntins. ***Muntins*** are small strips that divide the glass into smaller panes. The glass is sometimes called the *lite*.

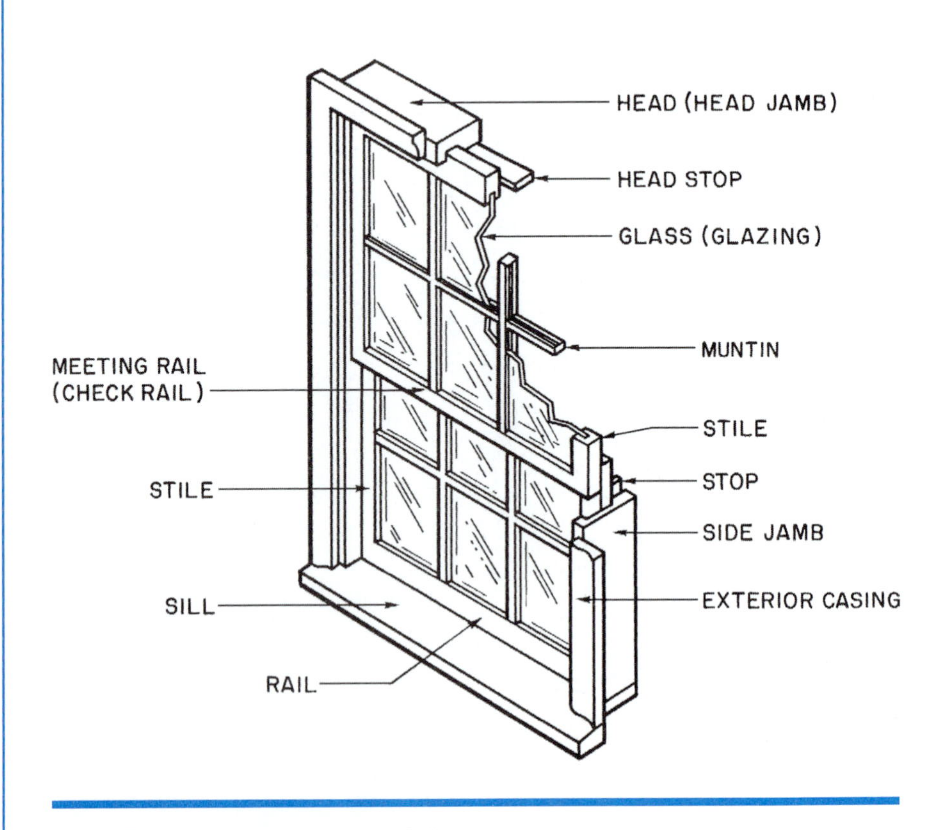

Figure 25–1. Parts of a double-hung window.

The window frame is made of the *side jambs*, the *head jamb*, and the ***sill***. Stop molding is applied to the inside of the jambs to hold the sash in place. Factory-built windows also come with the exterior casing installed. The ***casing*** is the molding that goes against the wall around the frame. The interior casing and the apron, if one is included, are applied after the window is installed.

Metal Windows

Many buildings have vinyl or metal windows. Improvements in the design of metal windows have made them competitive with wood windows in both cost and energy efficiency. The most important of these design improvements has been the development of thermal-break windows. ***Thermal-break windows*** use a combination of air spaces and materials that do not conduct heat easily to separate the exterior from the interior.

The basic parts of a metal or vinyl window are similar to those of a wood window. The sash consists of a stile, rails, and glazing. The frame is made up of side jambs, a head jamb, and a sill. However, the trim (casing) is not included as part of the window. Often the window frame itself is the only trim used on the exterior. The frame includes a nailing fin for attaching the window to the building framing.

Window Details

All windows include the parts discussed so far. However, to show the smaller parts, which vary from one window style to another, architects and manufacturers use detail drawings. The most common type of window detail is a section (see **Figure 25–2**). All the parts can be shown in section views of the head, sill, and one side jamb. These sections also usually show the wall framing around the window.

Some of the parts that can be found on window sections are defined here. Find each of the parts on the sections and illustrations in **Figures 25–3, 25–4,** and **25–5.**

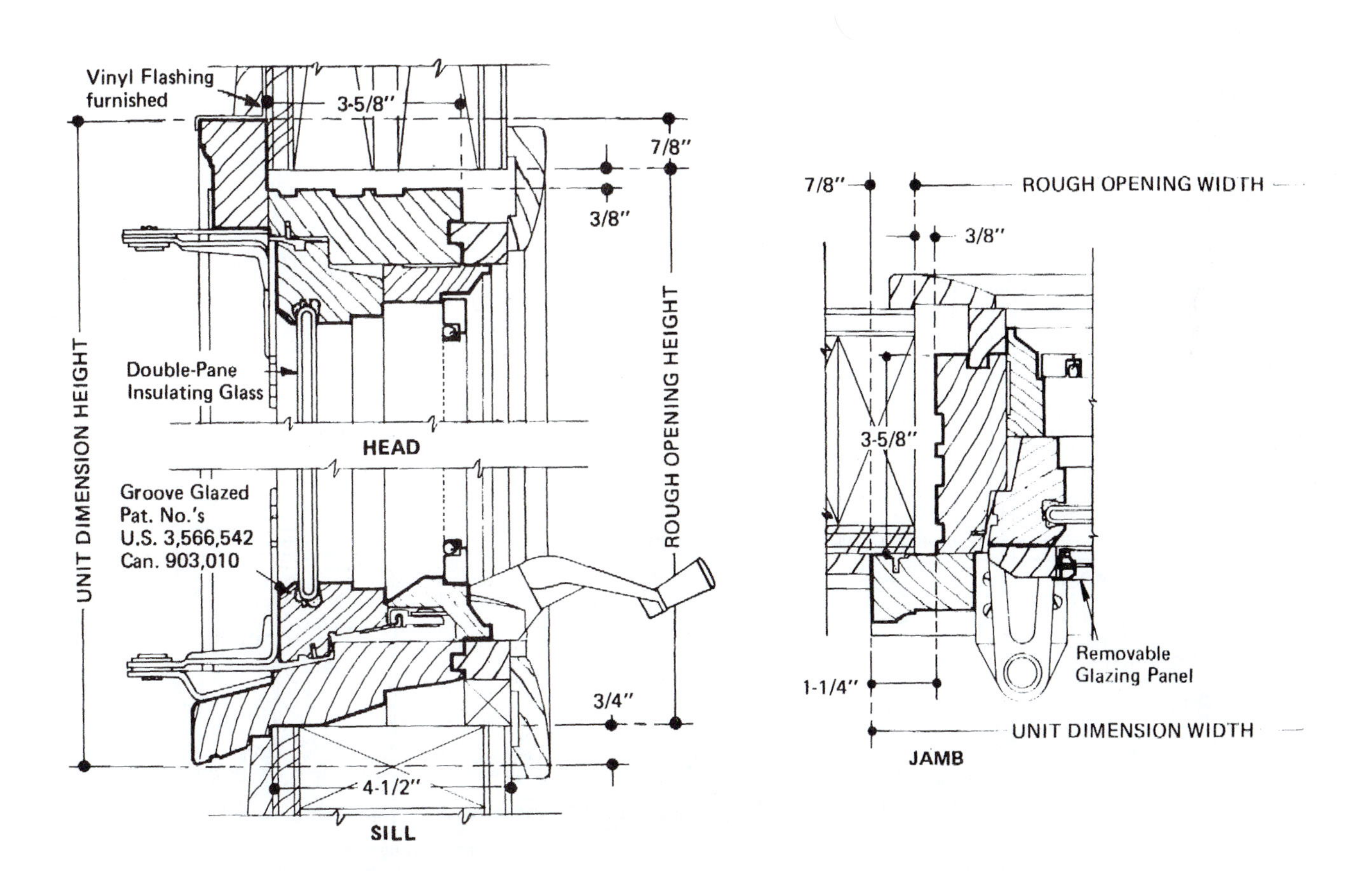

Figure 25–2. Typical casement window detail drawing.

GREEN NOTE

Fenestration is all of the openings in the building that are intended to allow air, daylight, people, animals, and vehicles to enter and exit the building. This is primarily windows, doors, and skylights. Even the best windows and doors tend to have poorer thermal insulating qualities than the surrounding wall. By eliminating all windows and doors, the heating and cooling load for the building could be greatly reduced, however, it would not be a pleasant home in which to live. Green home designers pay particular attention to the sizes and locations of all fenestration.

The thermal quality of a window is also greatly affected by its glazing. By using glass with low energy transmission properties (low-E glass) and filling the space between the layers of glass with inert gas, its heat transmission and solar properties can be improved.

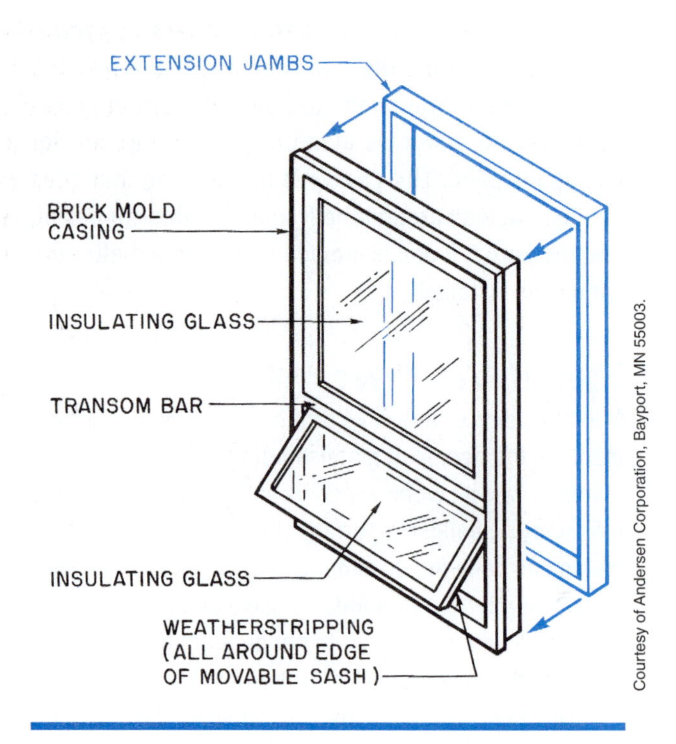

Figure 25–3. Window parts.

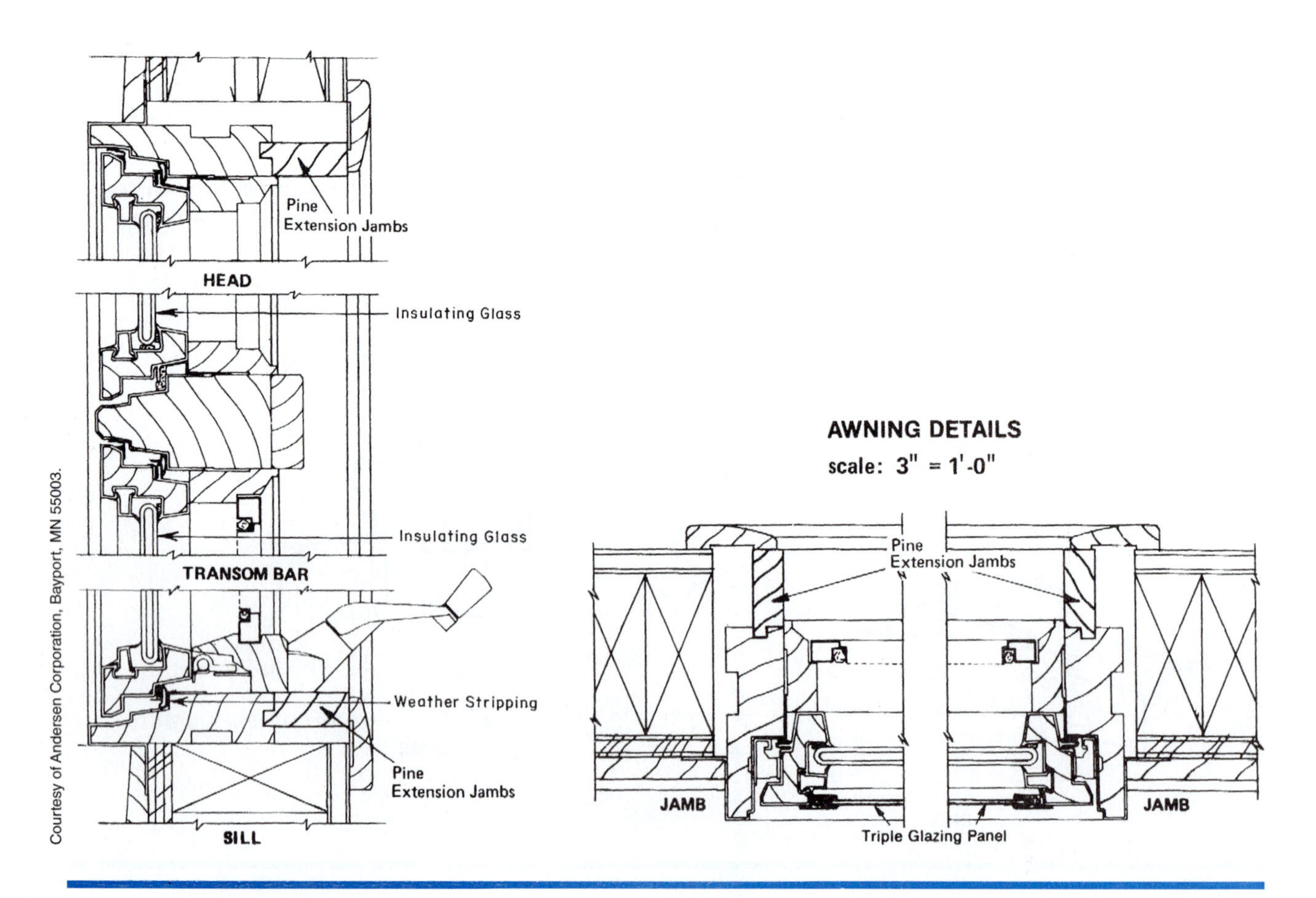

Figure 25–4. Section views of an awning window.

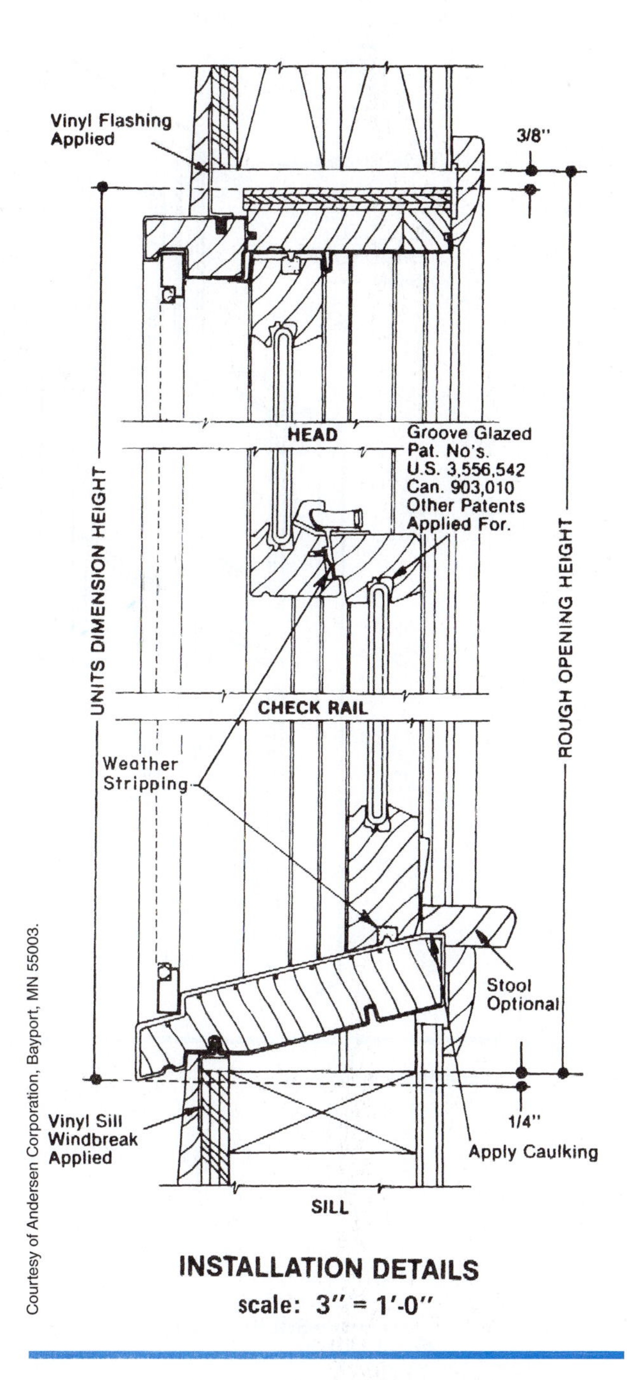

Figure 25–5. Section view of a double-hung window.

- *Weather stripping* is used on windows that open and close. It forms a weather-tight seal around the sash.
- The *transom bar* is the horizontal part of a window frame that separates the upper and lower sash when one is a fixed sash.
- *Meetings rails* or check rails are the rails that meet in the middle of a ***double-hung window***.
- *Insulating glass* is a double or triple layer of glass, creating a space between the layers. The space is filled with inert gas such as argon and acts as an insulator.
- *Extension jambs* are fastened to standard jambs when the window is installed in a thicker than normal wall.
- A ***mullion*** is a vertical section of the frame that separates side-by-side sash. If the mullion is formed by butting two windows together, it is called a *narrow mullion.* If the mullion is built around a stud or other structural support, it is called a support mullion.

Door Construction

Doors include many of the same basic parts as windows (see **Figure 25–6**). A door frame consists of side jambs and a head jamb with ***stop*** and casing. Exterior door frames also include a sill. Many doors are made of a framework with panels (see **Figure 25–7**). The parts of ***panel doors*** are named similarly to the parts of a window. The vertical parts are *stiles,* and the horizontal parts are *rails.* Doors with glass or louvers are variations of panel doors. The framework is made of rails and stiles, and the glass or louvers replace the

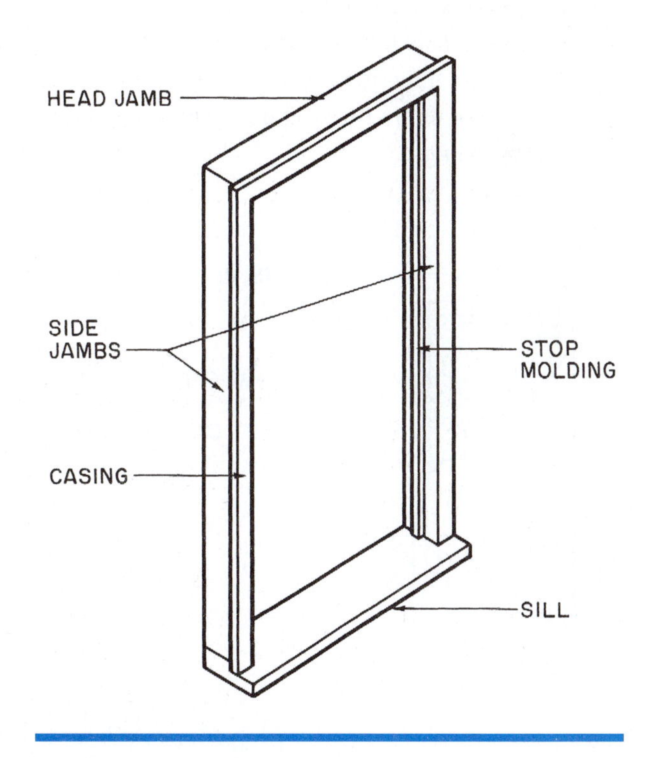

Figure 25–6. Parts of a door frame.

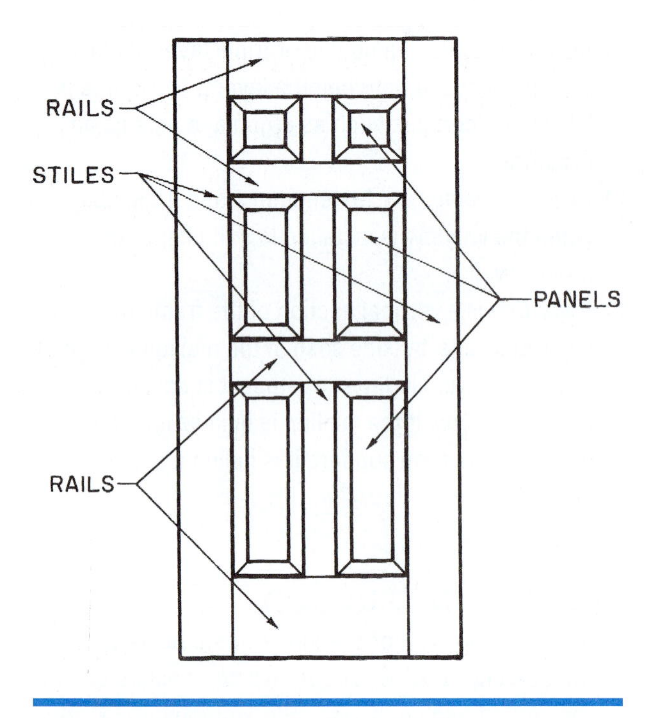

Figure 25–7. Construction of a panel door.

panels. Several manufacturers make molded doors. The most common type of molded door is made of hardboard for interior doors or steel for exterior doors which is manufactured in folds that contour the surface to look like panel doors (see **Figure 25–8**). *Hollow core doors* consist of an internal frame with "skin" applied to each side (see **Figure 25–9**). Exterior doors have insulation between the two outer steel skins. These insulated doors result in considerable heating and cooling savings.

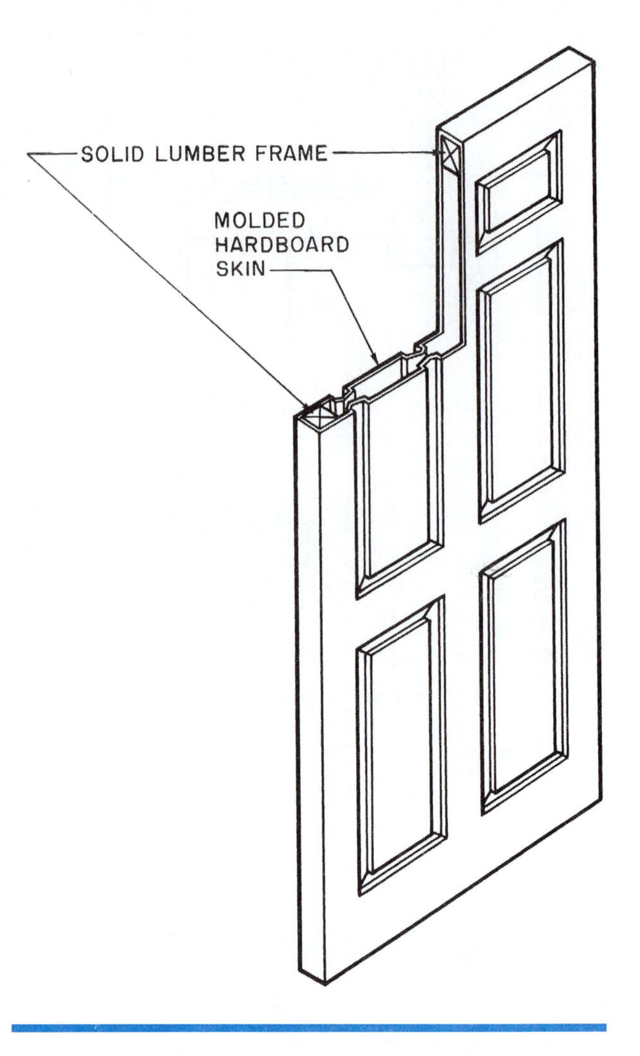

Figure 25–8. Molded hardboard door.

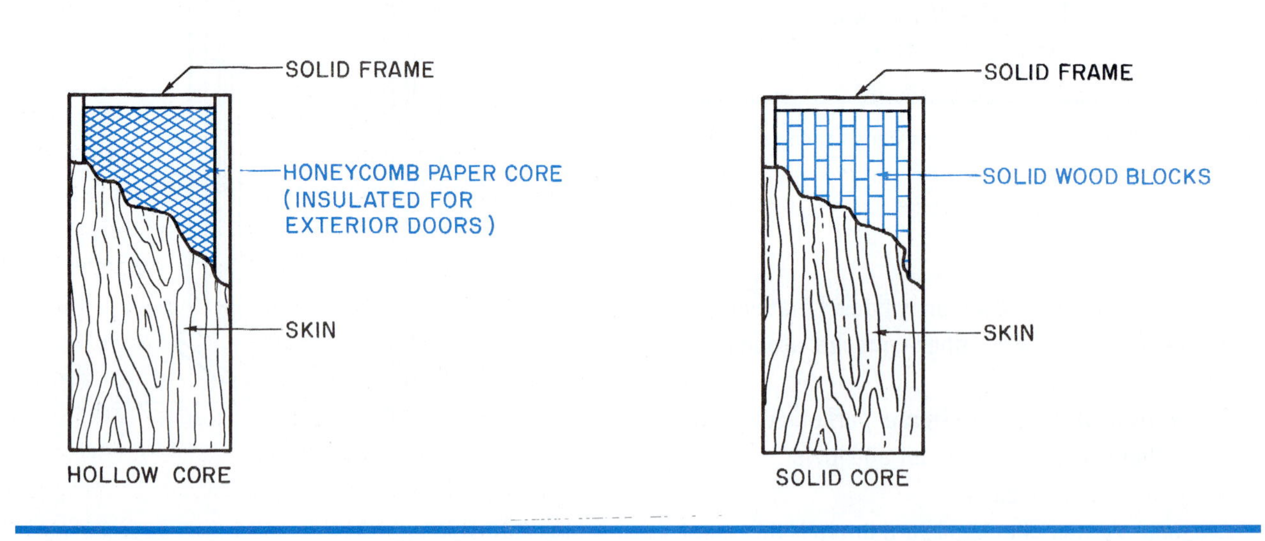

Figure 25–9. Flush doors.

Door Details

Door details are usually less complex than window details. Where security, fire alarm, electronic lock, or special systems must be run in door and window metal frames, the corresponding installers must familiarize themselves with the actual manufacturer details and coordinate their installation with the door and window installation schedule. Carpenters rarely make doors, so all that is needed are simple details of the door frame and its trim (see **Figure 25–10**).

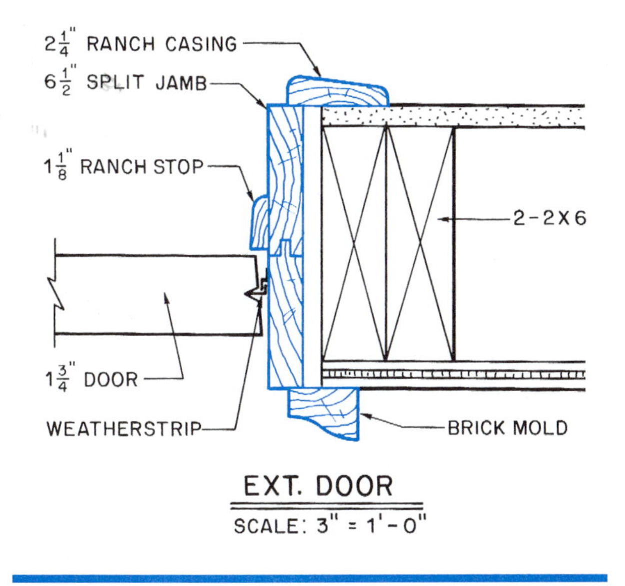

Figure 25–10. Typical jamb detail for an exterior door.

GREEN NOTE

Doors must insulate the conditioned space inside the home from the hot or cold weather outside and seal the opening against infiltration of air and water, sometimes in strong winds. If the door is glazed, that part of the door can be treated like a window and insulated with low-E glass and inert gas between the layers of glass. The rest of the door can be insulated by filling it with insulating material. To properly seal the opening requires a good weather seal on a good fitting door. If the weather seal is pinched or has gaps, air and water will enter at that point. If the door does not close evenly against the weather seal, that will cause leaks. The same principles apply to windows and skylights.

Many doors are sold as prehung units. In these units, the frame is assembled, including the trim, and the door is hung in the frame. A section view of the jambs shows how the door is installed. For example, the door detailed in **Figure 25–10** is made with two-piece jambs. These split jambs are pulled apart; each side is then slid into the opening for installation. The stop can be either applied or integral. *Applied stop* is molding that is applied to the jambs with finish nails. *Integral stop* is milled as a part of the jamb when the jamb is manufactured.

Reading Catalogs

It is often necessary to find specific information about windows or doors in the manufacturer's catalog or on its web site. Usually the catalog has a table of contents and web sites have a home page that lists the types of windows and doors available and illustrated. For each type of window or door, you will find some or all of the information listed below:

- A brief description of the window type and some of the features the manufacturer wants to highlight—a little advertising.
- Installation detail drawings.
- Sizes available—This information usually consists of drawings of the various sizes and arrangements, with dimensions for glass size, stud or rough opening, and unit dimensions. **Figure 25–11** shows typical window size Information reprinted from a manufacturer's catalog.
- Additional information, such as optional equipment available.

Each manufacturer uses its own design to show the information. It sometimes takes a minute of study to familiarize yourself with how the manufacturer's pages are designed. Also, each manufacturer may make slightly different sizes of stock windows. If the windows used in constructing a building are from a different manufacturer than the one the architect used to make the drawings, it may be necessary to find windows that are as close as possible to the sizes shown on the drawings. Of course, if the construction specifications call for a particular manufacturer, that specification must be followed unless a change is authorized by the architect or owner.

Table of Basic Unit Sizes Scale 1/8" = 1'-0" (1:96)

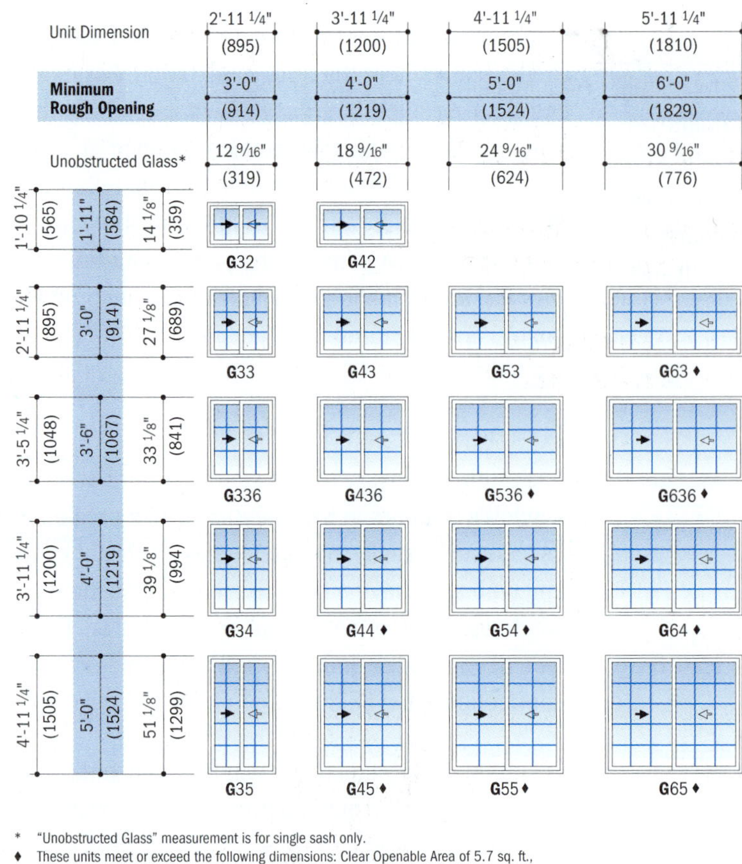

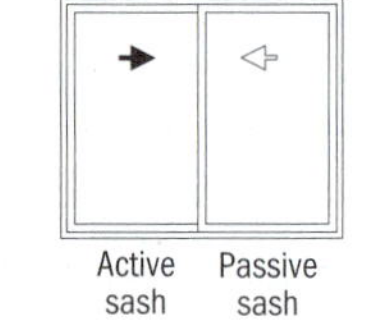

Venting Configuration

As viewed from the exterior. Passive sash will open after active sash has been opened.

* "Unobstructed Glass" measurement is for single sash only.
♦ These units meet or exceed the following dimensions: Clear Openable Area of 5.7 sq. ft., Clear Openable Width of 20" and Clear Openable Height of 24".

Rough opening dimensions may need to be increased to allow for use of building wraps, flashing, sill panning, brackets, fasteners or other items.

"Unit Dimension" always refers to outside frame to frame dimension.

Dimensions in parentheses are in millimeters.

When ordering, be sure to specify color desired: White, Sandtone or Terratone.®

Handle Locations — Operational Force = 8 lbs.

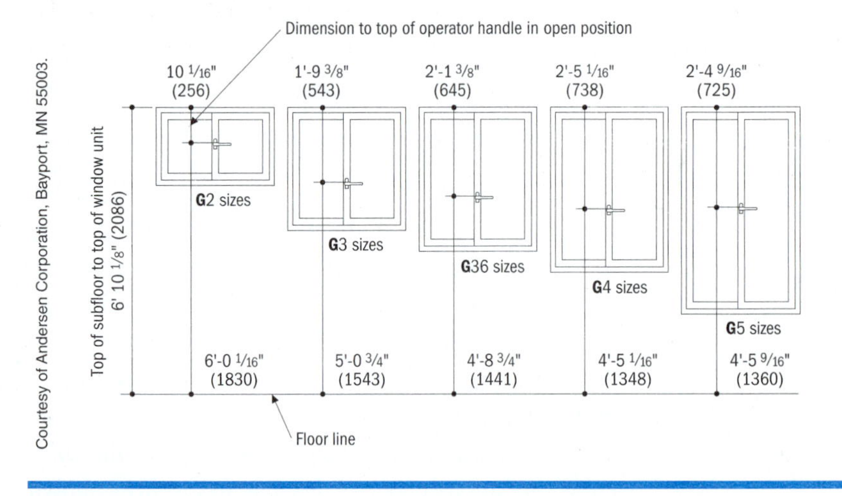

Figure 25–11. A typical manufacturer's catalog page showing window sizes.

USING WHAT YOU LEARNED

Sometimes drawings specify window sizes that are very close to but slightly different from the sizes that a particular manufacturer stocks. Custom-made windows can usually be ordered in any size requested, but when the stock sizes are very close, it is much less expensive to use a stock window that is close to the size shown on the drawings. For example, what size window shown in **Figure 25–11** might be used in the west wall of the dining room?

Refer to the Lake House example in your textbook packet. According to both the Upper Level Floor Plan and the West Elevation of the building, this is a "B" window on the Window Schedule. The Window Schedule on Sheet 5 tells us that a "B" window is 3′-0″ × 3′-0′. On the schedule of sizes in Figure 25–11, the unit dimensions are the top row and the far left column, so the closest size would be 2′-11¼″× 2′-11¼″. Coincidentally, the rough opening for that window is 3′-0″ × 3′-0″, shown in the second row of dimensions across the top and the second column of dimensions down the left side.

Assignment

Refer to the Lake House drawings when necessary to complete the assignment.

1. Name the lettered parts (a through f) in **Figure 25–12.**
2. What is the nominal size of the window in the south end of the Lake House dining room?
3. In the catalog sample shown in **Figure 25–11**, what is the width and height of the rough opening for a 3′-11¼″ × 2′-11¼″ window?
4. In the catalog sample, what are the rough opening dimensions for a window with a $24^9/_{16}$″ by $39^1/_8$″ glass size?
5. In the catalog sample in **Figure 25–11**, what is the glass size of the window in bedroom #2 of the Lake House?
6. In the catalog sample, what is the RO for the window in bedroom #2 of the Lake House?
7. Is the exterior door in the Lake House kitchen to be prehung or site hung?
8. What type and size are the doors in the Lake House playroom closet?

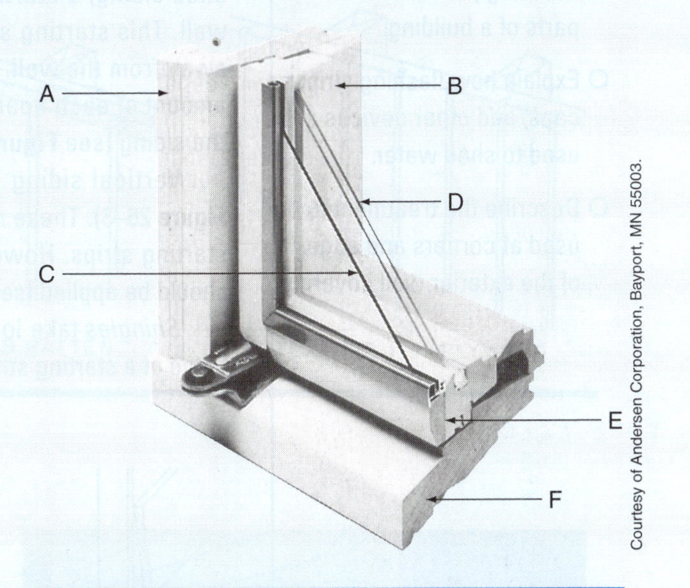

Figure 25–12.

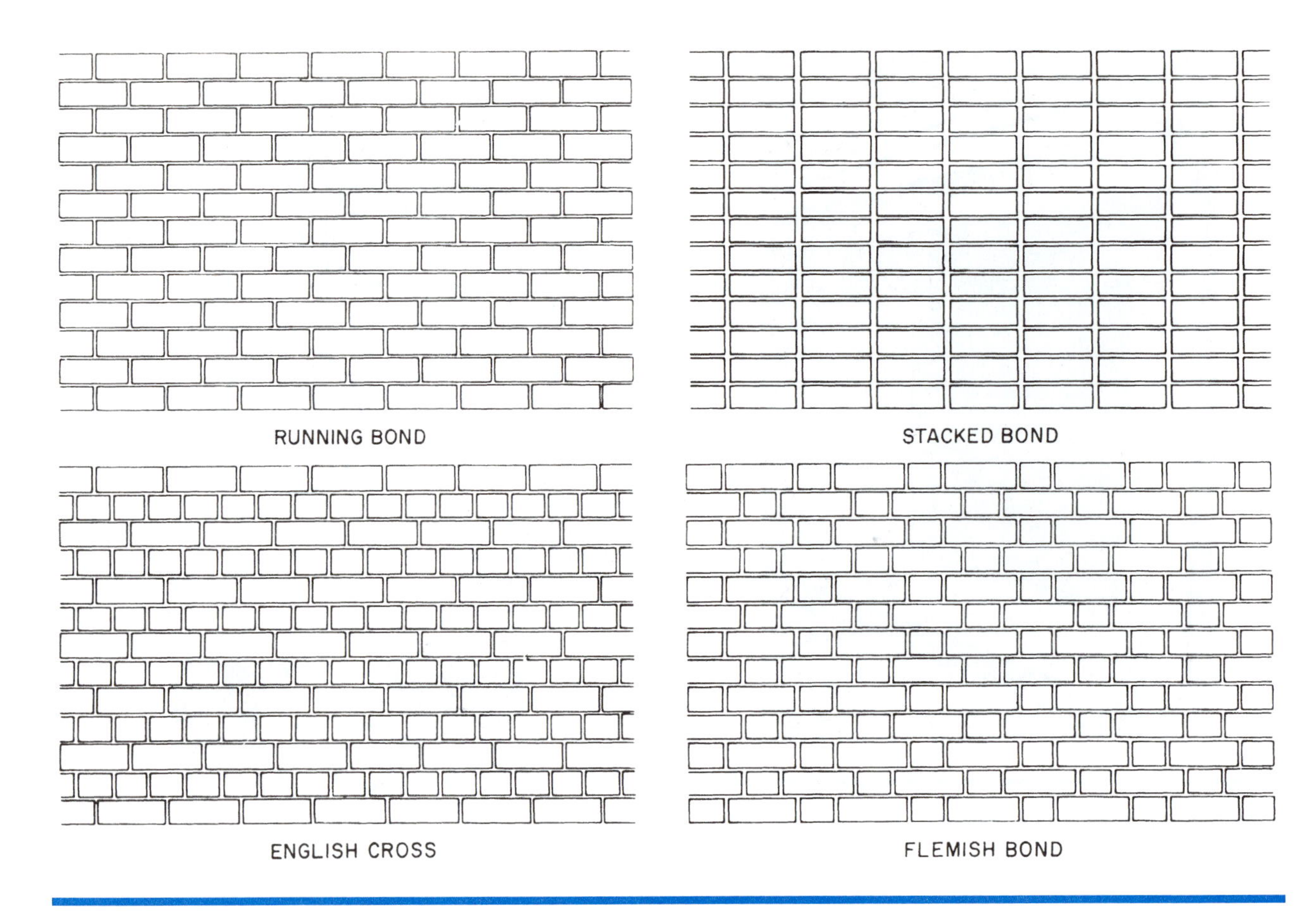

Figure 26–14. Frequently used bond patterns.

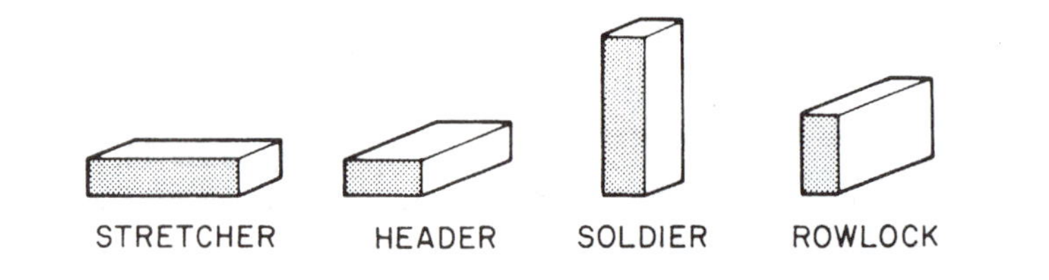

Figure 26–15. Brick positions.

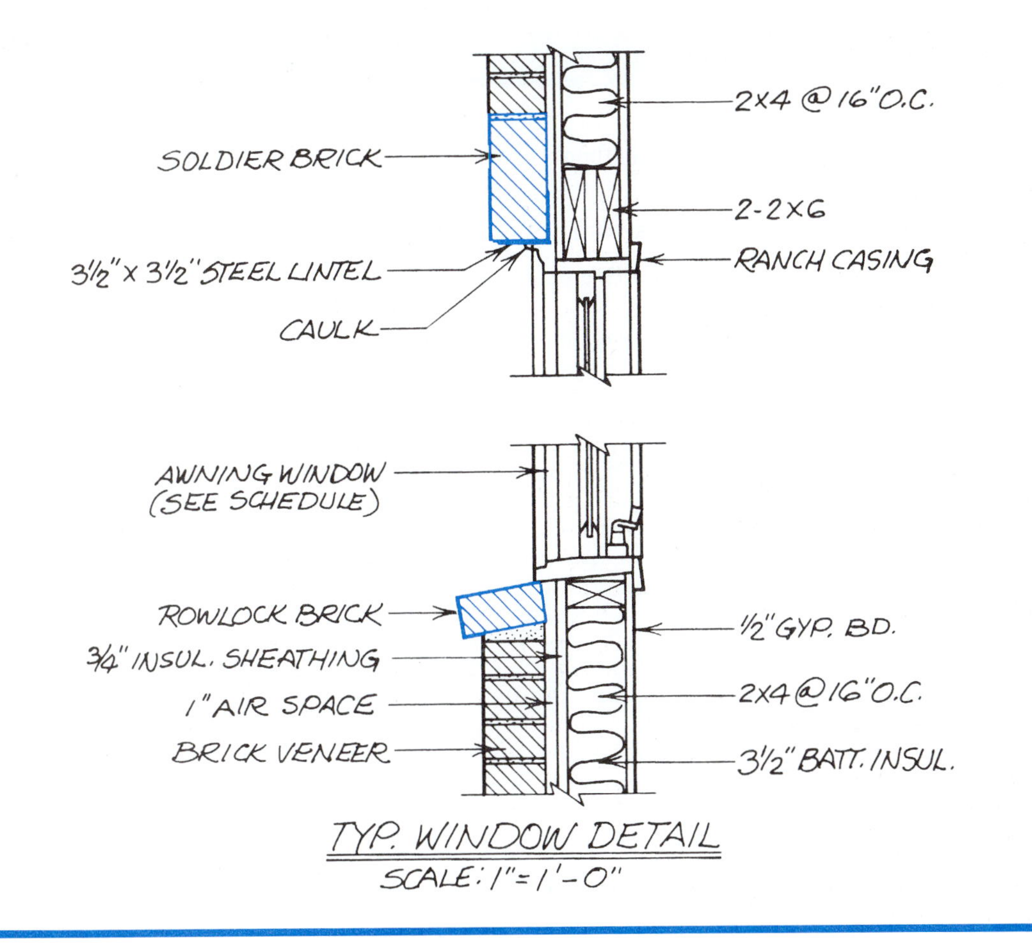

Figure 26–16. Notice the rowlock and soldier bricks on the window detail.

USING WHAT YOU LEARNED

Before the siding is applied, the carpenters should examine the drawings to know exactly how all of the edges are to be treated. Questions to consider include:

- Where is caulk to be used?
- Where are siding edges covered with trim?
- Exactly where do the edges fall? For example, where is the bottom edge of the siding relative to the top of the foundation? This is shown on the Lake House Wall Section 3/4. The bottom of the siding aligns with the top of the foundation wall.

Assignment

Refer to the Lake House drawings in your textbook packet to complete the assignment.

1. What material is used for the Lake House siding?
2. How are the outside corners of the Lake House siding finished?
3. Where is the bottom edge of the siding relative to the wall construction?
4. What prevents water from running under the siding at the heads of the Lake House windows?
5. Detail 1/7 of the Lake House shows aluminum screen nailed behind the top edge of the siding. What is the purpose of the opening covered by this screen?
6. Describe one use of aluminum flashing under the siding on the Lake House.

Decks

UNIT 27

Objectives

After completing this unit, you will be able to perform the following tasks:

- Explain how a deck is to be supported.
- Describe how a deck is to be anchored to the house.
- Locate the necessary information to build handrails on decks.

Wood decks are used to extend the living area of a house to the outdoors. A deck may be a single-level platform, or it may be a complex structure with several levels and shapes. However, nearly all wood decks are made of wood planks or synthetic planks made to simulate wood laid over joists or beams (see **Figure 27–1**). The planks are laid with a small space between them, so rainwater does not collect on the deck.

The same construction methods are used for decks and porches as for other parts of the house. The parts of deck construction that require special attention or that were not covered earlier in the text are discussed here.

Support

The deck must be supported by stable earth. The support must also extend below the frost line in cold climates. The most common method of support is by concrete columns with metal post anchors (see **Figure 27–2**). A typical metal post anchor is shown in **Figure 27–3**. Such anchors fasten posts to the concrete pier while keeping them from contacting the concrete in order to prevent decay. Indeed, all wood used in the construction of a deck should be pressure treated with a chemical to prevent such decay. Note that deck detail 3/6 for the Lake House includes a note that all wood is to be pressure treated.

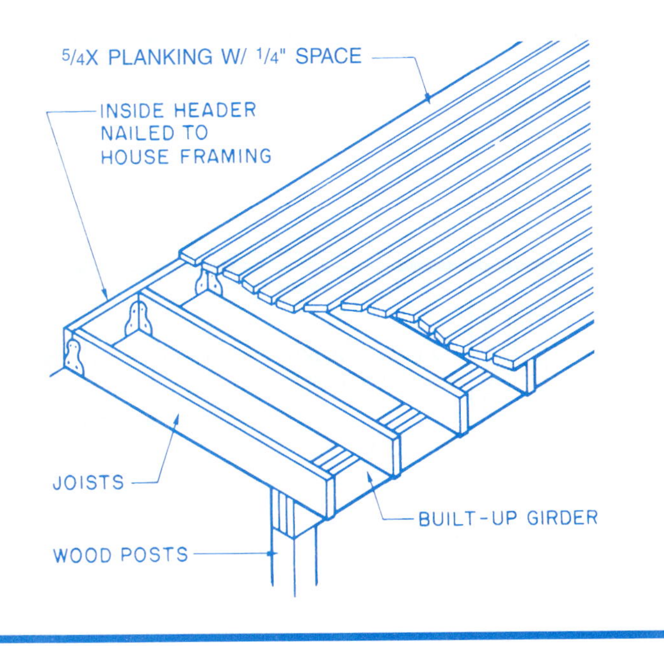

Figure 27–1. Typical deck construction.

> **GREEN NOTE**
>
> *Decks and outdoor structures are parts of the home. In the spirit of designing and building a green home, the same principles that apply to the house should be applied to decks. Design on a 2-foot module; avoid unnecessary corners, angles, and curves; select materials that require a minimum of maintenance and do not involve unnecessary embodied energy (the energy required to manufacture and transport a product); and always observe best practices—use quality workmanship.*

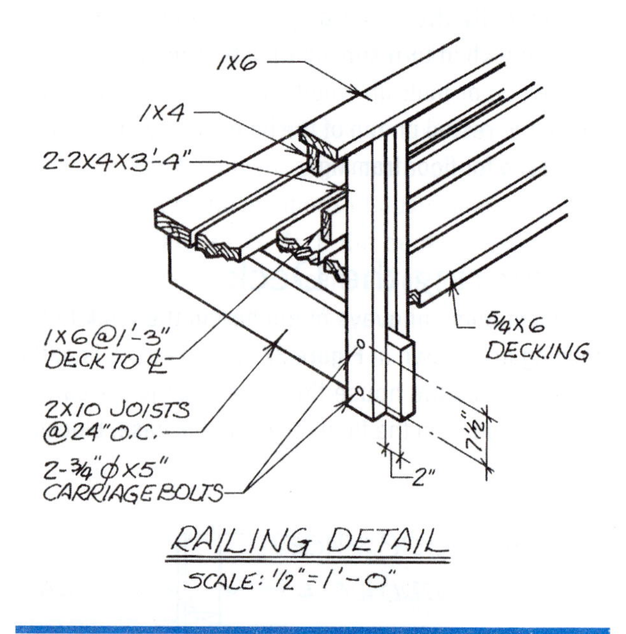

Figure 27–6. Railing detail.

Although all deck framing should be built of pressure-treated lumber, it is still important to prevent water from collecting on top of the header. This can be done by cementing flashing into a kerf (i.e., saw cut) in the concrete wall or with caulking.

Railings

Most decks have a railing because they are several feet from the ground. Although metal railings are available in ready-to-install form, the architectural style of most wood decks calls for a carpenter-built wood or composite railing. The simplest type of railing is made of uprights and two or three horizontal rails. The uprights are bolted to the deck frame, and the rails are bolted, screwed, or nailed to the uprights. The style of the railing and the hardware involved are usually indicated on a detail drawing (see **Figure 27–6**).

> **USING WHAT YOU LEARNED**
>
> To build a deck, it is necessary to know the sizes of all of the parts. The joists that support the deck planks must be of the correct material and cut to the correct length. On the Lake House, what are the dimensions of the joists that support the deck between the kitchen and the garage?
>
> The basic arrangement of these joists is shown in the floor framing plan 1/6. The joists are 2 × 8s, which span from the three 2 × 8s at the west end of the house to the east end. The upper level floor plan 2/2 shows the dimension from the west edge of the deck (and the side of the garage) to the west wall of bedroom #2 to be 13′-9″. Subtract 3 × 1½′ for the header at the west side end of the joists to find the length of the joists. Thus, 13′-9″-4½″ = 13′-4½″. The joists are 2 × 8 (1½″ × 7¼″) by 13′-4½″ long.

Assignment

Refer to the Lake House drawings in your textbook packet to complete the assignment.

1. What supports the south edge of the decks located outside the Lake House living and dining rooms?
2. How far from the outside of the house foundation is the centerline of these supports?
3. How far apart are these supports?
4. How many anchor bolts are required to fasten both of these decks to the Lake House foundation?
5. What is the purpose of the aluminum flashing shown on deck detail 3/6?
6. What material is used for the railings on the Lake House south decks?
7. How many lineal feet of horizontal rails are there on these decks? (Do not include the cap rail.)
8. What is the total rise from the lower deck to the higher deck? Which deck is higher?
9. What supports the west edge of the deck between the Lake House kitchen and the garage?

UNIT 28 Finishing Site Work

Objectives

After completing this unit, you will be able to perform the following tasks:

- Describe retaining walls, planters, and other constructed landscape features shown on a set of drawings.
- Find the dimensions of paved areas.
- Identify new plantings and other finished landscaping shown on a site plan or landscape plan.

As the exterior of the building is being finished or soon after it is finished, the masons, carpenters, and landscapers begin the finished landscape work. Any constructed features (called *site appurtenances*) are completed first. Then trees and shrubs are planted. Finally, lawns are planted.

Retaining Walls

Retaining walls are used where sudden changes in elevation are required (see **Figure 28–1)**. The retaining wall retains, or holds back, the earth. Where the height of the retaining wall is several feet, the earth may put considerable stress on the wall. Therefore, it is important to build the wall according to the plans of the designer. A section through the wall usually is included to show the thickness of the wall, its foundation, and any reinforcing steel

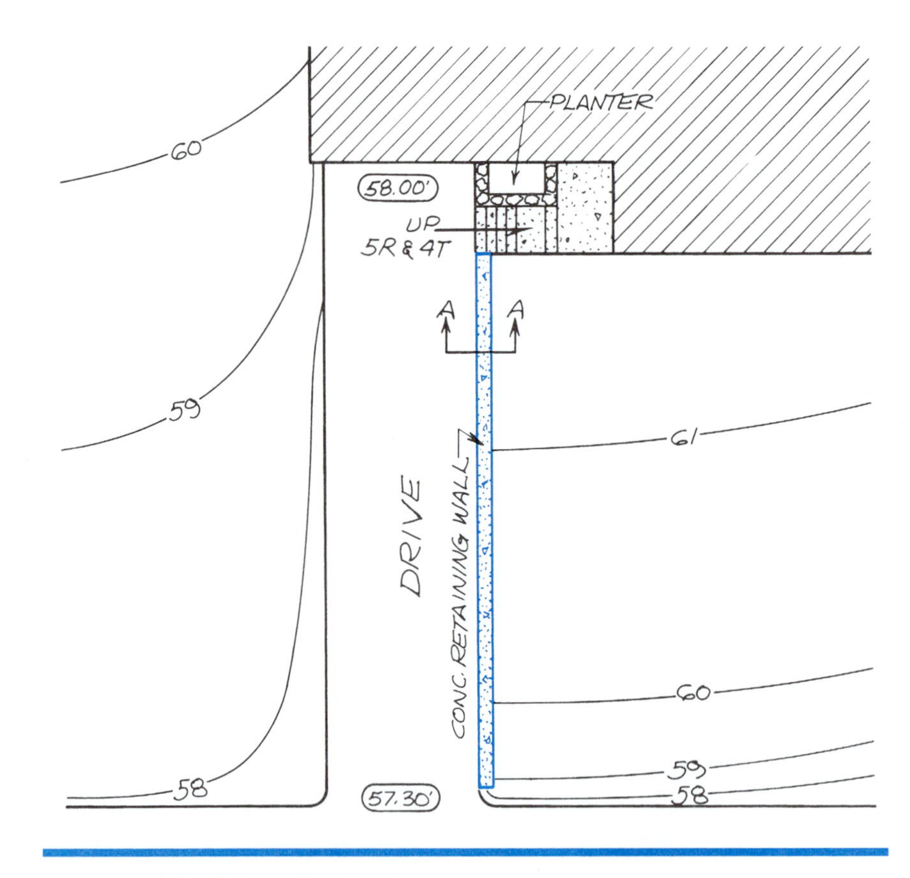

Figure 28–1. Retaining wall on site plan.

(see **Figure 28–2)**. For low retaining walls, the site plan may be the only drawing included.

A low retaining wall is sometimes built around the base of a tree when the finished grade is higher than the natural grade. This retaining wall forms a well around the tree, allowing the roots of the tree to "breathe." An example of a tree that will require a well can be seen in **Figure 28–3,** taken from the Lake House site plan. The 24-inch oak is at an elevation of approximately 333 feet, but the finished grade at this point is 336 feet. Therefore, a well 3-feet deep is required.

Planters

Planters are sometimes included in the construction of retaining walls or attached to the building. In these cases, the information needed to build the planter is included with the information for the building or retaining wall (see **Figure 28–4)**. The planter is built right along with the house or retaining wall. If a planter that is separate from other construction is included, it is usually shown with dimensions on the site plan. A special section may be included with the details and sections to show how the planter is constructed (see **Figure 28–5)**.

The planter should be lined with a waterproof membrane, such as polyethylene (common plastic sheeting). This keeps the acids and salts in the soil from seeping through the planter and staining it. The planter should also include some way for water to escape. This can be through the bottom or through weep holes. ***Weep holes*** are openings just above ground level. In cold climates, the planter may be lined with compressible plastic foam. This allows the earth in the planter to expand as it freezes, without cracking the planter. If the planter is to have landscape lighting, automatic watering, and so on, additional waterproofing may be required where these utilities penetrate the waterproof membrane.

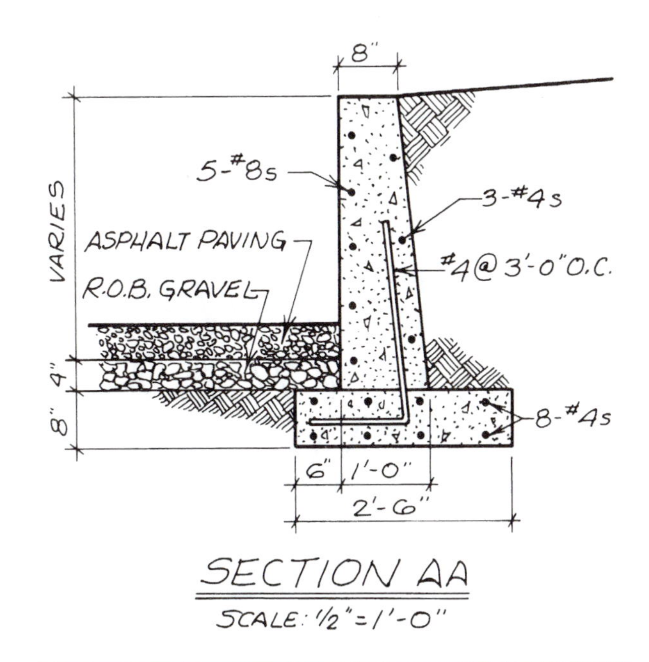

Figure 28–2. Typical retaining wall construction detail.

GREEN NOTE

Paved areas can be either impervious (water will not pass through) or pervious (water passes through to the earth below). For small areas such as walkways, the choice of pervious or impervious is not significant, but for large areas this can be an important design question. If ½ inch of rain falls on a paved area that is 100 feet by 20 feet, that is over 83 cubic feet or 617 gallons of water. If that much water drains at one point, it can do considerable damage to the landscape in that area. If impervious pavement, such as asphalt or concrete is to be used, provisions should be made for groundwater runoff.

Paved Areas

Paved areas on housing sites are drives, walks, and patios. Drives and walks are usually described most fully in the specifications for the project. However, the site plan includes dimensions and necessary grading information for paved areas, as shown in **Figure 28–3.** These dimensions are usually quite straightforward and easy to understand.

Patios are similar to drives and walks in that they are flat areas of paving with easy-to-follow dimensions. They may differ from drives and walks by having different paving materials, such as slate, brick, and flagstone, for example. Patios may also be made of a concrete slab with different surface material.

Assignment

Refer to **Figures 28–8** and **28–9** to complete the assignment.

1. What is the height of the retaining wall above the patio surface at A?
2. How long is the retaining wall?
3. Of what material is the retaining wall constructed?
4. What is the width and the length of the patio?
5. What materials are used in the construction of the patio?
6. Describe the weep holes in the planter.
7. How is the planter treated to prevent acids and salts from staining its surface?
8. How many deciduous trees are to be planted?
9. What is the area of the driveway?
10. Assuming that the driveway is 4 inches thick, how many cubic yards of asphalt does it require? (See Math Review 22 in Appendix B.)

Fireplaces

UNIT 29

Basic Construction and Theory of Operation of Wood-Burning Fireplaces

A fireplace can be divided into four major parts or zones: foundation, firebox, throat area, and chimney (see **Figure 29–1**). Each of these zones has a definite function. To understand the construction details, it is necessary to know how these zones work.

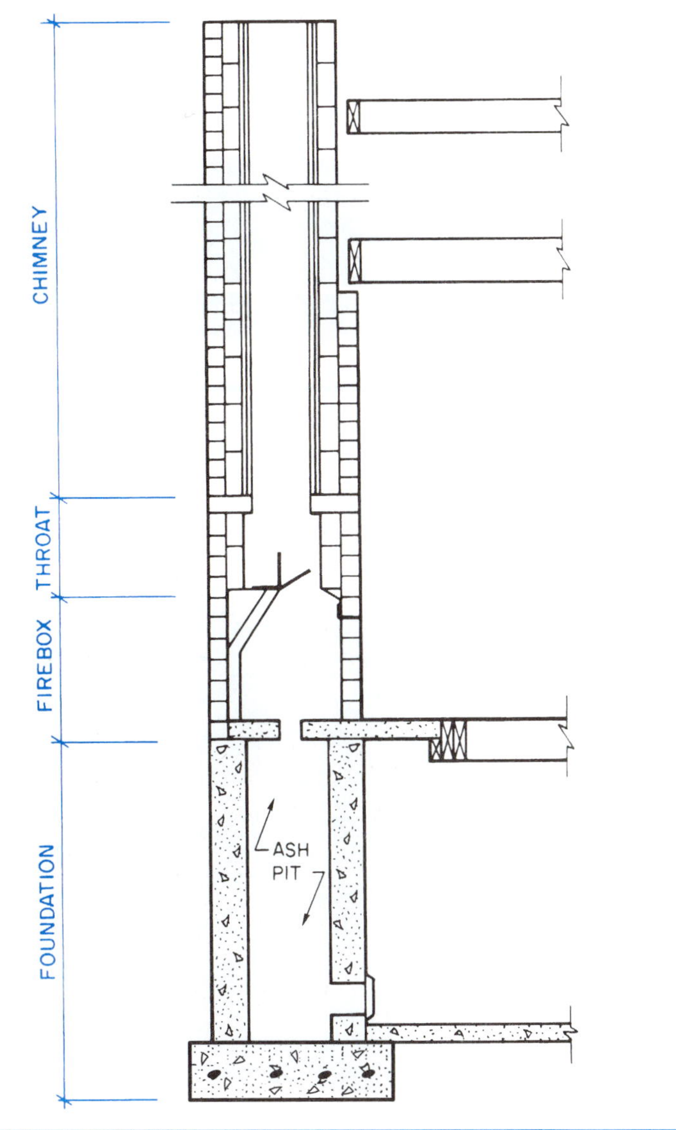

Figure 29–1. Four zones of a fireplace.

Objectives

After completing this unit, you will be able to perform the following tasks:

- Describe the foundation, firebox, throat, and chimney of a wood-burning fireplace using information from a set of construction drawings.
- Explain the finish of the exposed parts of the fireplace, using information from a set of construction drawings.
- Explain the methods used to vent a gas-burning fireplace.

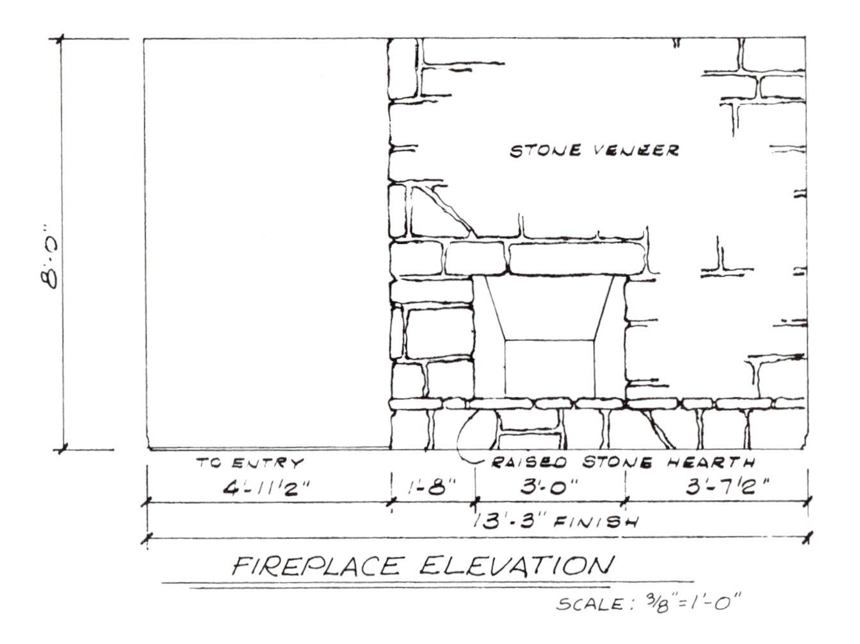

Figure 29–14. An elevation view is a good guide to the finished appearance of the fireplace.

USING WHAT YOU LEARNED

Natural stone products are usually ordered well in advance of the time they are needed on the site and cut to finished size by the supplier or a company that specializes in working with these materials. This means that the precise size of each cut piece must be specified when the piece is ordered. What is the width and length of the granite piece on top of the Lake House fireplace? Allow 4 inches for the width of the soldier-course bricks, including mortar. There are three fireplace details on sheet 7. Section 4/7 shows that the granite is to fit inside a soldier course of bricks (bricks laid in a vertical position.) The overall size of the fireplace, without the hearth, is shown in Fireplace Plan 5/7 as 3′-4″ by 7′-0″. Subtract 4 inches from each side (8 inches from the width and 8 inches from the length) for the soldier course bricks and the size of the granite is 2′-8″ by 6′-4″.

Assignment

Refer to the Lake House drawings in your textbook packet to complete the assignment.

1. What type of fireplace does the Lake House have?
2. How wide is the opening of the firebox?
3. How high is the opening of the firebox?
4. What is the opening next to the fireplace?
5. Determine the overall width and length of the fireplace, including the hearth.
6. Of what material is the hearth constructed?
7. What is used for a lintel over the firebox opening? (Include dimensions.)
8. Briefly describe the foundation of the fireplace.
9. How far above the highest point on the roof is the top of the chimney?
10. What is the total height from the playroom floor to the top of the chimney?
11. What is the overall height of the brickwork involved in the fireplace construction?
12. The top of the fireplace is covered with granite on ¾-inch plywood. How much clearance is there between that plywood and the chimney?
13. Where would you look for the minimum clearance around a gas fireplace vent where it passes through a wall?
14. Where would dimensions be found for the foundation of a wood-burning fireplace?

Stairs

UNIT 30

Stair Parts

In order to discuss the layout and construction of stairs, you need to know the parts of stairs described below and shown in **Figure 30–1:**

- *Stringers* are the main support members. The assembly made up of the stringers and vertical supports is called a ***stair carriage***.
- ***Treads*** are supported by the stringers. The treads are the surfaces one steps on.
- *Risers* are the vertical boards between the treads.
- A *landing* is a platform in the middle of the stairs. Landings are used in stairs that change directions or in very long flights of stairs.
- The *run* of the stairs is the horizontal distance covered by the stairs.
- The *rise* of the stairs is the total vertical dimension of the stairs.
- The ***nosing*** is the portion of the tread that projects beyond the riser.

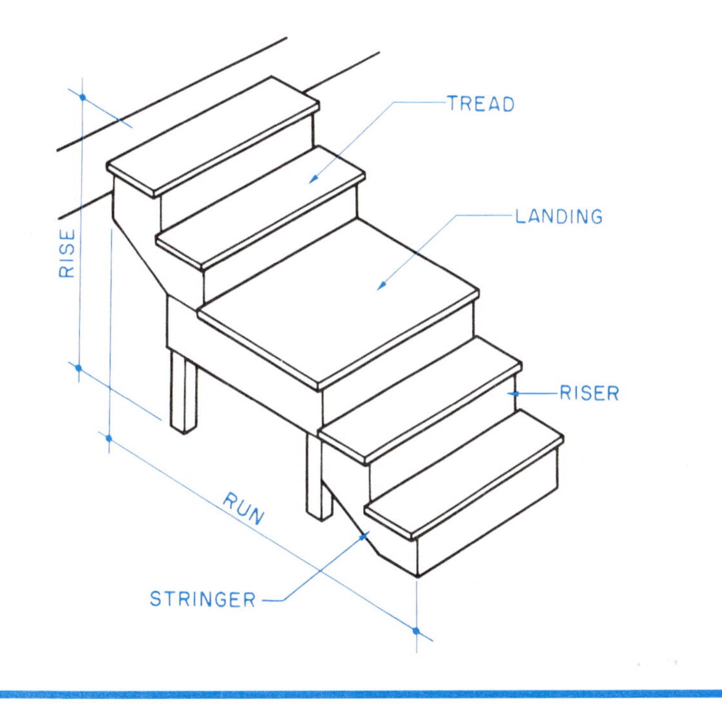

Figure 30–1. Basic stair parts.

Objectives

After completing this unit, you will be able to perform the following tasks:

- Identify the parts of stairs.
- Calculate tread size and riser size.

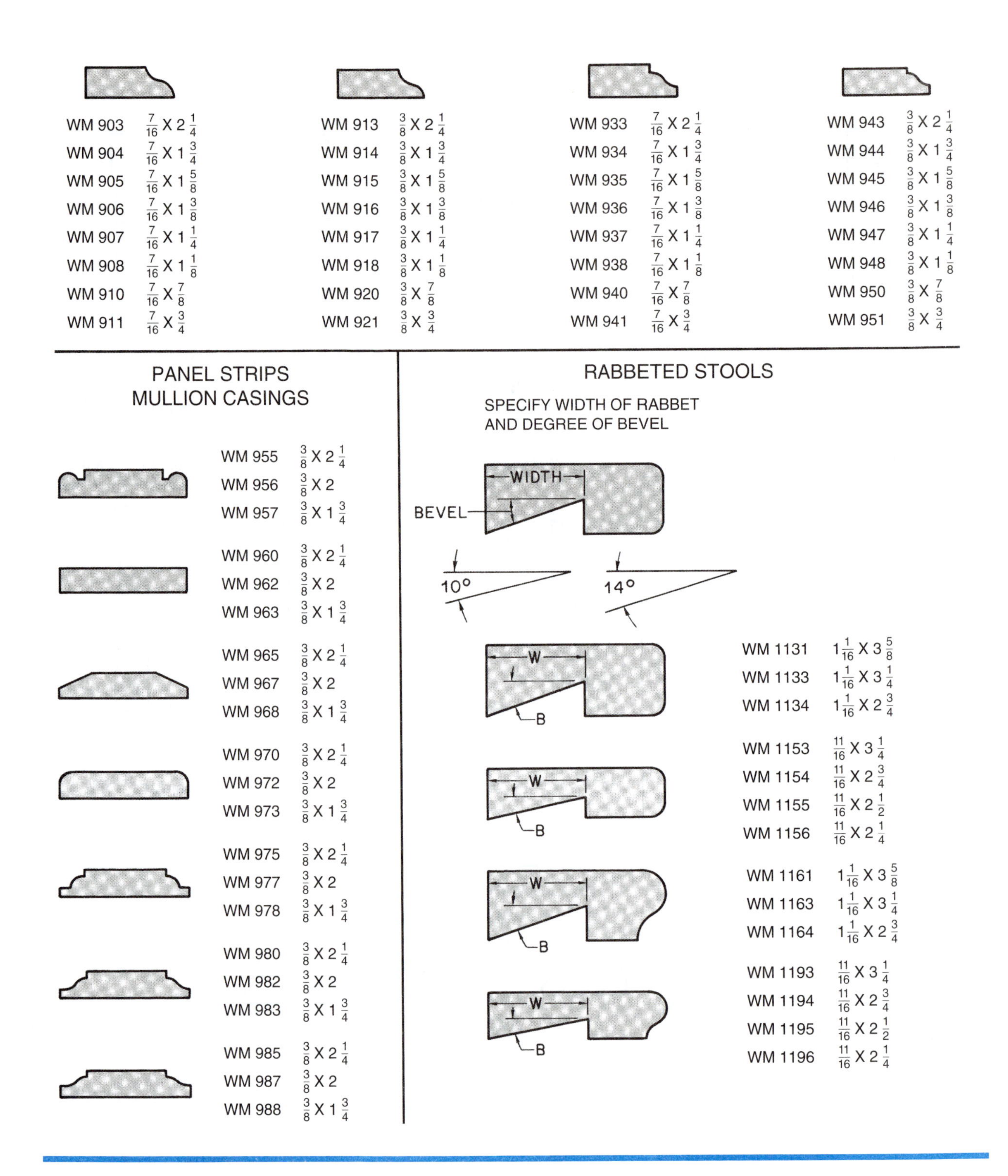

Figure 31–10. *(Continued)*

USING WHAT YOU LEARNED

It is important to search the construction drawings to determine what kind and size of material is used for interior trim throughout the house. In the Lake House, what kind and size of material is used as trim at the bottoms of the walls in the lower level, where the floors are concrete? Wall Section 3/4 shows 1″ × 4″ cellular PVC base trim at this location. Cellular PVC is not affected by the moisture that may be in the concrete.

Assignment

Refer to the Lake House drawings in your textbook packet to complete the assignment.

1. What size or rating and what kind of insulation are to be used in each of the following locations?
    a. Framed exterior walls
    b. Roof
    c. Under heat sink
    d. Masonry walls of playroom
2. What type of molding is to be used as casing around interior doors?
3. What type of molding is used at the bottom of interior walls and partitions?
4. What kind of trim is used to cover the lower edges of exposed LVL beams?
5. Describe the wall finish in the playroom including:
    a. What the wall finish material is fastened to
    b. The kind of material used for wall finish
6. What material is used for ceiling finish in the playroom?
7. What material is used for subflooring on typical framed floors?
8. What is the finished floor material at the heat sink?
9. What covers the interior faces of LVL beams?
10. What is the finished wall material in the bedrooms?

GENERAL CONDITIONS

The General Conditions of the Contract for Construction, AIA Document A 107, whether or not bound herein, are hereby incorporated into and made a part of this contract and these specifications.

01000 GENERAL REQUIREMENTS

A. ARCHITECT'S SUPERVISION
The architect will have continual supervisory responsibility for this job.

B. TEMPORARY CONVENIENCES /
The general contractor shall provide suitable temporary conveniences for the use of all workers on this job. Facilities shall be within a weathertight, painted enclosure complying with legal requirements. The general contractor shall maintain all temporary toilet facilities in a sanitary condition.

C. PUMPING
The general contractor shall keep the excavation and the basement free from water at all times and shall provide, maintain, and operate at his own expense such pumping equipment as shall be necessary.

D. PROTECTION
The general contractor shall protect all existing driveways, parking areas, side walks, curbs, and existing paved areas on, or adjacent to the owner's property.

E. GRADE LINES, LEVELS, AND SURVEYS
The owner shall establish the lot lines.
The general contractor shall:
1. Establish and maintain bench marks.
2. Verify all grade lines, levels, and dimensions as shown on the drawings, and report any errors or inconsistencies before commencing work.
3. Layout the building accurately under the supervision of the architect.

F. FINAL CLEANING
In addition to the general room cleaning, the general contractor shall do the following special cleaning upon completion of the work:
1. Wash and polish all glass and cabinets.
2. Clean and polish all granite.
3. Clean and polish all hardware.
4. Remove all marks, stains, fingerprints, and other soil or dirt from walls, woodwork, and floors.

G. GUARANTEES
The general contractor shall guarantee all work performed under the contract against faulty materials or workmanship. The guarantee shall be in writing with duplicate copies delivered to the architect. In case of work performed by subcontractors where guarantees are required, the general contractor shall secure written guarantees from these subcontractors. Copies of these guarantees shall be delivered to te architect upon completion of the work. Guarantees shall be signed by both the subcontractor and the general contractor.

H. FOREMAN
The general contractor shall have a responsible foreman at the building site from the start to the completion of construction. The foreman shall be on duty during all working hours.

I. FIRE INSURANCE
The owner shall effect and maintain builder's risk completed value on this job.

02000 SITE WORK

WORK INCLUDED
This work shall include, but shall not be limited by the following:
A. Clearing the site.
B. Excavating, backfilling, grading, and related items.
C. Removal of excess earth.
D. Protection of existing trees to remain on the site.
All excavation and backfilling required for heating, plumbing and electrical work will be done by the respective contractors and are not included under site work.
It is the contractor's responsibility to field inspect existing conditions to determine the scope of the work.

02100 CLEARING

A. Clean the area within the limits of the building of all trees, shrubs, or other obstructions as necessary.
B. Within the limits of grading work as shown on the drawings remove such trees, shrubs, or other obstructions as are indicated on the drawings to be removed, without injury to trunks, interfering branches, and roots of trees to remain. Do cutting and trimming only as directed. Box and protect all trees and shrubs in the construction area to remain; maintain boxing until finished grading is completed.
C. Remove all debris from the site; do not use it for fill.

02200 EXCAVATION

A. Carefully remove all sod and soil throughout the area of the building and where finish grade levels are changed. Pile on site where directed. This soil is to be used later for finish grading.
B. Do all excavation required for footings, piers, walls, trenches, areas, pits, and foundations. Remove all materials encountered in obtaining indicated lines and grades required. Beds for all foundations and footings must have solid, level, and undisturbed bed bottoms. No backfill will be allowed and all footings shall rest on unexcavated earth.
C. The contractor shall notify the architect when the excavation is complete so that he may inspect all soil before the concrete is placed.
D. Excavate to elevations and dimensions indicated, leaving sufficient space to permit erection concrete forms, walls, waterproofing, masonry, and inspection of foundations. Protect the bottom of the excavation from frost.

02260 BACKFILL

A. All outside walls shall be backfilled to within 6 inches of the finished grade with clean fill. Backfill shall be thoroughly compacted.
B. Unless otherwise directed by the architect, no backfill shall be placed before the first floor framing is in place. No backfill shall be placed until all walls have developed such strength to resist thrust due to filling operations.

02270 GRADING

A. Do all excavating, filling, and rough grading to bring the entire area outside of the building to levels shown on the drawings.
B. Where existing trees are to remain, if the new grade is lower than the natural grade under the trees, a sloping mound shall be left under the base of the tree extending out as far as the branches; if the grade is higher, a well shall be constructed around the base of the tree to provide the roots with air and moisture.
C. After rough grading has been completed and approved, spread topsoil evenly to the previously stripped area. Prepare the topsoil to receive grass seed by removing stone, debris, and unsuitable materials. Hand rake to remove water pockets and irregularities. Seeding will be done by the owner.
D. Furnish and place run of bank gravel as approved under all floor slabs.

03000 CONCRETE

WORK INCLUDED

Provide all materials, labor, equipment, and services necessary to furnish, deliver, and install all work of this section, as shown on the drawings, as specified herein, and/or as required by job conditions including, but not limited to the following:

A. Concrete for all footings and piers.
B. Concrete for all foundation walls.
C. Concrete for all slabs on ground.
D. Concrete for slab at heat sink.
E. Furnishing and installation of all required anchors.
F. Supplying, fabrication, and placement of all reinforcing bars and mesh and wire reinforcement for concrete where shown, called for, or required with proper supporting devices.
G. Erection of all forms required for concrete work and removal upon completion of the work.
H. The finishing of all concrete work as hereinafter specified.
I. Porous fill below slabs on ground.

03010 MATERIAL

A. Fine Aggregate
   Fine aggregates for concrete shall consist of natural sand having clean, hard, sharp, uncoated grains free from injurious amounts of dust, lumps, soft or flaky particles, shale, alkali, organic matter, loam, or other deleterious substances.
B. Coarse Aggregate (stone)
   Coarse aggregates shall consist of crushed stone or gravel having clean, hard, strong, durable, uncoated particles, free from injurious amounts of soft, friable, thin, elongated or laminated pieces, alkali, organic, or other deleterious matter.
C. Water
   All water used in connection with concrete work shall be clean and free from deleterious materials or shall be water used for drinking daily.
D. Portland Cement
   Portland cement shall be an approved domestic brand complying with Standard Specifications for Portland Cement, ASTM Designation C-150, Type 1. Only one brand of cement shall be used throughout the course of the work.
E. All concrete is to be machine mixed in an approved mixer with a water metering device. Concrete is to reach a compressive strength of 2500 psi after 28 days.
F. Reinforcement
   All reinforcing, unless otherwise shown or specified, shall conform to ASTM A-615, Grade 60. Wire mesh reinforcing shall have a minimum ultimate tensile strength of 70,000 psi, and shall conform to ASTM Specifications A-185, latest edition.

03320 INSPECTION & PLACING

A. All reinforcing shall be free of rust, scale, oil, or other coatings that tend to reduce the bond to concrete. All reinforcing is to be tied with 18 gauge wire at intersections and shall be held securely in position during the pouring of concrete.
B. The architect will inspect all footing beds, forms, and reinforcing, just prior to placing concrete for footings, foundation, and slabs.
C. All concrete shall be placed upon clean surfaces, and properly compacted fill, free from standing water. The concrete shall be compacted and worked into corners and around reinforcing.
D. All concrete to be true and level as indicated on drawings to within ± ¼ inch in 10 feet.

03330 FINISHING

A. Slabs in occupied spaces shall be troweled smooth and free of trowel marks.
B. Slabs in unoccupied spaces will have wood float finish.

04000 MASONRY

WORK INCLUDED

This work shall include but not be limited by the following:

A. Brickwork
B. Concrete block work
C. Mortar for brick and block work

04010 MATERIALS

A. Delivery and storage
   All material shall be delivered, stored, and handled so as to prevent the inclusion of foreign materials and the damage of the materials by water or breakage.

08010 PRODUCTS

A. Hollow metal doors to be manufactured by Pease, or approved equal.
B. Wood doors by Iroquois Millwork or approved equal. Birch plywood skin with phenolic impregnated Kraft core.
C. All door frames to be of clear pine in standard patterns as shown on the construction drawings. Side lites to be Iroquois, Weather Guard SL with 5/8″ insulating glass.
D. Windows to be Andersen, Series 200. Windows to include aluminum screens by the same manufacturer.
E. Skylight to be Skylight Concepts number CMDADE1850.
F. Door trim to be standard WM patterns milled from clear pine.
G. Contractor shall allow $1,500 for locksets, latches, bifold hardware, hinges, weather stripping, medicine cabinets, closet rods, and shower curtain rods.

08100 DOORS

A. Frames to be plumb and square with accurately fitted joints. Set exposed nails with a nail set.
B. Accurately align doors with frames and adjust hardware as necessary for smooth operation.
C. Install molding as shown on construction drawings with accurately mitered corners.

08500 WINDOWS & SKYLIGHTS

A. Install all windows true and plumb, and according to manufacturer's recommendations to produce a weathertight installation.
B. Install all hardware and accessories and check all moving sections for smooth operation.

08900 HARDWARE

A. Install all door and window hardware according to the manufacturer's recommendations and check for smooth operation.
B. Install a closet rod in each closet. Closet rods to be secured through wall finish to blocking installed with rough carpentry.
C. Install shower curtain rods over tub and shower stall. Shower curtain rods to be secured through wall finish to blocking installed with rough carpentry.
D. Install a medicine cabinet (by owner) over each lavatory.

09000 FINISHES

WORK INCLUDED

A. Gypsum wallboard
B. Ceramic tile in baths and toilet rooms
C. Quarry tile floors
D. Painting and varnishing

09250 WALLBOARD

MATERIAL

A. All wallboard material to be the product of one manufacturer; U.S. Gypsum, Flintkote, or approved equal. Drywall in baths, toilet rooms, and tub room to be moisture resistant.

INSTALLATION

A. Gypsum wallboard shall be installed with joints centered over framing or furring.
B. Fasten gypsum wallboard with power-driven drywall screw located not over 12″ O.C. at all edges and in the field.
C. Outside corners are to be protected with metal corner bead.
D. Finish all joints with a minimum of three coats of joint compound and standard gypsum board reinforcing tape in accordance with the manufacturer's printed instructions.
E. Dimples at screw heads shall receive three coats of compound.

09300 TILE

MATERIAL

A. Wall tile to be American Olean Tile Company standard grade bright glazed in a color selected. Bathroom accessories to be same manufacturer and color.
B. Floor tile in bath, toilet, and tub rooms to be American Olean unglazed 1′ × 1″ ceramic mosaic tile in color selected.
C. Quarry tile to be installed in this specification will be 6″× 6″ shale-and-clay tile provided by owner.
D. Marble thresholds at all doors adjacent to mosaic tile floors shall be Vermont Marble 7/8″×3 1/2″.

INSTALLATION

A. Layout ceramic tile on walls and floors so that no tiles less than half-size occur.
B. Cut and fit tile around toilets, tubs, and other abutting devices.
C. Install all floor tile by thin-set method in accordance with TCA recommendations.
D. Install wall tile in mastic cement conforming to the recommendations of the tile manufacturer.
E. Grout all tile work to completely fill joints.
F. Clean all tile surfaces to present a workmanlike job.
G. Install 12 bathroom accessories as follows:
   - 2 soap dishes
   - 3 toilet paper holders
   - 7 towel bars

09900 PAINTING

MATERIAL

A. Exterior stain: One coat Minwax exterior stain or approved equal, in color by owner.
B. Interior walls – flat: Two coats Martin Senour, Bright Life latex flat or approved equal, in colors by owner.
C. Interior walls – semi gloss: Two coats Martin Senour, Bright Life latex semi gloss or approved equal, in colors by owner.

GREEN NOTE

Volatile organic compounds (VOCs) are emitted as gases from certain solids or liquids. VOCs include a variety of chemicals, some of which may have short- and long-term adverse health effects. Many of the products used in the construction and maintenance of homes for decades emit VOCs: paint and varnish, carpets, drapes, and cleaning agents, to name a few. It has only been in recent years that we have been aware of these VOCs. The level of VOCs in some products is no regulated by law. Most of the products named above are now available with either very low or zero VOCs and it is common for construction specifications to specify the acceptable levels of VOCs in products to be used in construction.

D. Interior painted woodwork: Two coats Martin Senour, Bright Life latex semi gloss or approved equal, in colors by owner.
E. Metal: Two coats DeRusto rust resistant enamel or approved equal, colors by owner.
F. Polyurethane: Two coats United Gilsonite Laboratories, ZAR gloss.
G. Primers: All primer to be that recommended by manufacturer of top coat.

APPLICATION
A. Repair all minor defects by patching, puttying, or filling as normally performed by painting contractors.
B. Prime uncoated wood surfaces with tinted primer and touch up previously painted surfaces.
C. Sand all surfaces smooth before each coat of paint to produce a smooth and uniform job at completion.
D. Protect all adjacent surfaces and other work incorporated into project against damage or defacement.
E. Coat all surfaces according to the following painting schedule and as indicated on the drawings. All colors are to be selected by the owner.

Exterior Walls and Trim	Stain
Metal Railings	Rust-Resistant Enamel
Playroom	Walls Flat/Ceiling Flat
All Baths	Walls Semigloss/Ceilings Semigloss
Halls and closets	Walls Flat/Ceiling Flat
Living room	Walls Flat/Ceiling Flat
Dining room	Walls Flat/Ceiling Flat
Kitchen	Walls Semigloss/Ceilings Semigloss
Bedrooms	Walls Flat/Ceiling Flat
Loft	All Except Floor And Ladder: Flat
Loft floor and ladder	Polyurethane
Interior doors	Polyurethane
Interior stairs	Polyurethane
Unscheduled interior trim	Semigloss

10000 FIREPLACE

WORK INCLUDED
Provide and install fireplace, chimney, and accessories as indicated on the construction drawings. Related masonry and granite cap are not included in this section.

10310 EQUIPMENT

A. Majestic Company fireplace number SR36A.
B. Majestic Company chimney with 8″ flue.
C. All chimney accessories required to conform with printed instructions of Majestic Company.

10320 INSTALLATION

Fireplace and chimney are to be installed according to printed instructions of Majestic Company and as indicated on the construction drawings. All equipment is to be installed level and plumb and finished to provide a workmanlike appearance.

11000 EQUIPMENT

WORK INCLUDED
A. Provide and install kitchen cabinets. This section does not include kitchen sink or related plumbing.
B. Provide and install vanity cabinets. This section does not include lavatories or related plumbing.
C. Provide and install granite countertops, vanity tops, fireplace cap. Provide necessary cutouts for kitchen sink, by plumbing contractor; cooktop, by owner; oven, by owner; and chimney, by masonry contractor.

11910 MATERIAL

A. Cabinets and vanities shall be Merillat, Classic style.
B. Granite countertops and fireplace cap to be supplied by Granite Mountain Stone Design in color selected by owner from approved samples. Granite to be 3 cm thickness.

11920 INSTALLATION

A. Cabinets shall be installed level and true with no less than 4 screws per base unit and present a workmanlike appearance.
B. Granite shall be adhered to cabinet frames with a full bead of silicone.

C. All granite shall be installed with as few joints as possible. Joints shall be tight and filled with resinous material according to the manufacturer's instructions. Exposed edges shall be polished to match the finish of the top.
D. Provide necessary cutouts for installation of kitchen sink, cooktop, and grill.
E. Provide cutout for fireplace chimney with 1/8" clearance.

22000 PLUMBING

This division includes all plumbing. It is omitted here because these topics have not been covered earlier in the textbook.

23000 HEATING VENTILATING AND AIR CONDITIONING (HVAC)

This division includes heating, ventilating, and air conditioning. It is omitted here because these topics have not been covered earlier in the textbook.

26000 ELECTRICAL

This division covers all aspects of electrical work. It is omitted here because these topics have not been covered earlier in the textbook.

USING WHAT YOU LEARNED

The specifications for a construction project are part of the contract for the work to be performed, so it is very important to carefully read and fully understand what is written in the specs. The key to accurately applying the specs is to know where to find the information they contain. Look at the Table of Contents if there is one and find the division you are working on. Where in the lake house specifications would you find instructions for how to treat remaining trees that are above the natural grade when the site is being graded?

In Division 02000, Section 02270 is grading. Item B of that section says, "Where existing trees are to remain, if the new grade is lower than the natural grade under the trees, a sloping mound shall be left under the base of the tree extending out as far as the branches."

Assignment

Refer to the Lake House specifications to complete the assignment.

1. What division covers vapor barriers under slabs?
2. What is the material and thickness of the vapor barrier under the heat sink?
3. What material and rating is the thermal insulation under the heat sink?
4. What is the strength requirement for the concrete in the heat sink?
5. How is the concrete slab in the heat sink to be finished?
6. What make and model number is the skylight?
7. What division of the specs includes installing shower curtain rods?
8. What brand and type of paint is to be used on the kitchen walls?
9. Who is to choose the color of the paint for the kitchen walls?
10. What brand and type of paint is to be used on the trim around the skylight?

PART 2

Test 2

A. Refer to the Lake House drawings in your textbook packet to determine each of the following dimensions.

1. Northwest corner of the site to the high-water mark at the west boundary
2. Septic tank to the nearest property line
3. Width of the walk on the west side of the house
4. Width of the south end of the drive (in front of the garage)
5. Depth of the earth fill over the septic drainfield
6. Elevation at the butt of the most southerly maple tree to remain
7. Outside of the garage foundation (width × length)
8. Outside of the garage footing (width × length)
9. North to south spacing of the square steel columns
10. Length of the northwest square steel column
11. Length of the haunch under the slab between the north steel columns
12. East-west dimension inside the lower level bathroom at the widest end
13. Closet in bedroom #1 (width × length)
14. Inside of bedroom #1 (width × length)
15. Deck between the kitchen and the garage (width × length)
16. Length of the joists in the loft
17. Height of the foundation wall at the overhead garage door
18. Elevation at the bottom of the concrete piers for the kitchen deck
19. Difference in elevations at the bottom of the south garage footing and the bottom of the nearest house footing
20. Elevation at the bottom of the deepest excavation
21. Width of the cabinet over the refrigerator
22. Width of the cabinet closest to the living room stairs
23. Total thickness of a typical exterior frame wall (allow ½ inch for siding)
24. Lineal feet of soldier-course bricks in the fireplace
25. Lineal feet of 2 × 10 lumber in the eave of the corrugated roof
26. Finish floor to the top surface of the freestanding closet/cabinet unit in the kitchen (include surface material)
27. Length of the studs in the wall separating the playroom from the crawl space under the kitchen
28. Length of the steel dowels that anchor the foundation walls to the footings
29. Length of the C8 × 11½ structural steel beam
30. Outside surface of the foundation under the dining room to the centerline of the deck footings
31. Door in the south end of the garage (width × height × thickness)
32. Length of the 2 × 4 studs under the living room bench
33. Width of the treads (front to back excluding nose) in the stairs from the kitchen to the bedroom level
34. Rough opening for the door from the hall into bedroom #1 (width × height)
35. Thickness of the concrete at the haunches

B. Describe the material at each of the following locations in the Lake House. Include such considerations as the kind of material, nominal size of masonry units, and nominal size of lumber.

1. Tile field pipe
2. Reinforcement in the concrete slabs
3. Rungs in the ladder to loft
4. Hip rafter in the northeast corner of house
5. Roof insulation
6. Roof deck
7. Exterior wall studs
8. Rafter headers at skylight
9. Purlins under the corrugated roof
10. Finished surface on the west wall of playroom
11. Reinforcement in the footings
12. Anchor bolts in the wood sill
13. Wood sill
14. Vapor barrier under the concrete slab
15. Subfloor in the bedrooms
16. Floor joists in the kitchen
17. Floor joists in the bedrooms
18. Stair stringers between the decks
19. Top rail around the south decks
20. Top course of the foundation at heat sink
21. Expansion joint at the edge of heat sink
22. Rafters above the loft
23. Posts supporting the LVL beams
24. Girder under the west wall of bedroom #1
25. Base plates on the posts supporting LVL beams
26. Girder under the kitchen deck joists
27. Finished surface of the living room bench
28. Housed stringer on the north side of stair to bedrooms
29. Stair treads
30. Door casings (interior)
31. Door casings (exterior)
32. Window casings (interior)
33. Glazing (glass) in the window on south side of bedroom #1
34. Planks on the kitchen deck
35. Foundation at the west side of the kitchen deck

C. Answer each of these questions about the Lake House.

1. How many cubic yards of concrete are required for the concrete slab at elevation 337.00 feet?
2. What is the length of the concrete piers between the kitchen and the garage?
3. How many lineal feet of 1× pine trim are needed to cover the bottom of the LVL beam over the kitchen?
4. How long are the rafters in the corrugated roof? (Refer to the illustrated rafter table.)
5. How long is the hip rafter at the northeast corner of the roof? (Refer to the illustrated rafter table.)

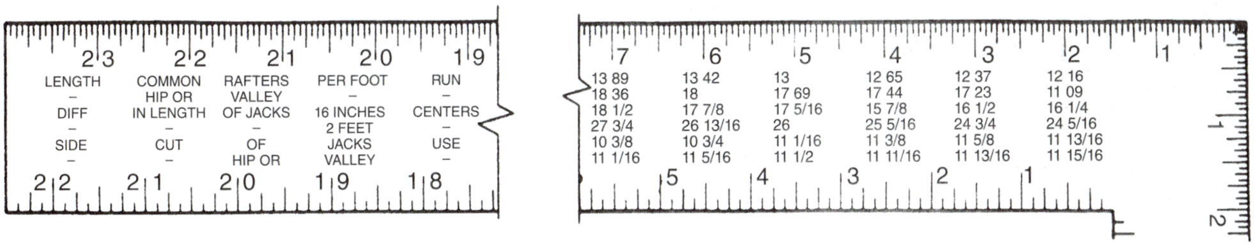

D. Answer the following questions.

1. Who should design roof trusses?
   a. architect
   b. carpenter
   c. engineer
   d. any of the above
2. Which drawing will include information about the slope of a roof?
   a. building elevation
   b. truss delivery sheet
   c. truss detail drawing
   d. all of the above
3. What is a piggyback truss?
   a. a truss that is designed to be used as the top half of a large truss
   b. a truss design that uses extra web members
   c. a truss that is to be placed against the side of another truss
   d. a particular brand of roof truss
4. Which sheet would be the easiest one on which to find the required number of trusses of a particular type?
   a. floor plan
   b. truss delivery sheet
   c. truss detail sheet
   d. building elevation
5. What is the most reliable place to find the spacing between trusses?
   a. truss layout
   b. floor plan
   c. truss detail sheet
   d. truss delivery sheet

Courtesy of Berkus-Group Architects.

Figure 34–4. This is a section of the site plan at the size it was drawn.

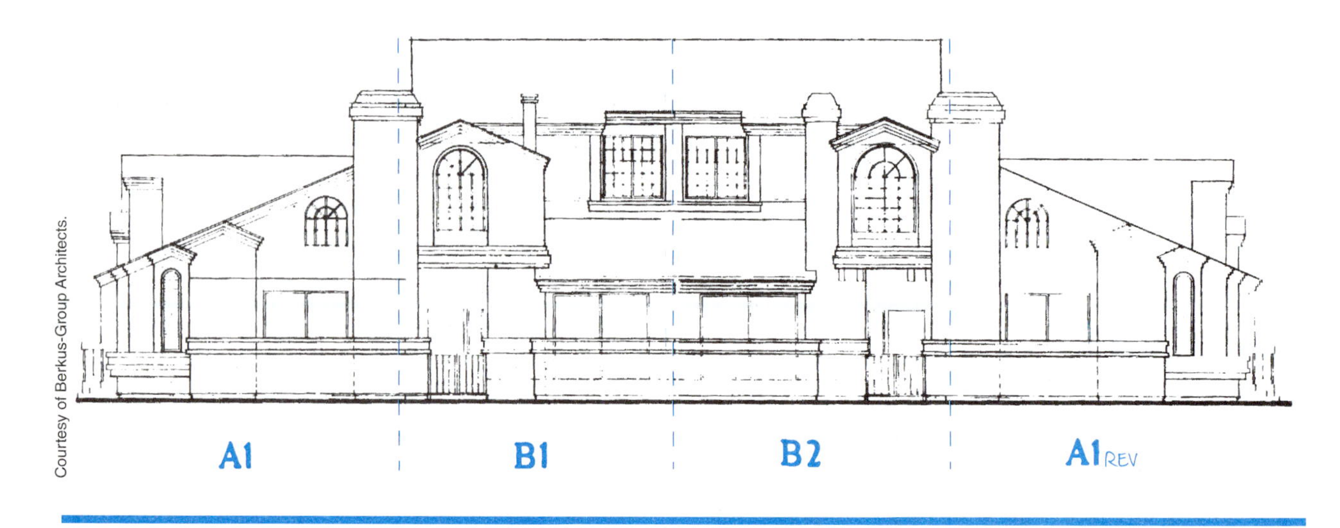

Courtesy of Berkus-Group Architects.

Figure 34–5. Building type II includes four housing units.

door. Inside the bathroom there is a toilet on the right, a vanity on the left, and a shower straight ahead.

Back out through the passageway and enter the kitchen area on your right. On the left is a storage closet. The kitchen has an island sink and cabinets and appliances on two walls. Beyond the kitchen is the living room, at the front of the building. Turn left at the front wall of the building into a passageway with the

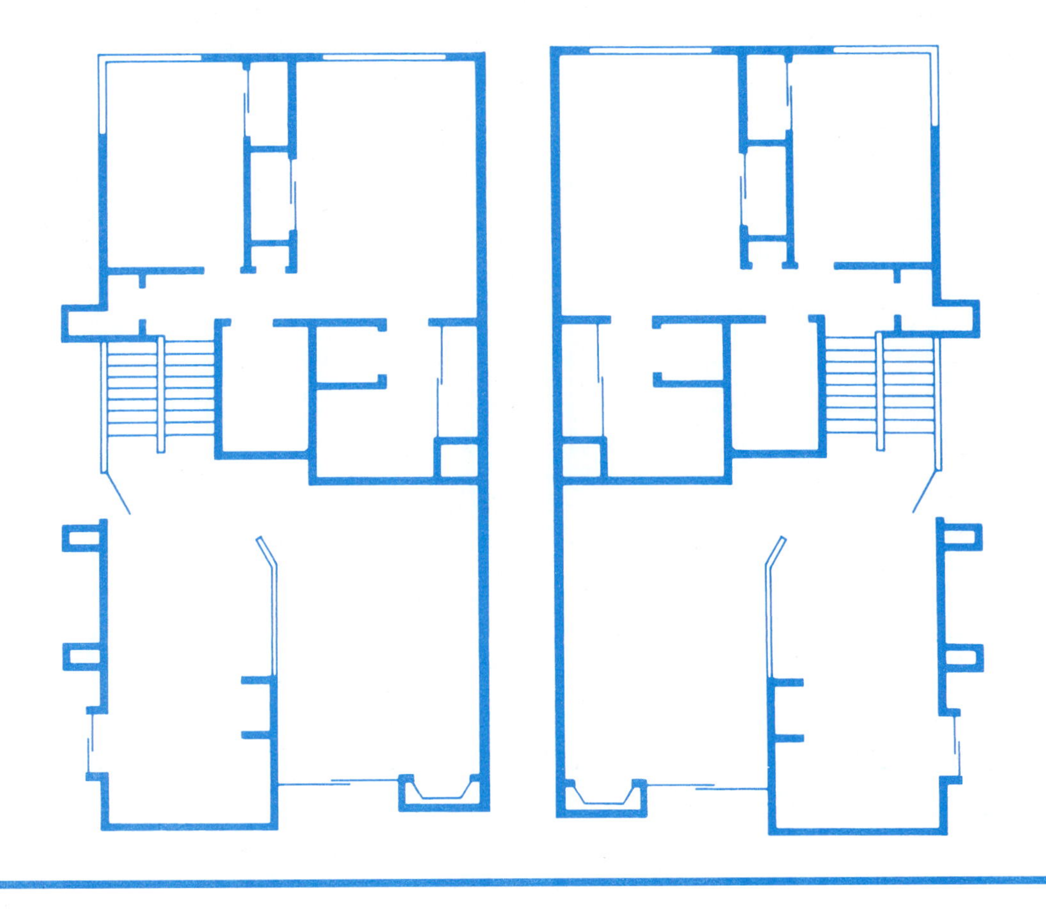

Figure 34–6. A reversed plan is similar to what would be seen by looking at the plan in a mirror.

laundry door on the left. Beyond the laundry room is a stair leading up.

Refer to the Fourth Floor Plan on Sheet A3.4. The stairs lead up to a loft, providing more living space. There is also a storage area. On the left wall of the loft is a sliding door onto the roof deck. The roof is sloped ¼″ in 1 foot, but that is gradual enough to allow for outdoor furniture. Two sides of the roof deck are open, but protected by a railing, described in Keynote 17.

USING WHAT YOU LEARNED

When similar style buildings are built in one development, it is desirable to make them appear to be different from one another. This can be done by using mirror images of the floor plans and elevations, by varying the architectural trim, and even by slight modifications in the buildings' exterior elevations. Some of these techniques are used in the Urban Courts. List all of the variations that are to be built in Block 3 (see **Figure 34–7**): Building 1, Elevation A; Building 1, Elevation B; Building 1, Elevation C; Building 2, Elevation A; Building 2, Elevation B; Building 2, Elevation C; and each of these reversed (mirror images).

Figure 34–7. Urban Courts sheet A1.2.

Assignment

Refer to the Urban Courts drawings in your textbook packet to complete this assignment.

1. Define the following abbreviations:
   a) SOFF
   b) TOS
   c) CR
   d) P/L
   e) PL
2. What sheet has details for exterior doors?
3. Sheets A5.01 through A5.052 have been changed since the originals were drawn. Who requested the changes?
4. **Figure 34-7** is Sheet A1.2 of the Urban Courts drawings. How many times is Building type 1, Elevation A, Reversed to be built in Block 3?
5. What wall surface material is to be used on the faces of the wall separating the two garages?
6. What fire protection requirements are stated for the doors that separate the garages from the living space?
7. List the types of windows and their sizes for the master bedroom of Unit A.
8. What size framing material is to be used for the wall separating the living room from the balcony in Unit A?
9. Between the coat closet and the master bedroom in Unit B there is a soffit overhead. What are the dimensions of the soffit, including the dimension from the finished floor to the surface of the soffit?
10. What type of wall construction is specified for the back wall of the storage closet on the third floor?

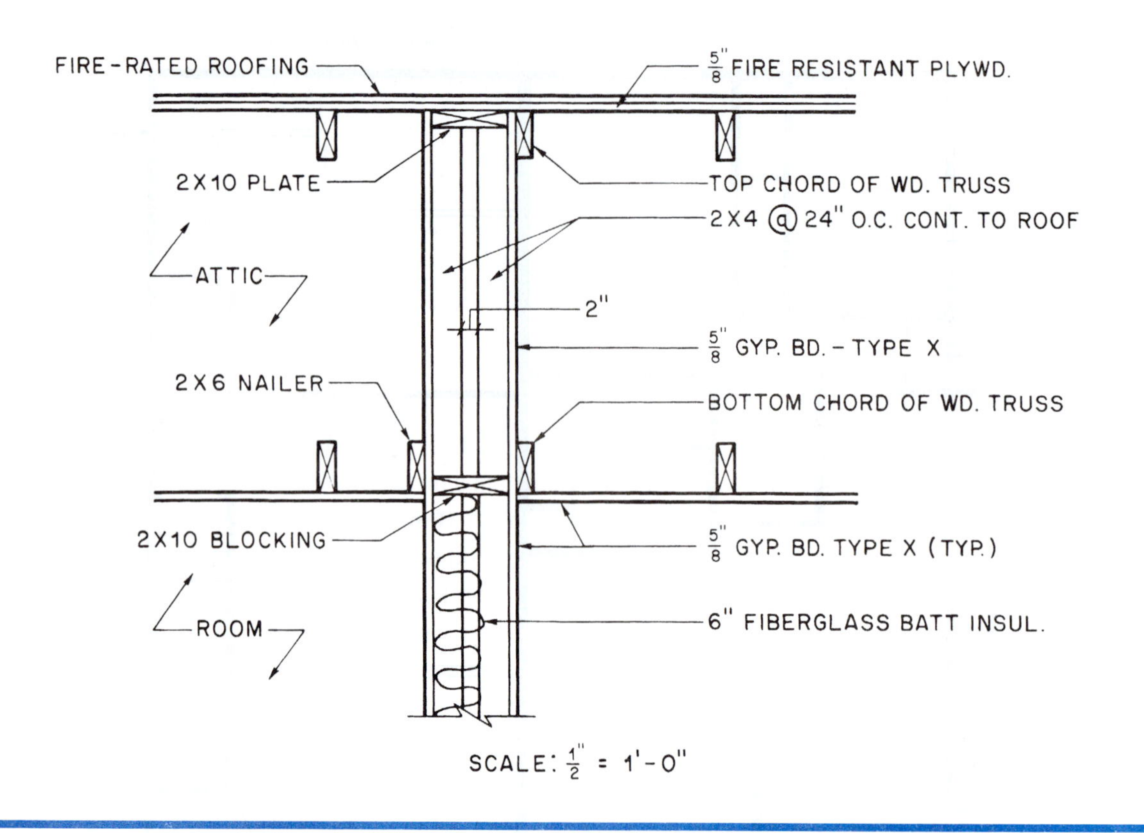

Figure 35–3. Fire-rated party wall in attic space.

10-feet high. The spaces between the studs are closed off with ***firestops*** at each level (see **Figure 35–4**). Instead of wood firestops, the wall cavity can be blocked off with fire-resistant fiberglass insulation as shown in **Figure 35–2**.

Openings are usually avoided in fire-rated walls. Where it is necessary to include a door, it is made of fire-resistant material. The fire rating of the door must comply with the building code. Doors in fire walls are equipped with a self-closing mechanism.

Sound Control in Walls

To provide privacy between the housing units, common walls and floors should not allow the sound from one unit to be heard in the next unit. The measurement of the capability of a building element to reduce the passage of sound is its *sound transmission classification* (STC) (see **Figure 35–5**).

The word *acoustic* is often used when talking about sound control. It simply means having to do with sound or hearing.

Sound is transmitted by vibrations in any material: solid, liquid, or gas. To slow the passage of sound, a party wall must reduce the flow of vibrations. The materials used in the construction of most walls vibrate relatively well. Also, they transmit these vibrations to the air inside the wall. Sound can bend around corners and find its way through the smallest of openings. The air carries the sound to the other side of the wall, where it is transmitted to the air on the opposite side of the wall. The drywall joint between the wall and the ceiling may require sound deadening putty to prevent sound for passing through that joint. The electrical installer may be required to utilize acoustical insulation or some other sound attenuation material around flush device boxes. Raceways may require installations that reduce sound transmission, such as expansion joints, flexible raceway, or raceway offsetting through the party wall.

The STC rating of a wall can be improved greatly by not allowing the studs to contact both surfaces. This may be accomplished in one of two ways. One method is to

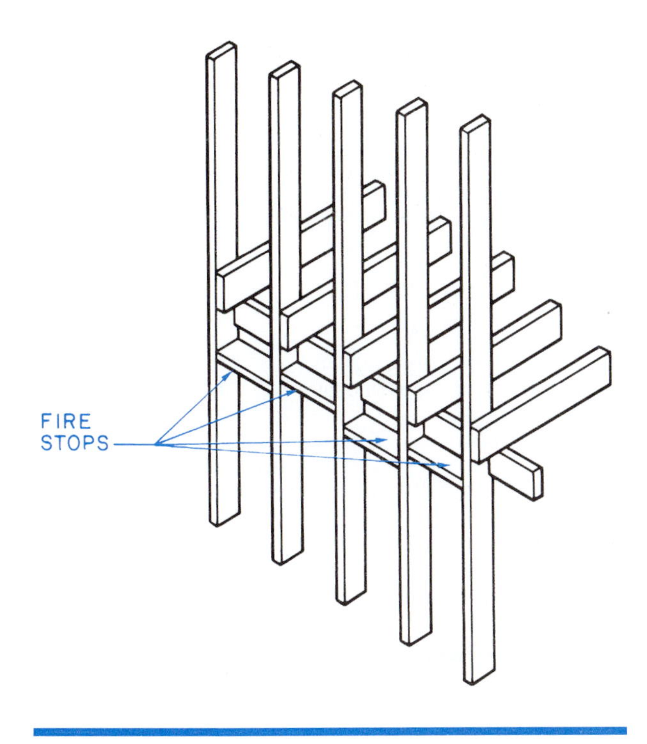

Figure 35–4. Firestops are installed between studs to prevent vertical drafts inside the wall.

STC RATING	EFFECTIVENESS
25	Normal speech can be understood quite easily.
35	Loud speech can be heard but not understood.
45	Must strain to hear loud speech.
48	Some loud speech can barely be heard.
50	Loud speech cannot be heard.

Figure 35–5. Sound transmission classes.

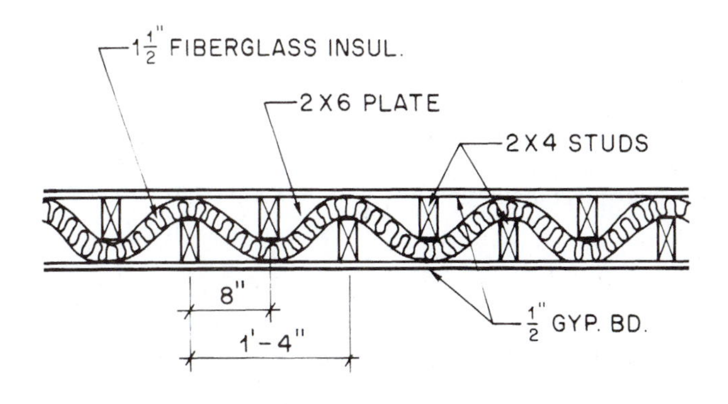

Figure 35–6. Typical sound-insulated wall.

attach clips made for sound insulation to the studs and then fasten the wallboard to these clips. The clips absorb the vibrations. Using the clips and sound-deadening fiberboard results in an STC rating of 52.

An STC of 45 is achieved without clips by using 2 × 4 studs and 2 × 6 plates. The studs are staggered on opposite sides of the wall so that no studs contact both surfaces (see **Figure 35–6**). The STC can be increased to 49 by including fiberglass insulation. Separating the two sides of the wall framing on separate plates and insulating both sides of the wall cavity yields an even higher STC rating (see **Figure 35–7**). There are many other products and designs that can be used to control the sound transmission through a wall. It is essential that those who assemble common walls understand and follow the architect's design.

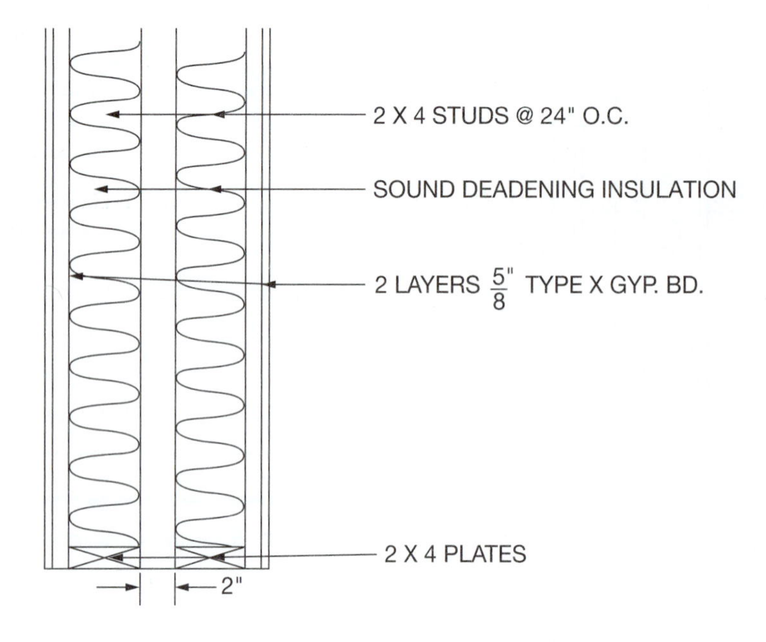

Figure 35–7. Double wall with insulation in both cavities yields an STC of 61.

Sound Control in Floor/Ceiling Assemblies

The principles that are used to stop the flow of sound through walls also apply to the floor/ceiling assembly, but wall construction is different from floor/ceiling construction. Some provision must be made to either absorb or block the vibrations that are sound. The IRC requires an STC rating of at least 45 for common walls and for floor/ceiling assemblies.

Another type of sound comes into play for floor/ceiling assemblies. When a person wearing hard-soled shoes walks across a wood floor the sound created by the impact of the footsteps can be a real annoyance to those living in the rooms below. The acoustics industry has devised a rating system for impact noise insulation. The Impact Isolation Class (IIC) is a rating of the floor/ceiling assembly's ability to isolate the sound from footfall and other impact sources. The IRC requires an IIC of 45 or more for common floor/ceiling assemblies between dwelling units. A 45 IIC rated floor would not be satisfactory to many people. A rating of 65 would satisfy most individuals.

There are so many options for floor surface materials, sound deadening products, and ceiling materials that it is not possible to list even the most common options here. For example, the floor/ceiling assembly shown in **Figure 35–8** would yield an IIC of 51 to 56 and an STC of 56 to 60.

Keynotes

On larger construction projects, the drawings might become very cluttered and confusing to read if all of the information is drawn and described on them. Also there can be multiple places where the same information applies. To avoid this confusion much of the information may be shown in keynotes. A numbered mark, in this case, a small rectangle containing a number, is placed on the drawing. A column of notes on the right side of the drawing sheet explains what each of those keynotes indicates (see **Figure 35–9**).

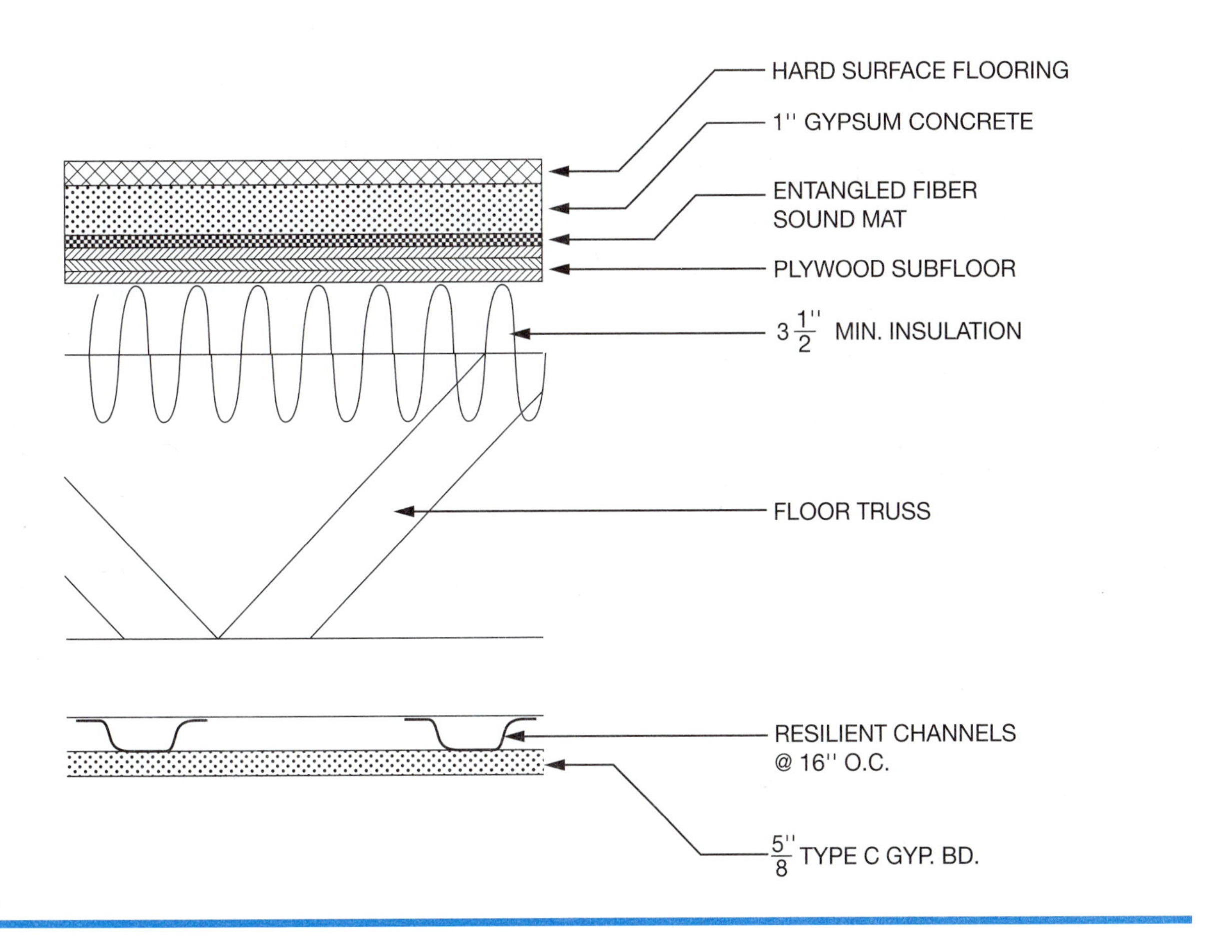

Figure 35–8. A typical floor/ceiling assembly rated IIC 51-56 and STC 56-60.

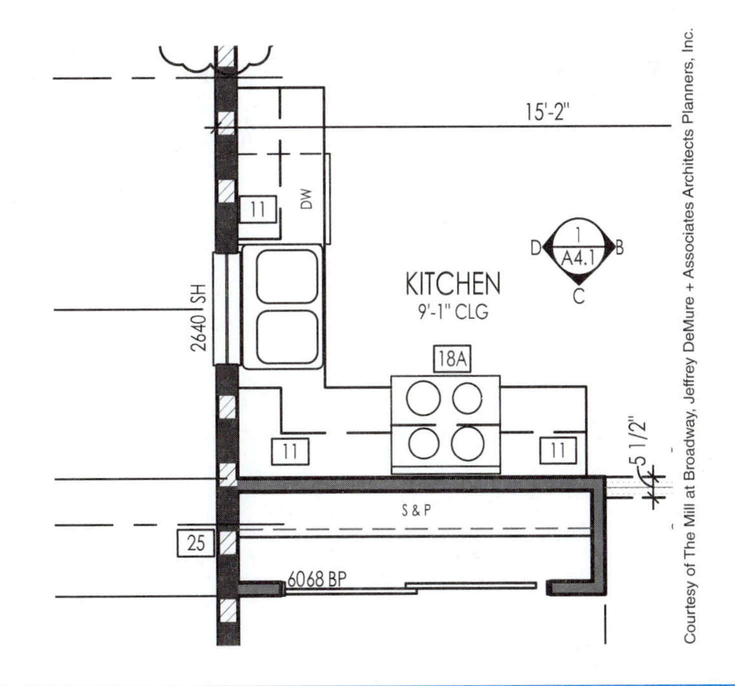

Figure 35–9. Floor plan keynote 11.

USING WHAT YOU LEARNED

The building code requires some provision for preventing the vertical spread of fire in the cavity of a frame wall that is more than one story high. The interior wall between the stairs and the living spaces is such a wall. What provision is made for resisting the vertical spread of fire in this wall?

The logical place to begin is by looking at the floor plan to identify the wall in question. The second floor plan shows both the stairs and the living space of Unit A. Keynote 26 (see the boxed number on the plan) identifies this as a 2-hour common wall and refers to Detail 9/FP1. That detail shows several features that provide fire resistance. Both sides of the wall are sheathed with 7/16″ Flameblock, a fire retardant OSB material, and both sides have 5/8″ Type X gypsum board. The interior of the wall has fire blocking at 10′ O.C. vertically and batt insulation, both of which resist the vertical spread of flames.

Assignment

Refer to the Urban Courts drawings in your textbook packet to complete the assignment.

1. List the type, thickness, and number of layers of gypsum wallboard for each of the following walls:
   a. wall separating the two garages
   b. wall between the entry area and Unit B garage
   c. wall between kitchen and master bedroom closet on second floor
   d. wall between the third floor kitchen storage room and the stairway
   e. wall on the right as you ascend the stairs to the fourth floor
2. List each of the components of the assembly that make up the floor of the third floor master bedroom and the ceiling of the second floor living room.
3. What is used to stop the vertical spread of fires inside the 1-hour exterior walls of the Urban Courts?
4. What is done to stop the transmission of sound through the floor/ceiling assembly between the third floor kitchen and the second floor kitchen?
5. What is the STC rating of the floor/ceiling assembly in Question 4?
6. What is the slope of the roof above the third floor living room?
7. What is the height of the railing on the second floor balcony?
8. How many lineal feet of standard 2 × 4 walls are on the third floor? Do not deduct for doors or windows.

Seismic Considerations

UNIT 36

Where Earthquakes Occur

In some areas of the United States, earthquakes are a potential hazard and buildings in those areas must resist these seismic forces.

Earthquakes are one of the three types of loads that buildings must support by providing load paths through the structure and back down to the ground:

- Dead loads—the weight of the building materials.
- Live loads—loads that change or move around the building, such as people and furniture.
- Dynamic loads—loads that produce varying amounts of force, such as snow, wind, or earthquakes (seismic loads).

Many of the methods used to resist earthquake damage will also prevent damage from strong winds. This chapter will concentrate on how buildings resist seismic forces and how to identify those design elements on construction drawings.

The Seismic Design Category map shown in four parts in **Figure 36–1** shows the seismic design standard for all areas of the country. The chart in **Figure 36–2** describes the potential risk for each category.

Notice that most areas of high seismic activity are in the western states including Alaska and Hawaii. California is particularly high in active seismic areas.

The Urban Courts drawings in your drawing packet show a multistory apartment complex in Sacramento, California, with a Seismic Design Category of D. This project requires extensive seismic-resisting elements. This unit discusses and explains those elements using selected drawings from the project's structural drawings.

Objectives

After completing this unit, you will be able to perform the following tasks:

- Identify the Seismic Design Category for a given location.
- Explain how earthquakes affect buildings.
- Find information about seismic considerations on construction drawings.

Forces Created By Earthquakes

The amount of damage done by a seismic event depends on several factors including distance from the earthquake's origin (epicenter), the strength and type of earthquake, site soil conditions, and the type of building construction. During a seismic event several types of energy waves are introduced into the structure. These include forces that shake the building back and forth, like moving a dish of jello rapidly left and right, and waves, much like ocean swells that lift the building up and then drop it. (See **Figure 36–3**.)

The side-to-side motion can come from any direction and causes the building to sway, particularly on the top floors. The ground waves can also tilt the

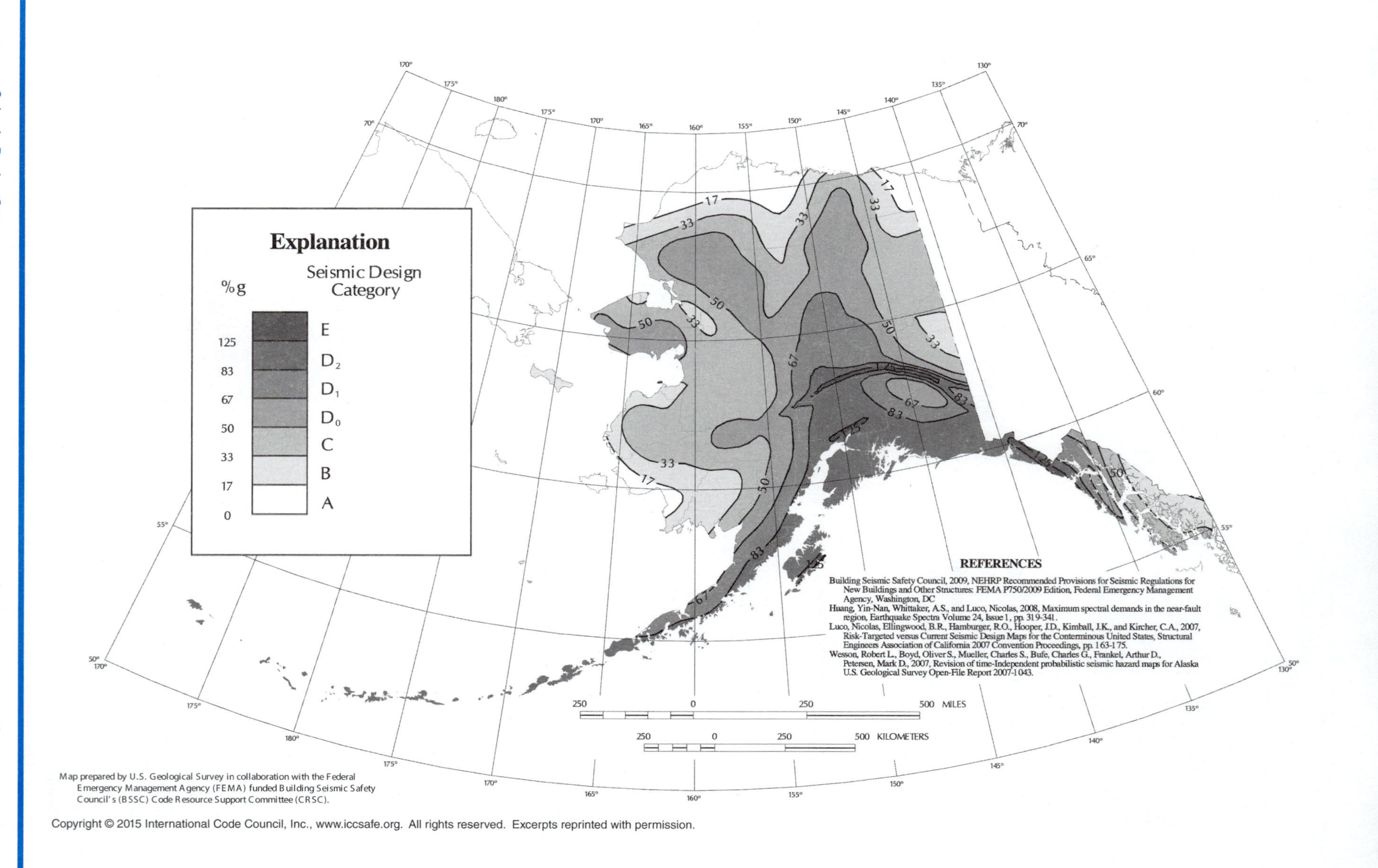

Figure 36–1. Seismic Design Category map.

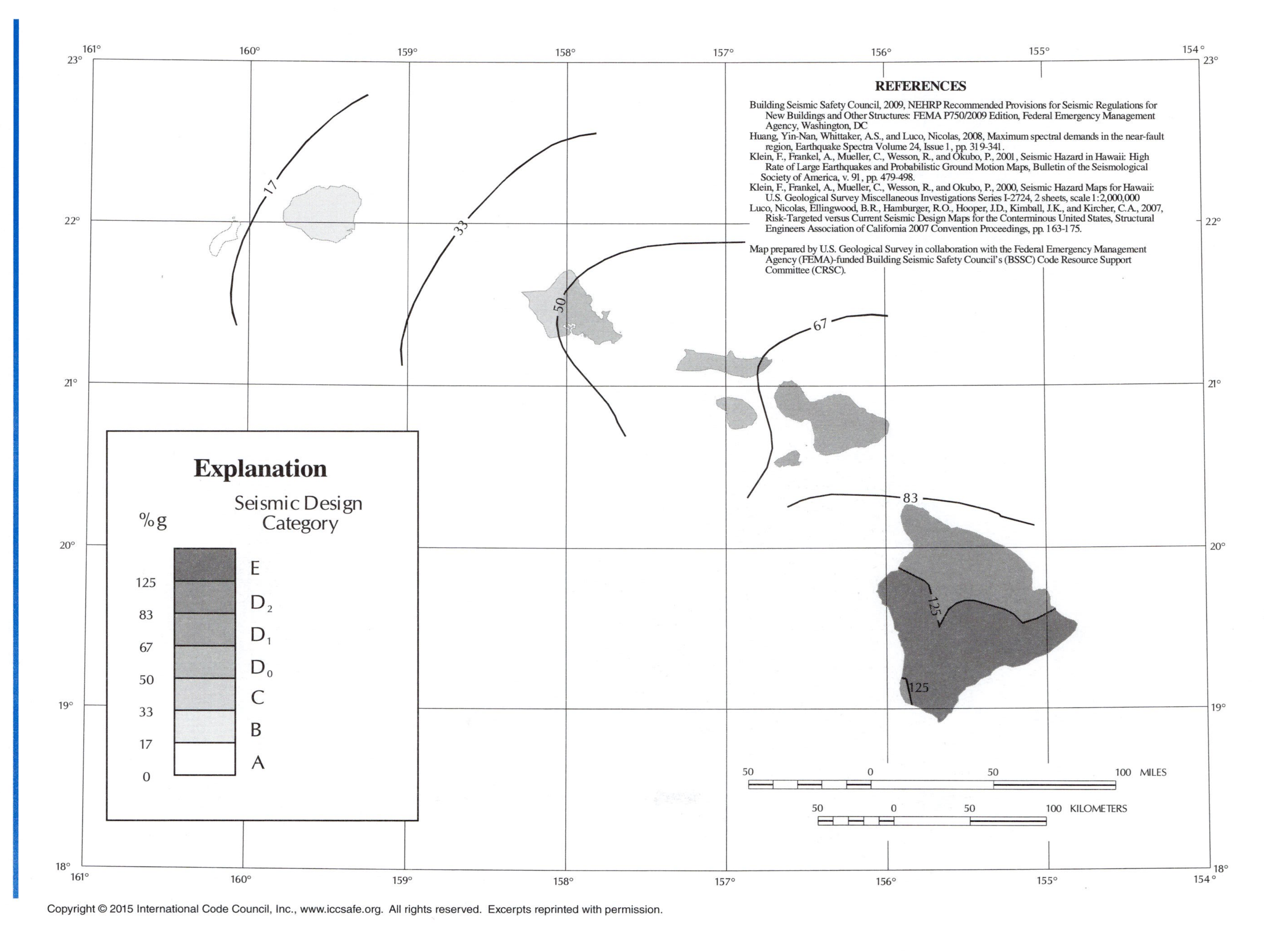

Figure 36–1. *(Continues)*

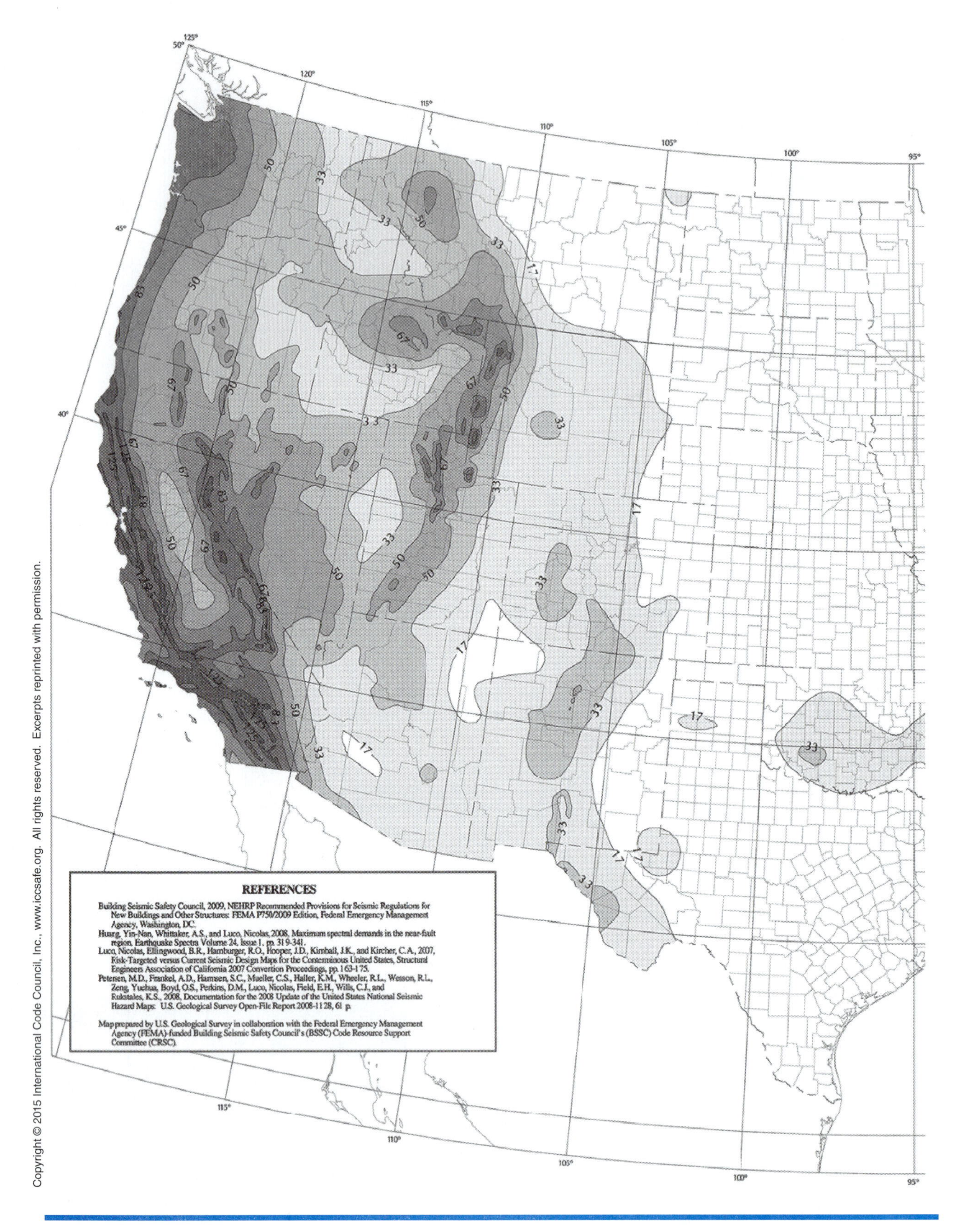

Figure 36–1. *(Continued)*

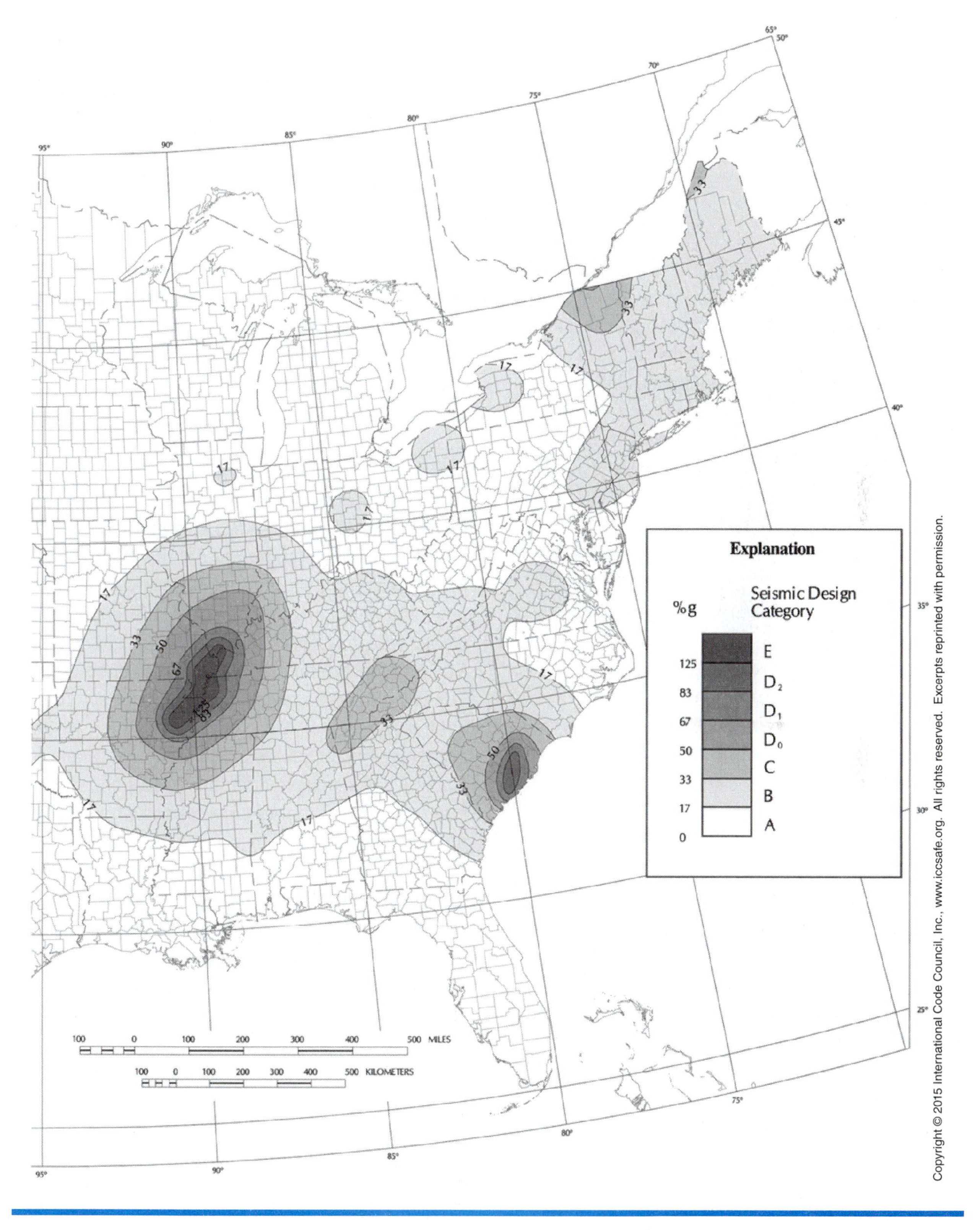

Figure 36–1. *(Continued)*

Seismic Design Category (SDC)	What Does It Mean?
A	Very small seismic vulnerability
B	Low to moderate seismic vulnerability
C	Moderate seismic vulnerability
D	High seismic vulnerability
E & F	Very high seismic vulnerability and near a major fault

Figure 36–2. Seismic Design Category risk.

building as well cause the building to lift off the foundation. Either of these forces can cause building floors to separate, the building to overturn, or the supporting walls to rack and fail. This racking would be similar to a free-standing row of books being pushed at the top edge until the row fell.

To resist this damage, the structural drawings for a building that might be subjected to seismic activity specify methods to ensure that

- the foundation design addresses specific site soil conditions.
- the foundation will hold together as one structural unit.
- the walls will be securely attached to the foundation.
- the walls will resist racking and overturning.
- the walls and floors of each story will be securely attached to the story above and below.
- the floor and roof will lend antiracking ability to the structure.

These structural design features allow the structure to absorb the energy of the earthquake and provide a load path for the energy to travel back into the ground while leaving the structure intact.

Soils Conditions

The site requires a geotechnical soils report, performed by a soils engineer. This provides information on soil types, the strength characteristics of the various soils present, landslide potential, and any other geologically related hazards.

Structures built on rock typically perform better than those built on soils, particularly loose soils. During earthquakes some soils are subject to liquefaction, where the soil acts like a liquid, offering the structure no support. The geotechnical report addresses all these concerns and provides information on preparing the site for the foundation. The report also provides the structural engineer the site and soil information needed to design the foundation.

Reading Structural Drawings

The Structural Drawings for Urban Courts that are included in your packet are two plan sheets, S7.1 and S7.2, and two sheets of details, SD1 and SD3. Familiarize yourself with the various views, schedules, and notes and how they are referenced in the plans. Note that most details on S7.1 will be found on SD1 and most details on S7.2 will be found on SD3. The foundation and framing plans have schedules, such as Shearwall Schedule; Notes, such as General Floor Framing Notes; and other miscellaneous Notes. They also have other plan views but are titled Elevations B or C not to be confused with architectural elevations.

Schedules use different shapes, such as △ or ⬡ to denote a reference to different schedules just as in architectural drawings.

The reference to details is the same as in the architectural drawings—a circle with the detail sheet number on the bottom and the detail number on the top.

Foundation Concrete is a very durable material and a good choice for building foundations, but it does have one drawback. While concrete is very strong in compression, it is weak in tension. Looking at the profile or elevation of a slab, the forces placed on the slab tend to make it deflect or bend causing a compression face and a tension face, **Figure 36–4**.

Figure 36–3. (A) When the first ground motion (from the left) arrives, the foundation and lower floors move to the right, but the upper floors remain in place momentarily. The upper floors eventually catch up to the bottom floors, but not before the ground moves to the right. Thus, shaking back and forth causes the building to sway, putting tremendous forces on the connections. (B) Surface waves roll under the building creating uplift that can move a building off its foundation or even overturn the structure.

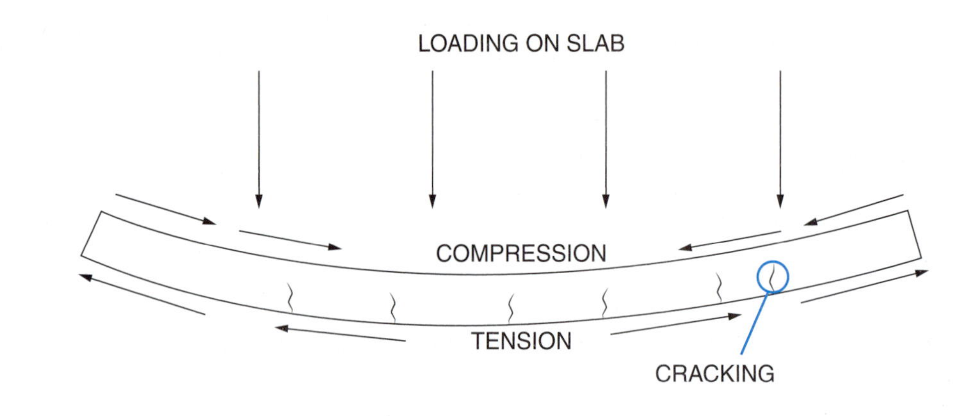

Figure 36–4. Concrete slab loaded and failing in tension. Rebar reinforcement and post tensioning are used to strengthen the slab.

Because steel has high tensile strength, steel rebar is added to the tension side of the slab, usually the bottom. This adds tension strength to the slab. To add even more strength to the slab on the Urban Courts project, a process called post tensioning is used in addition to using rebar.

Post tensioning involves the placement of a grid of plastic coated cables in the concrete forms prior to concrete placement. Both ends of each cable penetrate the form and extend past the edge of the slab after concrete placement. Once the concrete is placed the concrete is allowed to partially cure, usually about seven days or until a specific compressive strength is achieved. One end of the tensioning cable is then anchored and a hydraulic tensioner is used to stretch the cable inside the slab. The plastic coating allows the cable to stretch inside the now solid concrete. The cables are tensioned to approximately 33,000 pounds. Specifications for the spacing and how much the cable will be elongated once tensioned are shown on Urban Courts Sheet S7.1.

Post tensioning is similar to prestressed concrete beams used in highway bridges and the result is the same. The cables are pulling the sides of the concrete slab together and the slab gains a much higher resistance to tension stresses.

Sheet S7.1 has a holdown schedule, and **Figure 36–5** shows a typical holdown assembly. Holdowns are used to anchor the first-story walls to the

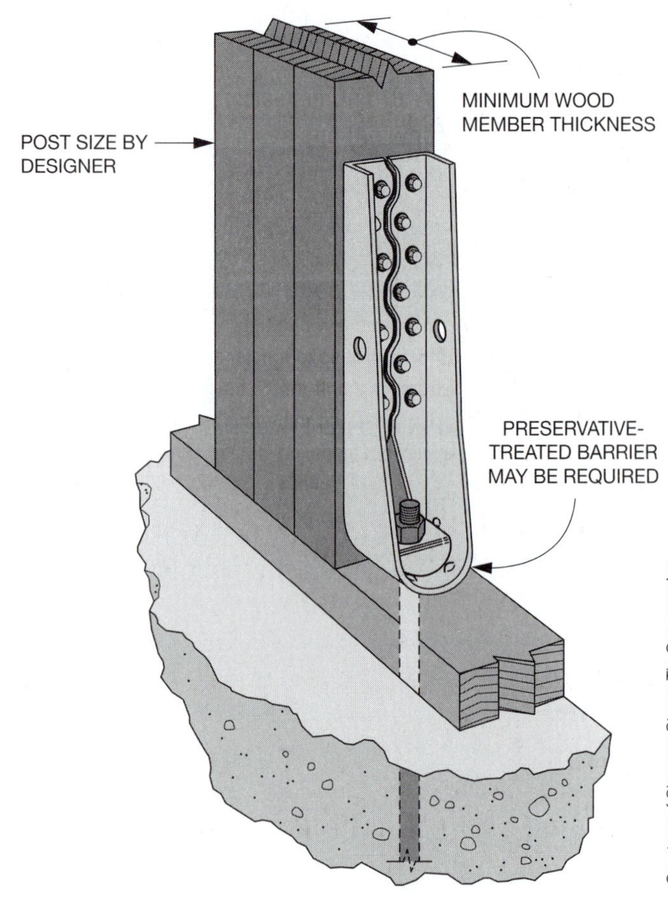

Figure 36–5. HDU holdowns connect to imbedded bolts and are one type of holdown used in the Urban Courts project to connect walls. HDUs resist overturning and shear forces.

foundation and resist the overturning forces discussed earlier. Holdowns are manufactured in a variety of configurations; they can attach to a bolt embedded in the concrete or be set directly in the concrete. Usually they bolt to a post or studs in the wall, but can also be nailed to the post or studs. They are a critical part of the shear wall design. The other anchoring system is a series of foundation anchor bolts embedded in the concrete that bolt the wall bottom plate to the concrete, this resists both uplift and lateral displacement.

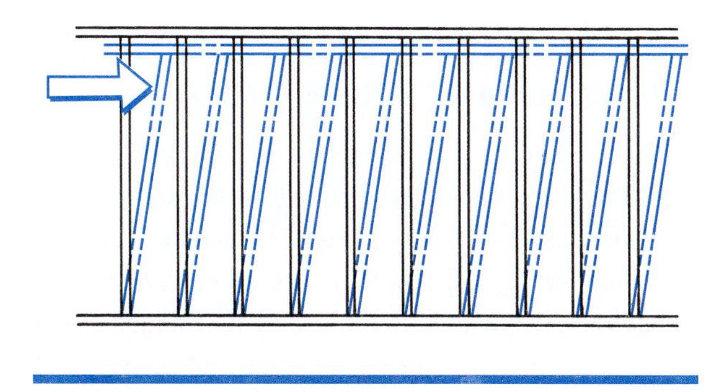

Figure 36–6. Wracking occurs when a frame wall is pushed sideways.

Walls

Urban Courts is a four-story wood frame building. Wood is a very forgiving material in earthquakes; it tends to bend and return to its original shape, but that same flexibility makes it too flexible for very large or tall buildings.

Any building in earthquake country needs to be braced to resist shear forces. Shear forces are those that move long walls or floors and distort the structure by racking. (See **Figure 36–6.**)

The standard shear resisting method used for walls is to apply structural sheathing such as plywood or OSB, nailed to the studs and plates in a prescribed manner. Because shear walls must resist the forces of an earthquake, the size, placement, and spacing of nails is especially important. In areas with very little seismic activity, it is not customary for construction drawings to specify nailing schedules, but in high-seismic areas nailing is typically specified by the structural engineer. Sheet S7.1 has a Shear Wall Schedule that includes information on sill plate thickness, nailing schedules, and whether structural panels are required on one or both faces of the wall. This schedule also calls out anchor bolt spacing and alternative connection methods.

Notice the garage door openings on the bottom floor. Since the shear panels cannot continue across those openings, the narrow wall on one side of the opening is panelled each side and has heavy hold downs tying it to the foundation (see ⟨J⟩ in Shear Wall Schedule). There is also shear transferred from floor two to the first-story wall. A critical design consideration is that the shear wall sections stack vertically, upper-level shear walls occurring directly above the shear walls below.

On Sheet S7.2, there are references to strap holdowns connecting each story to the story above, holding the building together vertically and transferring shear from floor to floor. Details of these connections, including the straps are included on Sheet SD2.

Floor and Roof

An uninterrupted expanse of floor or roof sheeting creates a strong horizontal shear-resisting diaphragm that not only prevents the building from twisting or racking horizontally but can also share its shear strength with the walls below. Shear transfer details are shown on Sheet SD3. Also detailed are Drag Collectors, reinforced floor and roof members that collect both tension and compression loads from walls and transfer them to the roof or floor. These are particularly necessary where there are reentrant (inside) corners or other discontinuous load paths in the walls.

USING WHAT YOU LEARNED

The size and spacing of nails in a shear wall is very important. What size nails are used to nail the edges of the structural panel to the outside of the shear wall beside the garage doors and what is the proper spacing of those nails? Find the shear wall in question on the structural plans. This identified as ⟨J⟩ on the Post Tension Foundation Plan, Sheet S7.1. The ⟨J⟩ indicates an entry on the Shear Wall Schedule. That schedule specifies that the ½″ plywood is to be nailed at its edges with 10d nails spaced 2″ O.C.

Assignment

Use the Urban Courts Structural and Architectural Drawings to answer the following questions.

1. Using the foundation plan, what does the ⑦ in the center of the west (left) garage door signify?
2. A wall separates Unit B garage from the storage and stair area. How is that wall secured to the floor?
3. What is the bottom plate material?
4. What is the minimum distance in height from the top of the footing to the adjacent grade?
5. How long will Post Tensioned Strand 3 be after tensioning?
6. What is the required sheer panel thickness on the longest exterior wall of Unit A garage?
7. What type of nails are used for the wall in Question 6 and how often is the edge nailed?
8. Along the same Unit A garage wall is a reserved space. What is the purpose of this space and why can't it be placed anywhere along the wall?
9. What details show the floor strap holdown?
10. On Sheet S7.2, what does [3] denote? Where does this detail occur?
11. What hanger is called out at the upper right corner of the Second Floor Framing Plan?

Refer to the Urban Courts Drawings in your textbook drawing packet to answer the following questions.

1. What is represented by the symbols shown here?

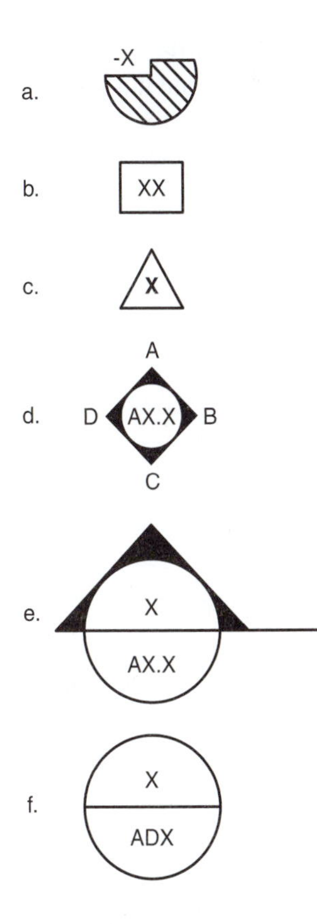

2. Define the abbreviations as used on the Urban Courts drawings.
   a. HDR
   b. UON
   c. AFF
   d. SLP
   e. CLR
3. On what sheet number would you find the following:
   a. details for exterior doors
   b. front building elevation for buildings with Elevation A
   c. foundation plan
4. How many times does Building 2 with Elevation C appear in phase 1, block 3? (See **Figure 34–7**.)
5. What is known about Building 57, based on the information shown on **Figure 34–7**?
6. What type of glass is indicated for the front entrance?
7. What framing material is to be used for the front wall on the first floor?

8. What is represented by the dashed line on the first floor plan in the area of the wall-bike storage?
9. Describe the window on the second floor in the stairway looking out onto the balcony.
10. Why is the floor in the first floor living room different from the floor in the second floor living room?
11. On the stair landing shown on the third floor plan, there is a short section of wall with only an outline and no internal symbol adjoining the back wall of the storage closet. Describe the construction of the short section of wall with no internal symbol.
12. How many lineal feet of standard 2×6 frame wall are there on the third floor?
13. List the types of wall construction shown on the fourth floor plan.
14. On the second floor plan, why is the wall between the living room and the balcony of different construction than the wall between the kitchen and the master bedroom?
15. What components of the exterior wall in the third floor kitchen are used to reduce the spread of fire?
16. List all of the components of the wall separating the third floor kitchen and the stairs starting at the kitchen side of the wall and ending at the stair side.
17. What is the STC rating of the common wall between the second floor kitchen and the stairs?
18. What is the fire resistance rating of the common wall between the second floor kitchen and the stairs?
19. What is the minimum allowable height for the nonabsorbent wall finish in the showers?
20. What provides fire resistance at the rim board on exterior walls?
21. For walls shown with the symbol below, is the insulation to be faced or unfaced?

22. What is the minimum embedded depth for an anchor bolt securing a sill to the slab?
23. What is the purpose of Detail A on Sheet S7.1 titled High Load Diaphragm?
24. What holdown is used to anchor the front wall to the foundation?
25. What is the post material that the holdown in Question 25 is attached to?
26. How far must the anchor bolt for the holdown at the right rear corner be embedded in the concrete?
27. What is the stud spacing for the wall between the two garages?
28. The Post Tensioned Foundation Plan on Sheet S7.1 has several boxed callouts referring to Maximum Cable Spacing. What are the cables the notes refer to?
29. On Sheet SD3, both Detail 1 and Detail 2 are both titled Floor Shear Transfer, and both have 2 alternates. What is the difference between the views on these two details?
30. Which shearwall is most often used on the second floor?
    See Second Floor Shearwall Plan

Multiple Choice

Select the best answer for the question.

31. Which type of gypsum board is most fire resistant?
    a. type X
    b. type M
    c. type F
    d. type L
32. Why is blocking paced between studs in walls over 10 feet tall?
    a. to prevent wracking
    b. to prevent vertical flame spread
    c. to resist shear forces
    d. all of the above

33. What is STC an abbreviation for?
    a. sound tested construction
    b. seismic tensile control
    c. sound transmission category
    d. sound transmission classification
34. Which is a rating of a floor/ceiling assembly's ability to control the sound of footfalls?
    a. FFI
    b. CRC
    c. IIC
    d. ASTM
35. What structural panel(s) is(are) required on the exterior side wall of the Unit A garage?
    a. 3/8″ plywood on the exterior only
    b. 3/8″ plywood on both sides
    c. 1/2″ CDX on the exterior only
    d. No structural panels are required
36. For the wall separating the Unit A garage and the entry area, how deep must the anchor bolts be embedded in the concrete slab?
    a. 7″
    b. 10″
    c. 12″
    d. 17″
37. What material is to be used for the post at the right rear corner of the garage?
    a. 2 × 4
    b. two 2 × 4s
    c. 6 × 6
    d. 8 × 8
38. How many and what size fasteners are used to attach the holdown to the post at the right rear corner of the garage?
    a. 8; 3/8 × 5 lag screws
    b. 12; 10d nails
    c. 12; 1/4 × 2 1/2 wood screws
    d. 36; 1/4 × 2 1/2 wood screws
39. What is the required thickness of a sill plate in a shearwall marked ⟨F⟩?
    a. 2 inches
    b. 3 inches
    c. 4 inches
    d. The thickness cannot be determined without more information.
40. What secures beam FB12 to the post that supports it at the end nearest the front wall?
    a. 4 16d nails
    b. HUC416 hanger
    c. LTP4 clips
    d. 2 CS16 straps

HEAVY COMMERCIAL CONSTRUCTION: SCHOOL ADDITION

Part 4 presents a thorough examination of the information found on prints for heavy commercial construction. The materials and methods used for large buildings are different from those used in light-frame construction. Also, the drawing set for a commercial project usually includes many more sheets than are found in the drawing sets for smaller buildings. To understand these drawings and make practical use of them, you need to understand the organization of the drawings and how the heavier materials are described on the drawings.

Part 4 also discusses the drawings used for air conditioning and heating, plumbing, and electrical systems. For those who work in the mechanical or electrical trades, the importance of understanding the drawings for these systems is obvious. However, all the trades work in the same spaces, and the work of one trade affects the other trades. Estimators, superintendents, inspectors, and many other construction professions also require an understanding of all the construction trades.

The School Addition project in your textbook packet drawings is an excellent building for inclusion in this book. It uses structural steel framing, reinforced concrete foundations, and a varied assortment of other materials typically

FILLET	PLUG OR SLOT	SPOT OR PROJEC-TION	STUD	SEAM	BACK OR BACKING	SUR-FACING	SCARF (FOR BRAZED JOINT)

FLANGE		GROOVE						
EDGE	COR-NER	SQUARE	V	BEVEL	U	J	FLARE-V	FLARE-BEVEL

NOTE: _ _ _ _ REPRESENTS REFERENCE LINE WHICH IS NOT SHOWN AS A SOLID LINE FOR THIS PURPOSE.

Figure 37–6. Weld symbols used to show various types of welds.

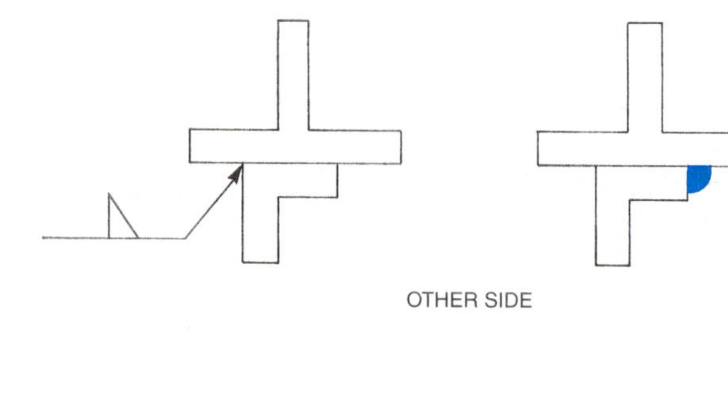

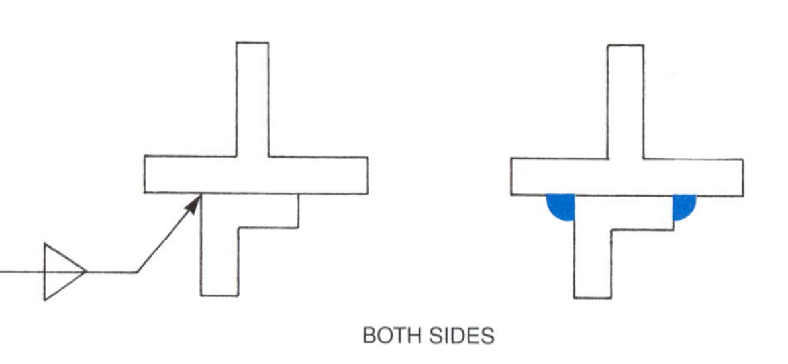

Figure 37–7. Arrow-side, other-side significance.

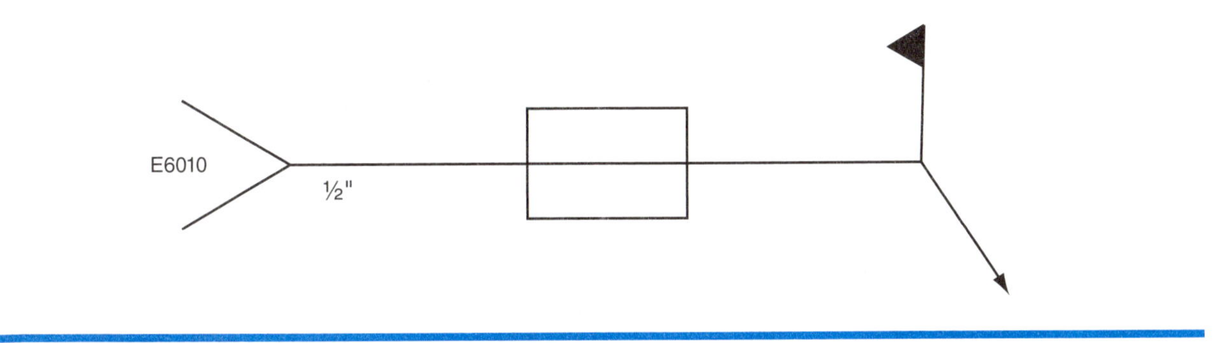

Figure 37–8. This can be recognized as a field weld symbol by the flag. It is a ½-inch plug weld on both sides of the part and is to be done with E6010 electrode or filler metal.

and will probably require more and larger-diameter reinforcing steel. The overall layout of the foundation is shown on a foundation plan (see **Figure 37–9**). The dimensions and the reinforcement of the footings and walls are planned by engineers to carry the necessary loads, so it is likely that there will be several detail drawings to further describe the foundation systems at different points in the building (see **Figure 37–10**). Some of the reinforcement may be bars that protrude only a short distance out of the footing and into the wall. Reinforcing bars used in this way are called *dowels* and are used to keep the wall from shifting on the footing.

Column footings are located by the column centerlines. All these footings must be centered under the columns they support unless specifically noted otherwise. The outline of the footing is shown on the foundation plan using dashed lines. This indicates that the footing is actually hidden from view by the concrete slab. Column footing dimensions are usually shown in a schedule (see **Figure 37–11**). There are usually several column footings of the same size. The mark-schedule system allows the drafter to show all the similar footings with a single entry on the schedule.

Piles are long poles or steel members that are driven into the earth to support columns or other grade beams. Pile foundations support their loads by means of several piles in a cluster, topped with a pile cap made of reinforced concrete. Pile foundations are not as common as spread footings and continuous walls. *Grade beams* are reinforced concrete.

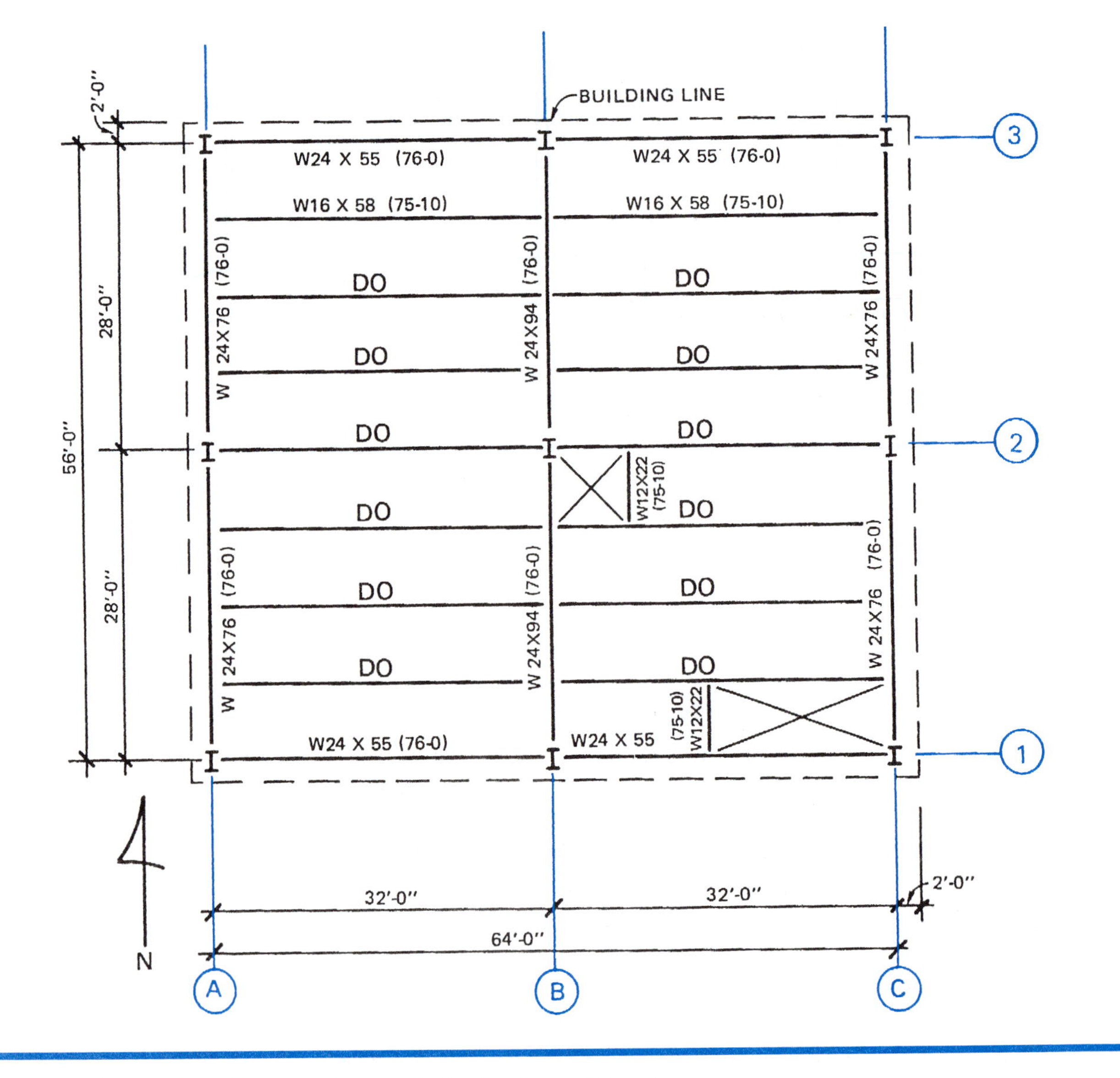

Figure 39–3. Structural coordinates are shown on the framing plans and floor plans.

The abbreviation *DO,* which stands for *ditto,* indicates that the specification for the first member in a series is to be repeated for all members. A number in parentheses at the end of the designation is used to indicate the elevation of the top of the member. This may be the elevation or the distance above or below the floor line (see **Figure 39–4**).

Joists are frequently *open-web steel joists,* sometimes called *bar joists* (see **Figure 39–5**). Bar joists are manufactured in H, J, and K series, depending on the grade of steel used and their strength requirements. K series are stronger than H series, and H series are stronger than J series. A bar joist designation includes a number to indicate depth, a letter to indicate strength series, and a number to indicate chord size. For example, the designation 16H6 indicates a 16-inch-deep H series joist with number 6 chords.

The actual lengths of members are not shown on the general contract drawings. This information is shown on shop drawings that are drawn by the steel fabricator some time after the construction drawings are completed. It is an easy matter, however, to find the span of the member by looking at the framing plan. Connections are shown on details and sections.

Lintels are sometimes categorized as *loose steel.* In masonry construction, a steel lintel is placed above each opening to support the weight of the masonry above the opening (see **Figure 39–6**). These lintels are

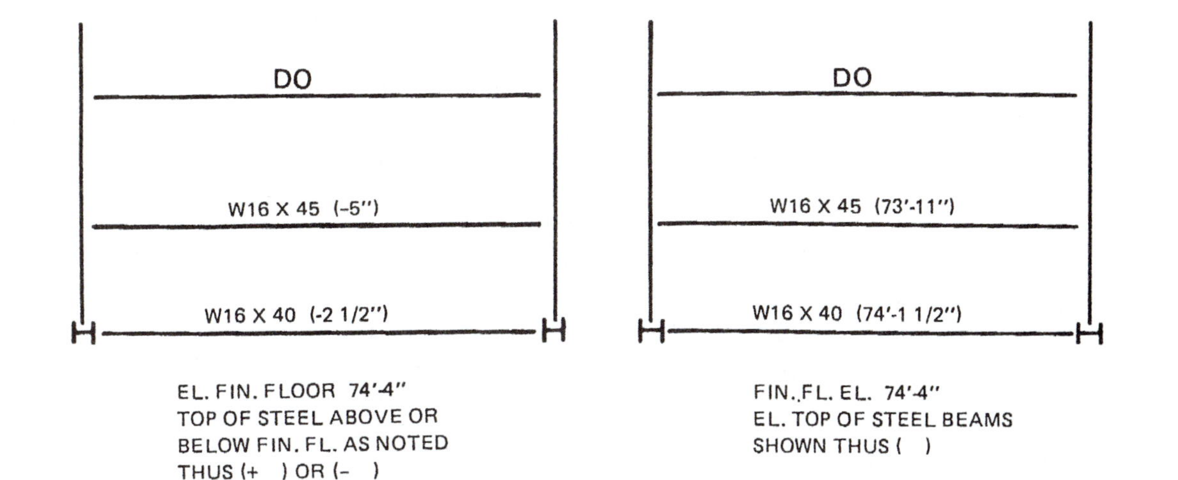

Figure 39–4. Notations on the drawings indicate the relative elevations of beams.

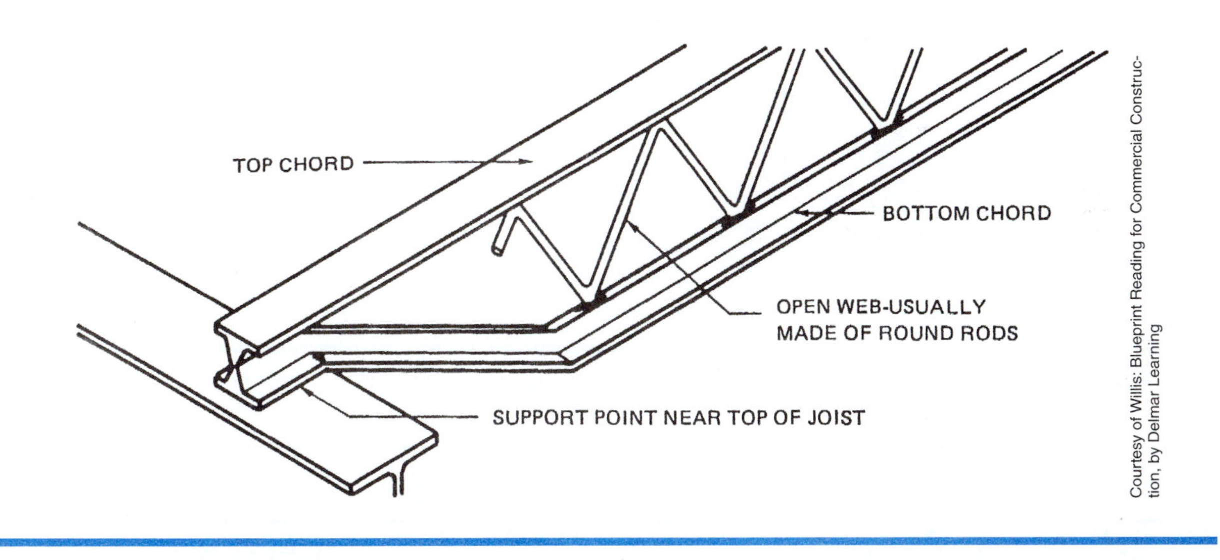

Figure 39–5. Open-web steel joists (often called bar joists).

not attached to the steel building frame, which is why they are called loose steel. If there are several lintels alike, they are normally shown on the plans by a symbol and are then described more fully in a lintel schedule (see **Figure 39–7**).

Masonry Reinforcement

Masonry materials have great compressive strength. That is, they resist crushing quite well, but they do not have good tensile strength. The ***mortar*** in masonry joints is especially poor at resisting the forces that tend to pull it apart, such as a force against the side of a wall or a tendency for the wall to topple. Masonry joint reinforcement is done by embedding specially made welded-wire reinforcement in the joints (see **Figure 39–8**). Greater strength can be achieved by building the masonry wall with reinforcement bars in the cores of the masonry units, then filling those cores with concrete. Concrete used for this purpose is called *grout.* Reinforcing steel is also embedded in masonry *walls* to tie structural elements together. For example, it is quite common for rebars to protrude out of the foundation and into the exterior masonry walls. The strength of a masonry wall can be increased considerably by the use of bond beams (see **Figure 39–9**). A *bond beam* is made by placing a course of U-shaped masonry units at the top of the wall. Reinforcing steel is placed in the channel formed by the U shape; then the channel is filled with grout. The result is a reinforced beam at the top of the wall.

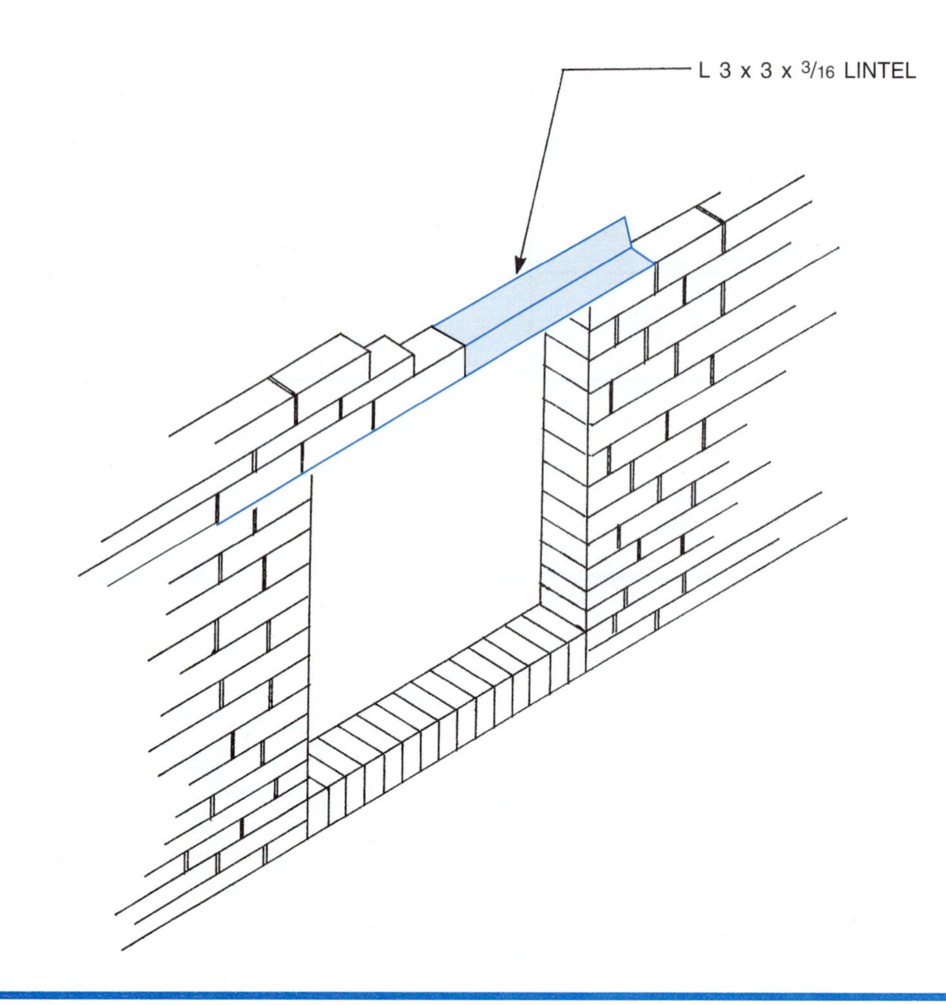

Figure 39–6. A lintel may be considered loose steel if it is not attached to the frame.

LINTEL SCHEDULE				
MARK	**MATERIAL**	**TYPE**	**MAS. OPNG**	**REMARKS**
L1	WT8 × 13	⊥	SEE ARCH.	ATTACH ONE END TO COL. SEE 1/S200
L2	WT8 × 13	⊥	"	END BEARING BOTH ENDS
L3	WT8 × 13	⊥	"	CONTINUOUS W/CENTER SUPPORT ON T.S. 4×4
L4	WT4 × 9	⊥	"	AT ALL BELOW WINDOW UNIT VENTILATOR OPENINGS
L5	(2) ∠ 6×4×5/16 W/ 1/4"×9 1/2 CONT. BOT. PL	JL	"	LINTEL IN FIRE WALL WELD BOT. PL TO ANGLES.
L6	(2) ∠ 6×4×5/16	JL	"	IN EXIST. BLDG. WALL
L7	(3) ∠ 4×3 1/2×5/16	JJL	"	AT CAFETERIA HVAC UNIT WALL OPENINGS COORDINATE WITH G-7/H3 AND A103 (2-LOCATIONS)

Figure 39–7. Lintel schedule.

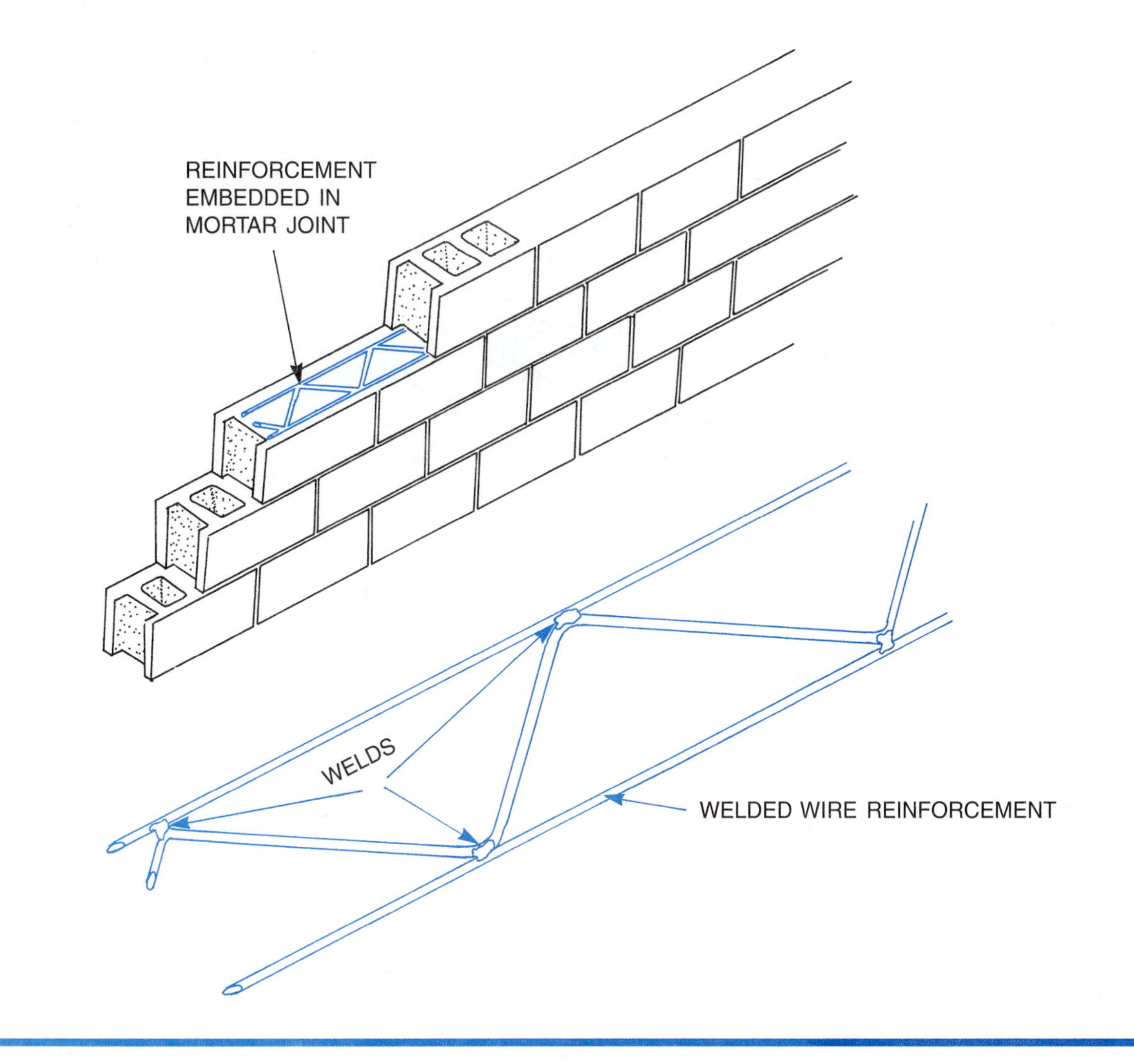

Figure 39–8. Masonry joint reinforcement.

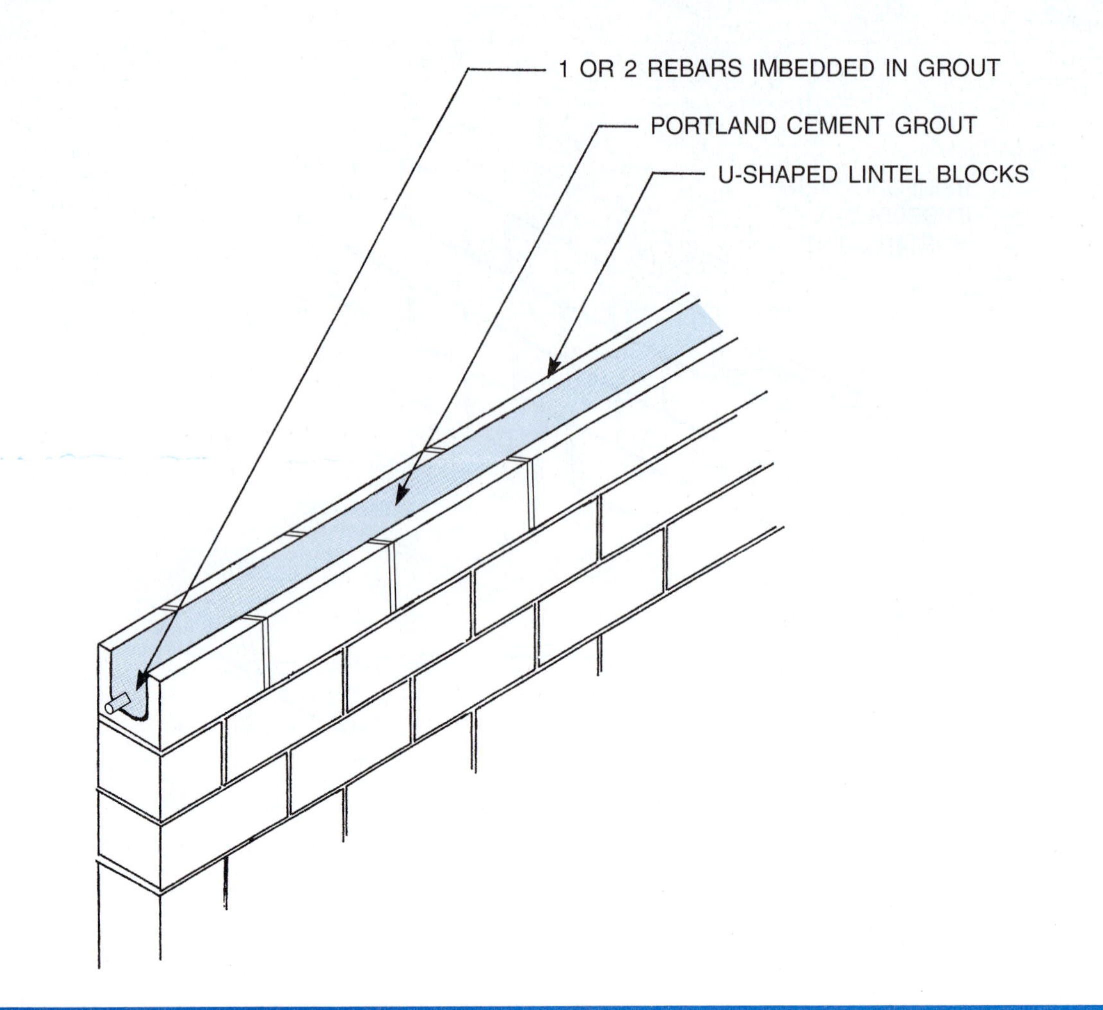

Figure 39–9. A bond beam is used to strengthen the top of the wall.

USING WHAT YOU LEARNED

The structural drawings for the School Addition in your textbook packet contain information about how the new addition is to be anchored to the existing building. If this anchoring is not done correctly, the risk that the buildings will separate over time increases. At the ground level, there will be a ramp down from the existing building to the addition. How is that ramp anchored to the existing building?

Being at the ground level, the ramp is shown on the Foundation Plan on Sheet S100. Section View 11/S101 shows a cross section of the ramp. At the far right of that section view is a callout that reads #4 × 1′-6″ DOWELS @12″ O.C. DRILLED & GROUTED INTO EXIST. SLAB 6.

Assignment

Refer to the drawings of the School Addition in your textbook packet to complete this assignment.

1. What are the dimensions of the footing for the northwestern most column?
2. What is the width and depth of the footing at the west end of the addition?
3. What is the elevation of the top of the floor in the elevator pit?
4. How many lineal feet of #5 reinforcement bars are needed for the footing under the east wall of the addition?
5. What size and kind of material is used to prevent the foundation wall from moving on the footings? How much of this material is needed for the west end of the addition?
6. How many pieces of what size reinforcing steel are to be used in the footing for the column between the entrances to classrooms 103 and 104?
7. How closely is the vertical reinforcement spaced in a typical section of the foundation wall?
8. What is the overall length of each piece of rebar used to secure the interior masonry partitions to the concrete slab?
9. What is the spacing of the dowels used to secure the exterior masonry walls to the foundation?
10. Describe how the corridor walls are secured to the columns.
11. How many pieces of W18 × 50 steel are used in the construction of the addition?
12. What size and shape structural steel supports the north ends of the floor joists under classroom 203?
13. What size and shape structural steel supports the ends of the girder in the second floor at the front (side with the door) of the elevator shaft?
14. What is the nominal depth of the joist in the second floor corridor?
15. What is the elevation of the top of the second floor girder between D5 and D6, relative to the second finish floor elevation of 110′-3″?
16. What is the shape and size of the member that supports the masonry above the windows in classroom 108?

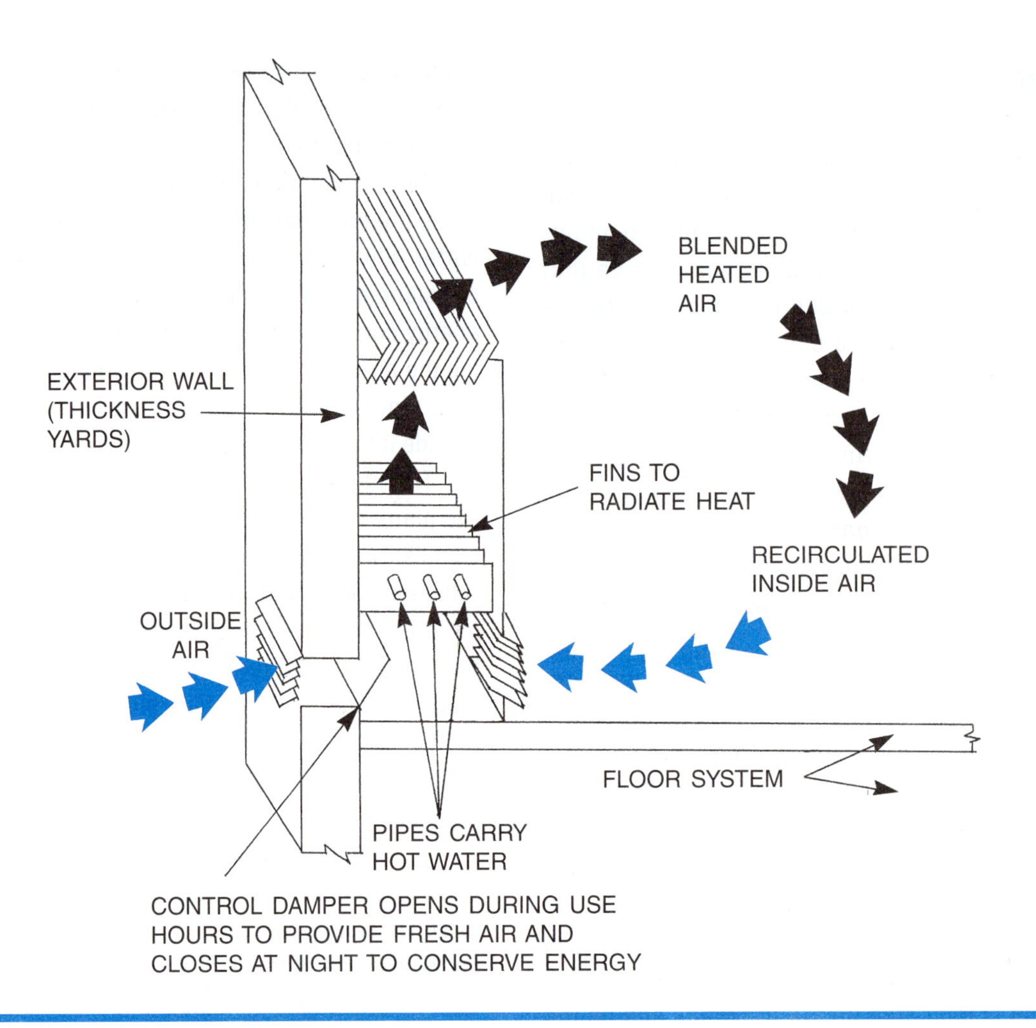

Figure 40–2. Operation of a unit ventilator.

CABINET UNIT HEATER SCHEDULE

MARK	MAKE	MODEL	STYLE & ARRANGEMENT	STEAM		WATER				CFM	RPM	EAT	MBH	ELECTRICAL	
				PSIG	IB/HR	EWT	ΔT	GPM	ΔP*					V/φ/HZ	HP
CUH-1	STERLING	RWI-1130-04	RECESSED WALL, INVERTER			180	34	2.0	0.44	420	1050	60	33.8	120/1/60	1/10

UNIT HEATER SCHEDULE

MARK	MAKE	MODEL	TYPE	WATER			MBH	CFM	EAT	LAT	MOTOR		
				ENT.	GFM	ΔP					RPM	H.P.	V/φ/HZ
UH-1	MODINE	HS-18L	HORIZONTAL	180	1.1	0.4	9.4	364	60	84	1550	16MHP	120/1/60

FAN COIL UNIT SCHEDULE

MARK	MAKE	MODEL	CFM & TOT./F.A.	EXT. S.P.	ELECTRICAL		HEATING				
					WATTS	VOLTS/PH/HZ	EAT	EWT	GPM	MBH	ΔP
FCU-1	AAF	SFG-TFA-3000	350	0	110	120/1/60	50	180	1.0	18	0.1

Figure 40–3. Schedules of cabinet unit heaters, unit heaters, and fan cabinet unit heaters.

Heating Piping

Once the basic type of heating system has been determined, the next step is to understand the piping that carries water to and from the heating units. It is easy to spot the pipes on the heating plan. A solid line indicates supply pipes, and a dashed line indicates return pipes. Pipe sizes are indicated by callouts throughout the plans. Fittings such as valves, elbows, and tees are represented by symbols. **Figure 40–4** shows the most common symbols for mechanical systems. There are standard piping symbols, but each drafter uses a few special representations of his or her own, so a symbol legend is included with the drawings.

Pipes are easier than ducts to coordinate with structural elements because the direction of the pipe can be changed easily and frequently. Pipes might run horizontally in a ceiling for some distance, then loop around a column and drop down to a unit ventilator. It is not practical to show the layout of all pipes with separate plan and riser views. The common method for showing pipes where both horizontal and vertical layout is involved is to use isometric schematics. Plumbing isometrics are discussed in Unit 35. Sheet H-1 of the School Addition drawings combines plan views and isometric schematics. The long runs of pipes are shown in plan view to indicate where the pipes are in relation to the walls. Where the pipes need to drop down from the ceiling to the level of the ventilators, they are shown in isometric views (see **Figure 40–5**).

For the sake of explanation, the following discussion traces the hot-water supply piping (broken line) in room 109, the storage room on the first floor, E-7 on Sheet H-1. Most of what is discussed here is also shown in **Figure 40–5**, but we will start closer to the center of the building, where the branch tees to the right (north) off the main supply running from top to bottom of the drawing. See the symbol legend A-11 on Sheet H-1 to interpret the symbols as you trace the piping. It is recommended that you use a colored pencil to trace the piping as you read this.

The first thing we encounter is a valve, an elbow toward the top of the drawing, and then another elbow back to the right. Construction Note 13 indicates that the pipes pass through a structural steel member. The line is broken for clarity through the area where ductwork is shown that would make reading the piping drawing difficult. Pick the line up again in the storage room 109. Next we come to a tee where the pipe drops down to the level of the unit ventilator. If we continued on to the right or north, we would come to Construction Note 2, which tells us that the symbol indicates a riser to the second floor. The riser is not shown in the wall, where it would actually be positioned, because doing so would bury the riser symbol in the wall with material symbols and other lines. The pipe going down is actually shown at an angle up and to the right. This is the isometric portion of the drawing. At the level of the unit ventilators, the pipes are arranged in a special way to eliminate noises that might result from expansion and contraction as the water temperature changes (see **Figure 40–6**). The ends of the pipes are shown with break lines, which indicate that the pipes actually continue to their obvious destinations: fittings on the unit ventilators.

Air-Handling Equipment

The unit ventilators introduce fresh air into the School Addition. A separate ventilation system removes stale air from each room of the school. All the rooms in the addition except the smallest closets and the stairways are provided with a system of ducts and a fan to ensure air circulation. A *louvered grille* in the room receives the air and channels it into a duct that carries it to the roof, where a fan exhausts it to the outside.

Figure 40–7 explains the designations for outlet and inlet grilles. The ducts that carry the air to the roof vent are shown by their outlines and do not indicate their sizes. A square or rectangle with a diagonal line through it indicates where the duct rises to the next floor or the roof. Several notes indicate that ducts rise up to EF-# on the roof. *EF* is an abbreviation for exhaust fan. The number indicates which of the several exhaust fans on the roof the duct is connected to. Some of the exhaust fans occupy several square feet of space on the roof, so they cannot be placed too close together.

Trace the flow of air as it is exhausted from the janitor's closet on the first floor. The path starts with a louver marked RB (return)—8 × 6 (cross-sectional size in inches)—75 (cubic-feet-per-minute nominal airflow). Between the janitor's closet and the storage room, the air enters a 6 × 6 duct, which goes up through a chase (space between two walls) to the second floor. At the second floor it is joined by the exhaust air from the second floor janitor's closet. On the Second Floor Plan

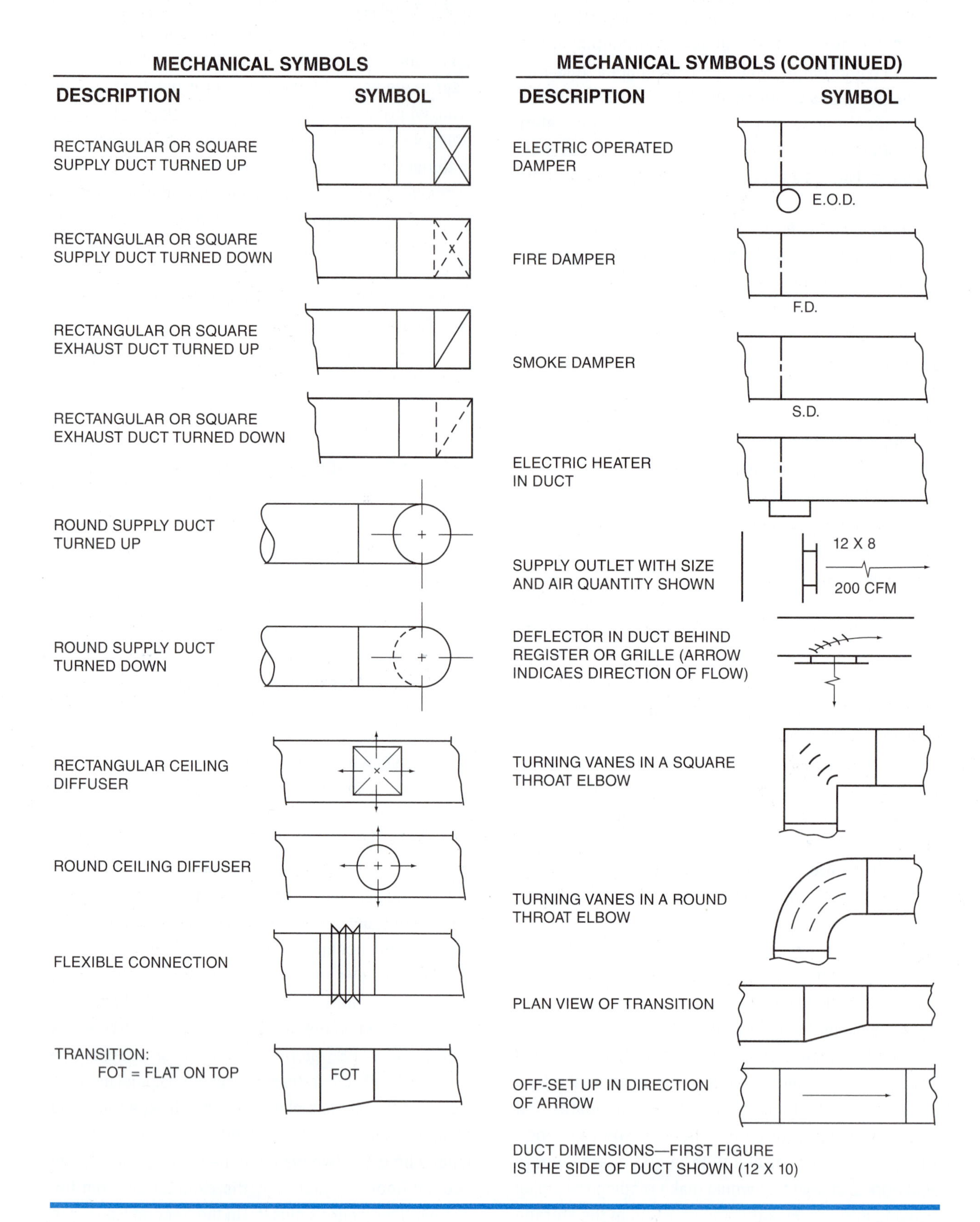

Figure 40–4. Mechanical and plumbing symbols.

MECHANICAL SYMBOLS (CONTINUED)	
DESCRIPTION	**SYMBOL**
ACCOUSTICAL LINING INSIDE INSULATION	
BRANCH TAP IN DUCT	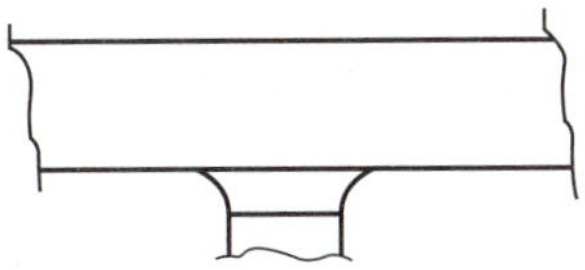
SPLITTER FITTING WITH DAMPER	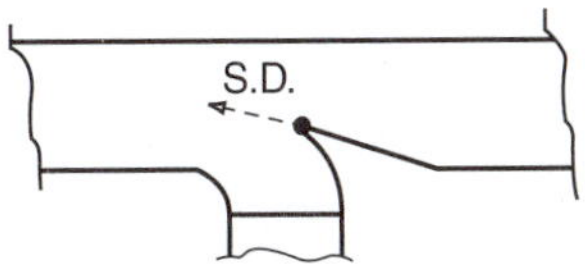 S.D.
VOLUME-DAMPER	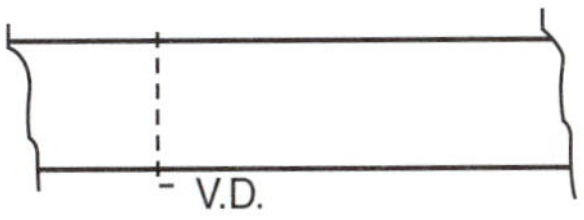 V.D.
BACKDRAFT DAMPER	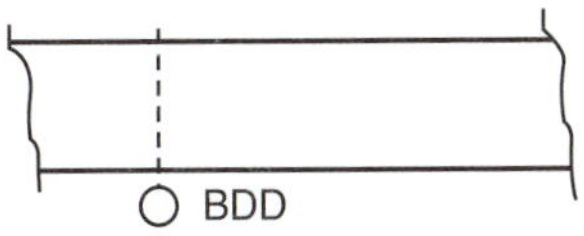 BDD
ACCESS DOOR IN DUCT 10" X 10" SIZE	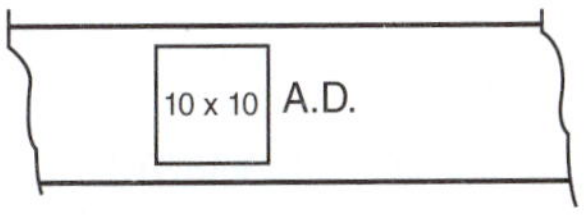 10 x 10 A.D.
PNEUMATIC OPERATED DAMPER	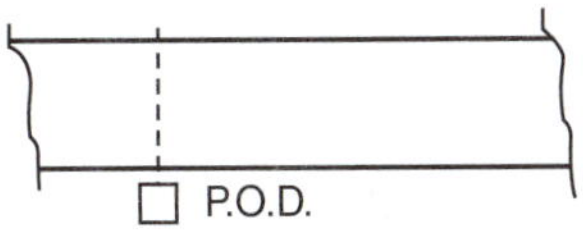 P.O.D.
THREE-WAY VALVE	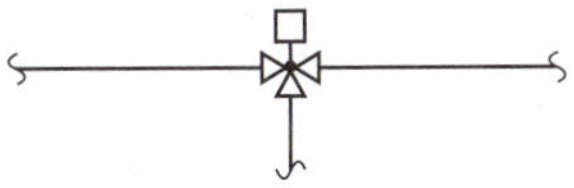
PRESSURE REDUCING VALVE	
PRESSURE RELIEF VALVE OR SAFETY VALVE	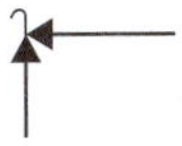
SOLENOID VALVE	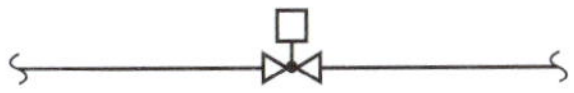
PIPE TURNED UP (ELBOW)	
PIPE TURNED DOWN (ELBOW)	
TEE (OUTLET UP)	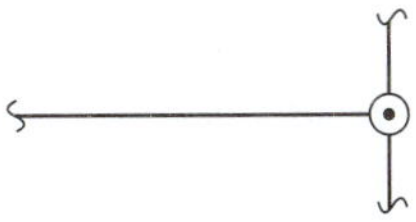
TEE (OUTLET DOWN)	

MECHANICAL SYMBOLS (CONTINUED)	
DESCRIPTION	**SYMBOL**
BACKFLOW PREVENTER	 B.F.P.
UNION	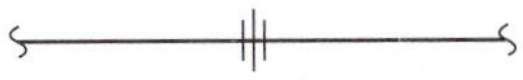
REDUCER	
CHECK VALVE	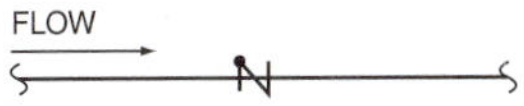 FLOW
GATE VALVE	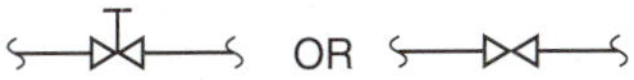 OR
GLOBE VALVE	 OR
BALL VALVE	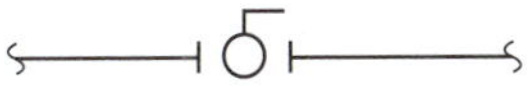
BUTTERFLY VALVE	
DIAPHRAGM VALVE	
ANGLE GATE VALVE	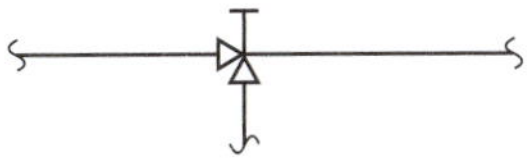
ANGLE GLOBE VALVE	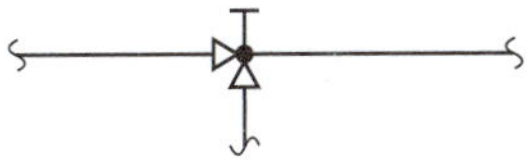
PLUG VALVE	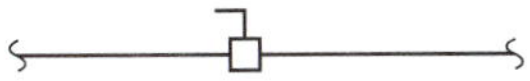
LOW PRESSURE STEAM	 LPS
LOW PRESSURE CONDENSATE	 LPC
PUMPED CONDENSATE	 PC
FUEL OIL SUPPLY	 FOS
FUEL OIL RETURN	 FOR
HOT WATER SUPPLY	 HWS
HOT WATER RETURN	 HWR
COMPRESSED AIR	 A
REFRIGERANT SUCTION	 RS

Figure 40–4. *(Continues)*

MECHANICAL SYMBOLS (CONTINUED)

DESCRIPTION	SYMBOL
REFRIGERANT LIQUID	RL
REFRIGERANT HOT GAS	RHG
CONDENSATE DRAIN	CD
FUEL GAS	G
CHILLED WATER SUPPLY	CWS
CHILLED WATER RETURN	CWR

PLUMBING SYMBOLS

DESCRIPTION	SYMBOL
METER	M
SPRINKLER PIPING	S
SPRINKLER HEAD	
FLOOR DRAIN	F.D.
CLEAN-OUT	C.O.
TUB	
TANK-TYPE WATER CLOSET	
WALL-MOUNTED LAVATORY	
URINAL	
SHOWER	
WATER HEATER	WH
MANHOLE	MH

PLUMBING SYMBOLS (CONTINUED)

DESCRIPTION	SYMBOL
WALL HYDRANT	
YARD HYDRANT	Y.H.
FLUSH VALUE WATER CLOSET	
COUNTER-TYPE LAVATORY	
KITCHEN SINK (DOUBLE BOWL)	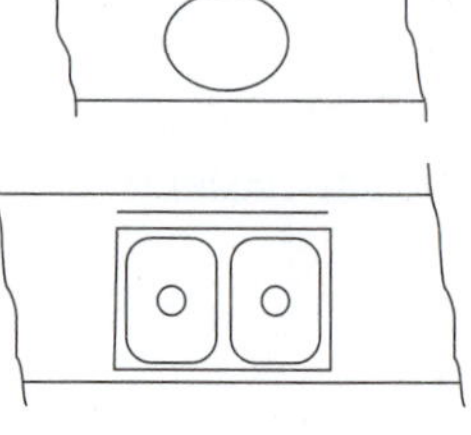
SOIL, WASTE OR DRAIN LINE	
PLUMBING VENT LINE	
COLD WATER (DOMESTIC)	
HOT WATER (DOMESTIC)	
HOT WATER RETURN (DOMESTIC)	
FIRE LINE	F
FUEL GAS LINE	G
ACID WASTE LINE	AW
VACUUM LINE	V
COMPRESSED AIR LINE	A
BACKFLOW PREVENTER	BFP
GATE VALVE	OR
GLOBE VALVE	OR
CHECK VALVE (ARROW INDICATES DIRECTION OF FLOW)	

Figure 40–4. *(Continued)*

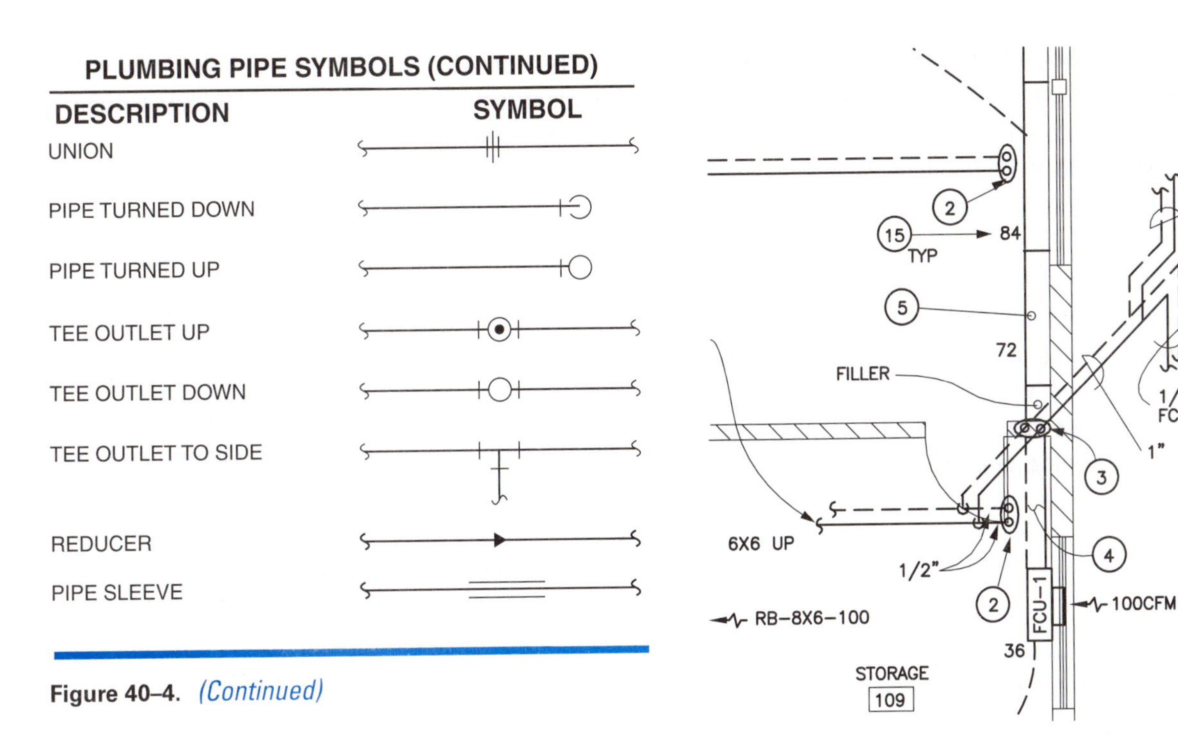

PLUMBING PIPE SYMBOLS (CONTINUED)

DESCRIPTION	SYMBOL
UNION	
PIPE TURNED DOWN	
PIPE TURNED UP	
TEE OUTLET UP	
TEE OUTLET DOWN	
TEE OUTLET TO SIDE	
REDUCER	
PIPE SLEEVE	

Figure 40–4. *(Continued)*

Cataldo, Waters, and Griffith Architects, P.C.

Figure 40–5. This isometric drawing shows how the heating pipes drop down from the ceiling to the level of the ventilators.

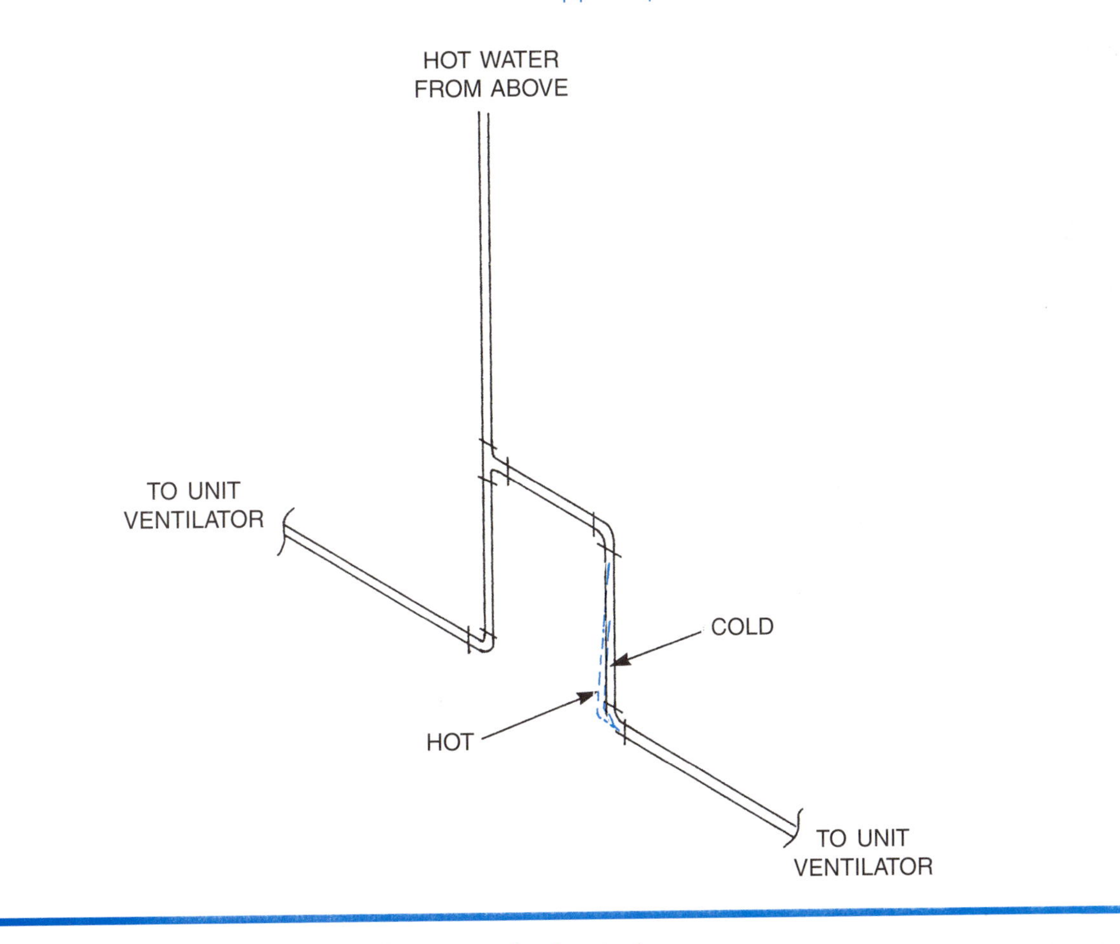

Figure 40–6. Noises are eliminated by allowing pipes to expand and contract.

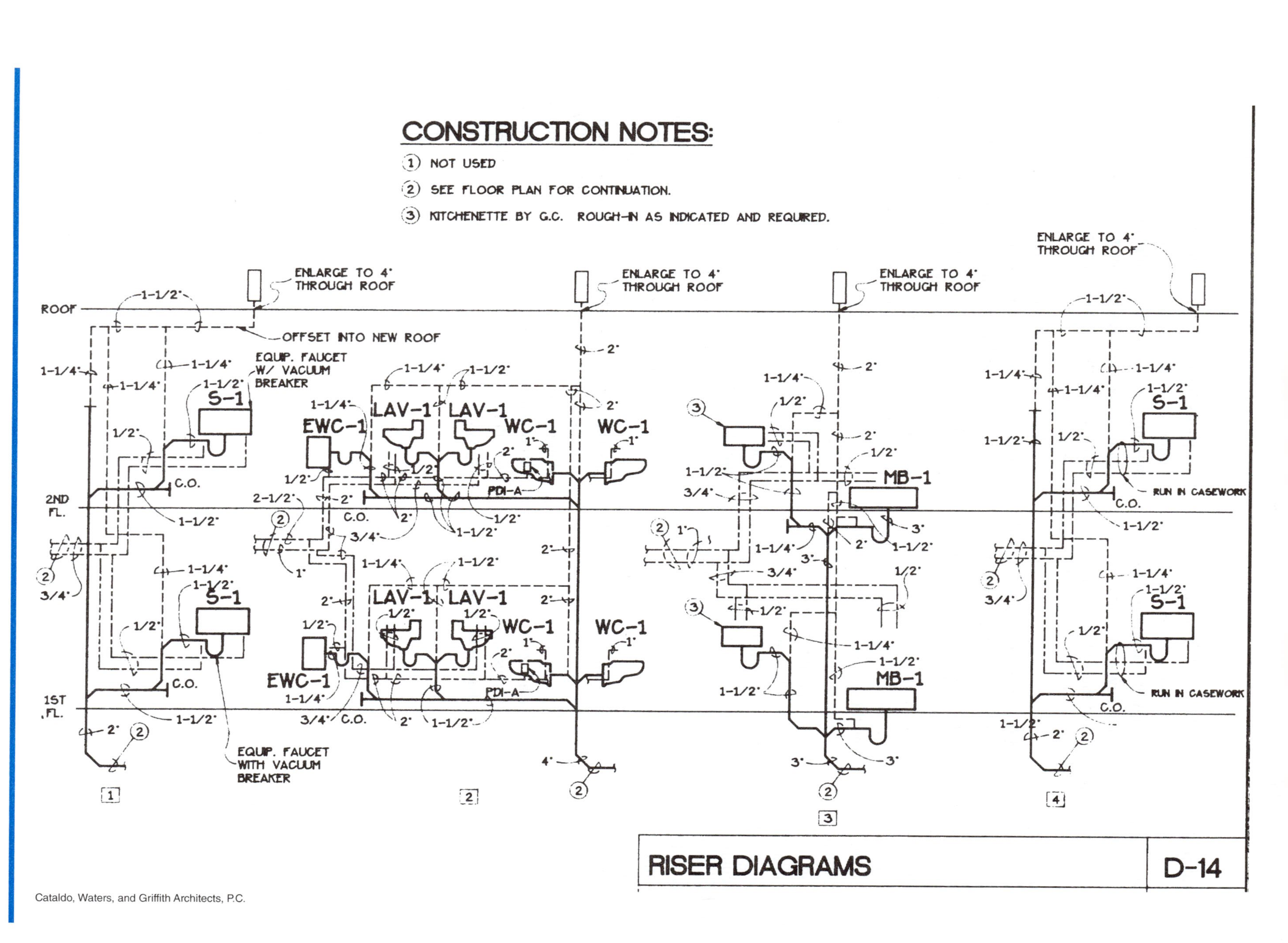

Figure 40–9. Riser diagrams for school fixtures.

USING WHAT YOU LEARNED

Even for workers not directly involved in installing the HVAC system in a building, it is useful to understand where piping, ductwork, and other mechanical equipment will be installed. For example, iron workers who install the structural steel need to know where pipes run through bar joists or around steel beams. As an exercise in tracing piping, list all the pipe sizes through which hot water passes from the time it enters the School Addition until it reaches the unit ventilator in Classroom 104. The easiest way to do this is to start at the unit ventilator and work back to where the piping enters the building. Find the Unit Ventilator (UV-2) in Classroom 104 on the First Floor Plan E-7. There are actually two UVs in this classroom, but both have the same size supply pipes. The pipes at this point are near the top of the classroom and labeled as ¾″. Follow the pipes to where they turn east (notice that the North Arrow point to the right) to follow the hall toward the existing building. At this point, they are labeled as 1″ pipes. After they pass the tees to Classroom 107, they are 1¼″ pipes. Beyond the tees, entering Classroom 103, they are 1½″ pipes, and, after the next tees, entering storage room 101, they are 2″. They are 2″ all the way to where they enter the addition. So, in the order the water passes through the pipes, they are 2″, 1½″, 1¼″, 1″, and ¾″.

Assignment

Refer to the drawings of the School Addition in your textbook packet to complete this assignment.

1. How many unit ventilators are shown on the first floor?
2. How is the machinery room on the first floor behind the elevator heated?
3. List in order the sizes of pipe that the water flows through, starting from the point where the water enters the addition to get to the heating unit in Classroom 210 and then back to the point where it leaves the addition.
4. What are the dimensions of the grille where stale air is vented out of Classroom 209?
5. List in order the sizes of the ductwork that exhaust air passes through from the point where the air leaves Classroom 209 until it is outside the building.
6. List in order the sizes of the ductwork that exhaust air passes through from the point where the air leaves the girls' toilet room on the first floor until it exits the building.
7. How many gallons per minute of water are expected to flow through the heating unit in stair #2?
8. Why are two roof drains shown on the First Floor Plumbing Plan of this two-story building?
9. What is designated as EWC-1 on the plumbing plans?
10. What is the vertical distance between the storm drain and the sanitary drain at the point where they cross?
11. What is the diameter of the pipe that is used for the storm drain where it goes from the second floor to the first floor?
12. What size pipe is used for the domestic hot-water return line?
13. Where does the domestic hot-water return line tee out of the domestic hot-water supply line?
14. Where would you shut off the cold-water supply to the kitchenette unit in Room 101 without shutting off the cold water to the toilet rooms on the same floor?

UNIT 41 Electrical Drawings

Objectives

After completing this unit, you will be able to perform the following tasks:

- Explain the information found on a lighting plan.
- List the equipment served by an individual branch circuit using electrical plans, riser drawings, and schedules.
- Explain the information on a schematic diagram.

Electrical Drawings

Electrical drawings deal predominantly with circuits. Knowledge and understanding of the basic circuits used in buildings is necessary for those working with and interpreting electrical drawings. The four basic methods of showing electrical circuits are:

1. Plan views
2. Single-line diagrams
3. Riser diagrams
4. Schematic diagrams

Plan Views

The electrical floor plan of a building shows all the exterior walls, interior partitions, windows, doors, stairs, cabinets, and so on along with the location of the electrical items and their circuitry.

Power Circuits

The *power circuit* electrical floor plan shows electrical outlets and devices and includes duplex outlets, specialty outlets, telephone, fire alarm, and the like. The conventional method of showing power circuits is to use long dash lines when specified to be installed in the slab or underground and to use solid lines when concealed in ceilings and walls. Short dash lines indicate exposed wiring. Solid lines are often used in commercial installations when the raceway (conduit) is to be exposed (surface mounted). Slash marks through the circuit lines are used to indicate the number of conductors. Full slash marks are the circuit conductors, a longer full slash is the neutral, and half slash marks are the ground wires. When only two wires are required, no slash marks are used. The typical wire size for commercial construction is No. 12 AWG (American wire gauge). The arrows indicate "home runs" to the designated panel. The panel and circuit number designations are adjacent to the arrowheads. **Figure 41–1** lists the typical circuiting symbols. The electrical installer will normally group home runs in a raceway with combinations similar to the following:

- Three-phase systems with three circuit wires and a common neutral in a raceway
- Single-phase systems with two circuit wires and a common neutral in a raceway

CIRCUITING

(arrows 1 2)	BRANCH-CIRCUIT HOME RUN TO PANEL*
(///)	THREE WIRES IN CABLE OR RACEWAY
(////)	FOUR WIRES IN CABLE OR RACEWAY, ETC.
(///)	SOME DRAWINGS SHOW THIS METHOD OF CONDUCTOR IDENTIFICATION: EQUIPMENT GROUNDING CONDUCTOR: LONG LINE WITH DOT. NEUTRAL CONDUCTOR: LONG LINE. PHASE CONDUCTOR WITH SWITCH LEGS: SHORT LINE.
	WIRING CONCEALED IN CEILING OR WALL
	WIRING CONCEALED IN FLOOR
	WIRING EXPOSED
	WIRING TURNED UP
	WIRING TURNED DOWN
CO	CONDUIT ONLY (EMPTY)
	SWITCH LEG INDICATION. CONNECTS OUTLETS WITH CONTROL POINTS.

* AN ARROW INDICATES A BRANCH-CIRCUIT HOME RUN TO PANEL.

THE NUMBER OF ARROWS INDICATES THE NUMBER OF CIRCUITS.

IF THERE ARE NO CROSSHATCHES, THEN IT IS ASSUMED THAT THE RACEWAY CONTAINS TWO WIRES.

Figure 41–1. Circuiting symbols.

Note: To reduce the harmonic problems caused by solid-state devices (nonlinear loads), many electrical designers are requiring separate neutrals with each phase conductor.

Some equipment, such as television, fire alarm, clock, PA, and sound, might not be connected with wiring lines on the plan. This indicates that wiring for that equipment is not part of the general circuit wiring and is probably not to be fed through the same panels as the rest of the electrical equipment. For example, on the School Addition the PA system and the clocks are fed from the sound system console and do not involve the general circuit wiring.

Lighting Circuits

There is usually a *lighting circuit* electrical floor plan for each level or major space within a building. This floor plan shows light fixtures, emergency lighting, security lighting, and special lighting control (photocell, motion detector, etc.). Typically included is a reflected ceiling plan (see **Figure 41–2**). That is, it is a plan view but shows what is on the ceiling as though it were reflected onto the floor plan (see **Figure 41–2**). The reflected ceiling plan shows each light fixture with a circle, square, or rectangle that approximates the shape of the fixture. The lighting circuitry on the School Addition is shown with three types of wiring circuit lines, solid for unswitched, dotted for switched, and line-dash-line for the motion sensor circuits. The conventional method of showing lighting circuits is the same as previously explained in the paragraph above on power circuits.

Motion sensors are an interesting feature of the lighting design for the School Addition. The motion sensors are shown as small circles containing the letter *M.* A motion sensor detects any movement in the area of the sensor, so that in a room controlled by a motion sensor, if there is no movement in the room for a period of time, the sensor opens its contacts, like turning a switch off. A motion detector can be used together with a switch so that the lights are controlled normally by the switch. But if someone forgets to turn the lights off, the motion detector will.

A low-voltage lighting control system requires a *relay* that is activated by the light switch. The relay opens or closes the higher-voltage fixture (see **Figure 41–3**). The relays on the School Addition are indicated by the letter *R* in a small square. Notice that every lighting switch is connected to a relay.

Symbols

Electrical symbols are used to simplify the drafting and later the interpreting of the drawings. Electrical symbols *are not standardized* throughout the industry. Most drawings have a symbol legend or list. You must be knowledgeable of the symbols specifically used on each project since designers modify basic symbols to

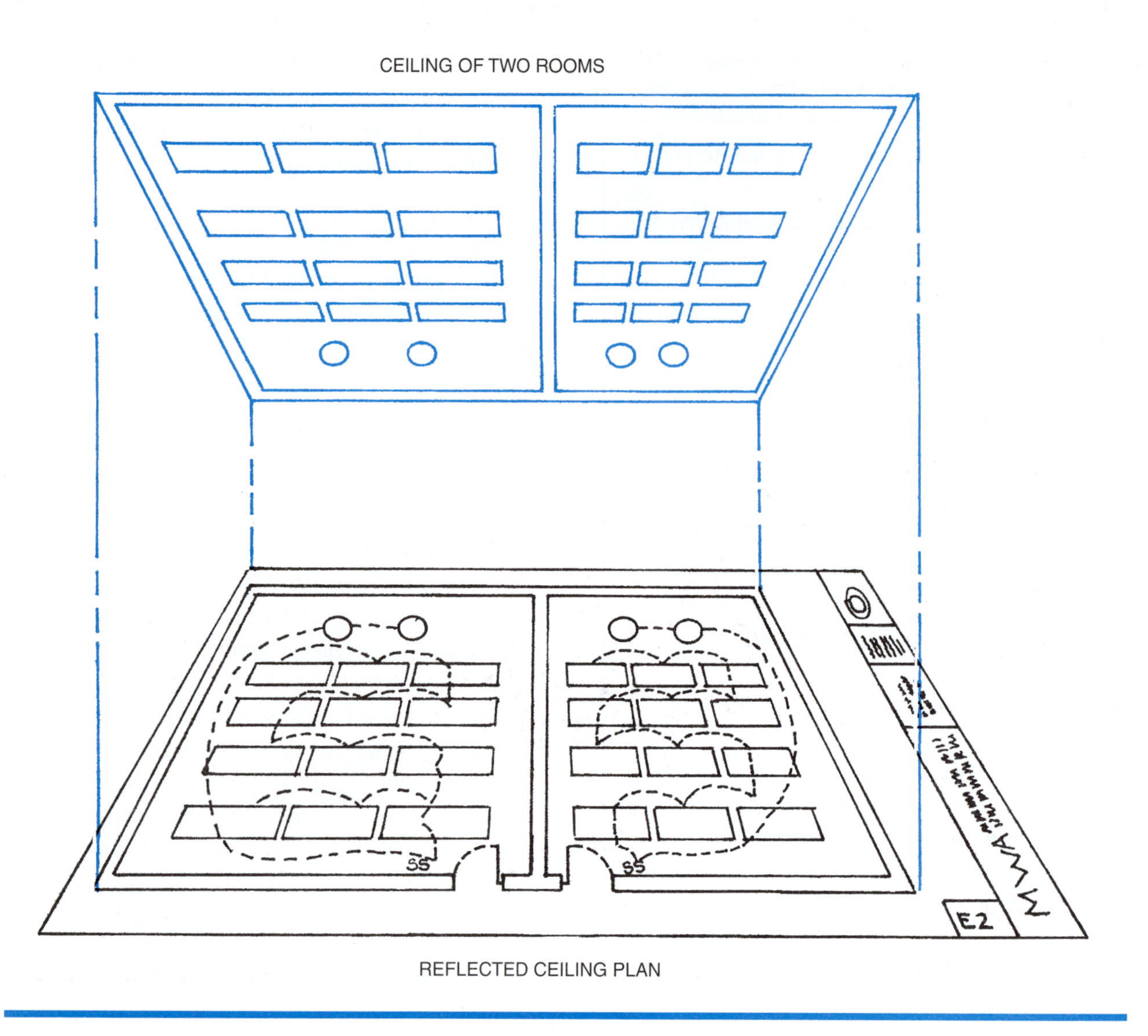

Figure 41–2. Reflected ceiling plan.

suit their own needs. Many symbols are similar(circle, square, etc.). The addition of a line, dot, shading, letters, numbers, and so forth gives the specific meaning to the symbol. Learning the basic form of the various symbols is the best starting point in developing the ability to interpret the drawings and the irrelated symbol meanings. **Figure 41–4** shows common electrical symbols and abbreviations.

The School Addition drawing E-1, symbol legend A-4, contains an electrical symbol list for the project. From the symbol list, it can be seen that there are duplex receptacles, switches, telephone outlets, special-purpose outlets, and fire alarm devices mounted at various heights. Then there is the General Note in the symbol legend that specifies that all mounting heights are to be verified and "modified as directed." This is an example of why the installer must become familiar with the drawings and specifications far in advance of the scheduled time for the installation. The installer must request clarification or direction and give the designers reasonable time to clarify the questionable specified instructions.

Single-Line Diagrams

The electrical service *single-line diagram* on the School Addition is shown on drawing E-2, Single-Line Diagram E-17. The electric power is brought into the school by way of the *service entrance section (SES)* The note "coordinate with the local utility requirements prior to rough-in" indicates that the electrical contractor

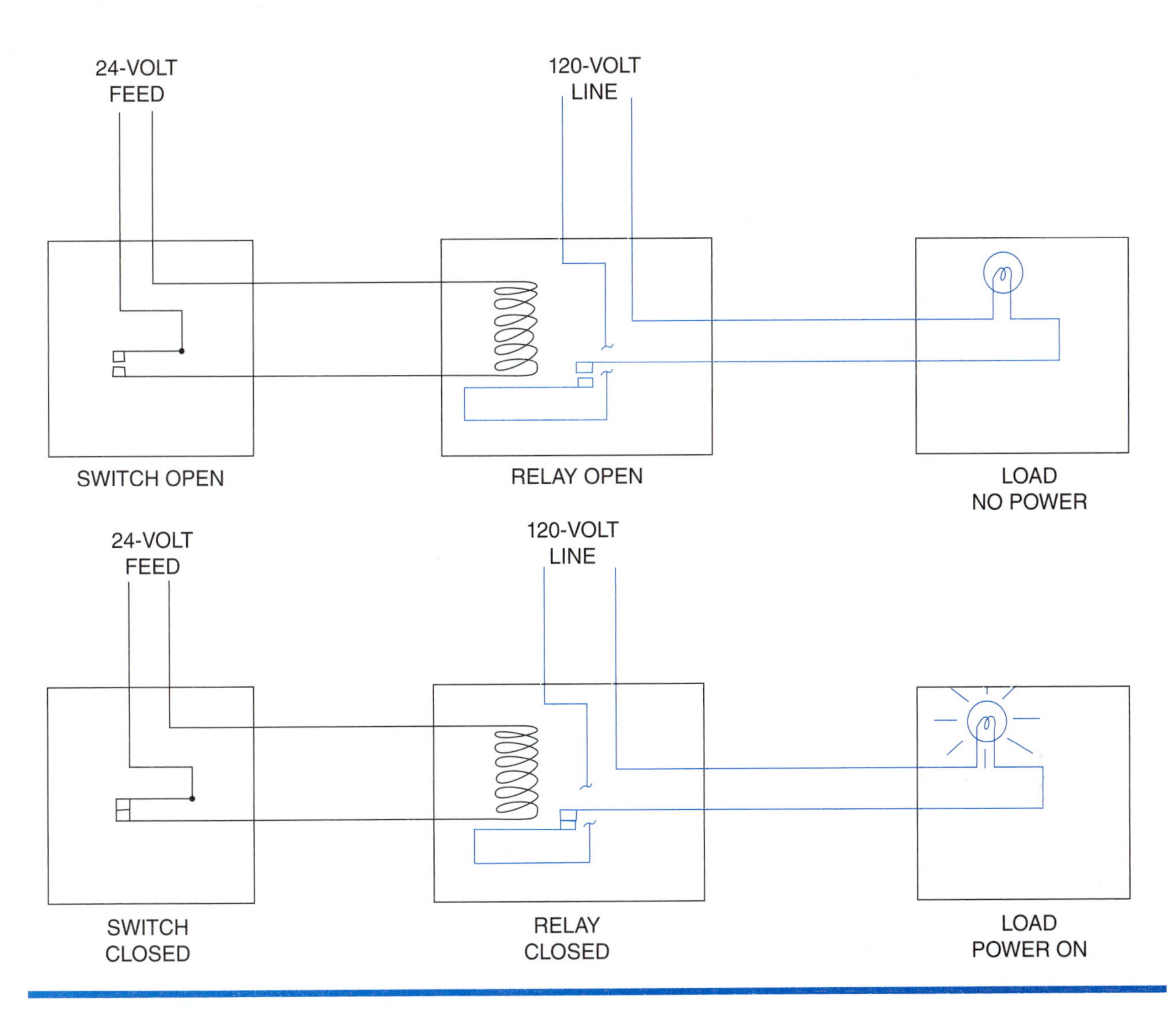

Figure 41–3. A relay allows a low-voltage circuit to start and stop a higher-voltage circuit.

is to furnish and install the three 4-inch conduits and that the utility company wants to inspect the conduit installation prior to backfilling the trench where the conduits are installed; however, this should be verified at a preconstruction meeting with the utility company. There are times when an electrical service, or electrical distribution raceway, would require being concrete encased for protection. This type of concrete-encased underground raceway system is called a *duct bank* (see **Figure 41–5**).

The note does not specifically indicate who furnishes and installs the utility feeders in the three 4-inch conduits. This could be an expensive item and should be clarified with the utility company prior to estimating and bidding the project.

To comply with the *National Electrical Code*® *(NEC®)*, the service entrance section may have up to, but must not exceed, six main disconnecting units (switches or circuit breakers). Four main fused switches are shown in the existing portion of the single-line drawing. Note that this single-line drawing shows only a portion of the complete existing service entrance section. The service entrance section indicates that it is located outside by the NEMA 3R ENCLOSURE. National Electric Manufacturers Association (NEMA) generates specifications that are recognized as design standards for electrical boxes, devices, and equipment. NEMA 3R is a rain-tight designation. The new portion of the single-line drawing shows the electrical distribution to be installed in this School Addition.

Electrical Reference Symbols

ELECTRICAL ABBREVIATIONS
(Apply only when adjacent to an electrical symbol.)

Central Switch Panel	CSP
Dimmer Control Panel	DCP
Dust Tight	DT
Emergency Switch Panel	ESP
Empty	MT
Explosion Proof	EP
Grounded	G
Night Light	NL
Pull Chain	PC
Rain Tight	RT
Recessed	R
Transfer	XFER
Transformer	XFRMR
Vapor Tight	VT
Water Tight	WT
Weather Proof	WP

ELECTRICAL SYMBOLS

Switch Outlets

Single-Pole Switch	S
Double-Pole Switch	S_2
Three-Way Switch	S_3
Four-Way Switch	S_4
Key-Operated Switch	S_K
Switch and Fusestat Holder	S_{FH}
Switch and Pilot Lamp	S_P
Fan Switch	S_F
Switch for Low-Voltage Switching System	S_L
Master Switch for Low-Voltage Switching System	S_{LM}
Switch and Single Receptacle	S
Switch and Duplex Receptacle	S
Door Switch	S_D
Time Switch	S_T
Momentary Contact Switch	S_{MC}
Ceiling Pull Switch	S
"Hand-Off-Auto" Control Switch	HOA
Multi-Speed Control Switch	M
Push Button	

Receptacle Outlets

Where weather proof, explosion proof, or other specific types of devices are to be required, use the upper-case subscript letters. For example, weather proof single or duplex receptacles would have the uppercase WP subscript letters noted alongside of the symbol. All outlets should be grounded.

Single Receptacle Outlet
Duplex Receptacle Outlet
Triplex Receptacle Outlet
Quadruplex Receptacle Outlet

Figure 41–4. Recommended electrical symbols.

Duplex Receptacle Outlet
(Split Wired)

Triplex Receptacle Outlet
(Split Wired)

250-V Receptacle Single Phase. Use Subscript Letter to Indicate Function (DW—Dishwasher, RA—Range, CD—Clothes Dryer) or Numeral (with explanation in symbol schedule).

250-V Receptacle Three Phase

Clock Receptacle — C

Fan Receptacle — F

Floor Single Receptacle Outlet

Floor Duplex Receptacle Outlet

Floor Special-Purpose Outlet *

Floor Telephone Outlet (Public)

Floor Telephone Outlet (Private)

Example of the use of several floor outlet symbols to identify a 2, 3, or more gang flow outlet.

Underfloor duct and junction box for triple, double, or single duct system as indicated by the number of parallel lines.

Example of use of various symbols to identify location of different types of outlets or connections for underfloor duct or cellular floor systems.

J

Cellular Floor
Header Duct

*Use numeral keyed to explanation in drawing list of symbols to indicate usage.

Circuiting

Wiring Exposed (Not in Conduit) — E —

Wiring Concealed in Ceiling or Wall

Wiring Concealed in Floor

Wiring Existing*

Wiring Turned Up

Wiring Turned Down

Branch Circuit Home Run to Panel Board — 2 1

Number of arrows indicates number of circuits. (A number of each arrow may be used to identify circuit number.)‡

Bus Ducts and Wireways

Trolley Duct‡ — T T

Busway (Service, Feeder, or Plug-in)‡ — B B

Cable Trough Ladder or Channels‡ — C C

Wireway‡ — W W

Panelboards, Switchboards, and Related Equipment

Flush-Mounted Panelboard and Cabinet‡

Surface-Mounted Panelboard and Cabinet‡

Switchboard, Power Control Canter, Unit Substations (Should be drawn to scale.)‡

Flush-Mounted Terminal Cabinet (In small scale drawings the TC may be indicated alongside the symbol.)‡ — TC

Surface-Mounted Terminal Cabinet (In small scale drawings the TC may be indicated alongside the symbol.)‡ — TC

Figure 41–4. *(Continues)*

Pull Box (Identify in Relation to Wiring System Section and Size.)	
Motor or Other Power Controller (May Be a Starter or Contactor.)‡	
Externally-Operated Disconnection Switch	
Combination Controller and Disconnection Means	

Power Equipment

Electric Motor (hp As Indicated)	1/4
Power Transformer	
Pothead (Cable Termination)	
Circuit Element (e.g., Circuit Breaker)	CB
Circuit Breaker	
Fusible Element	
Single-Throw Knife Switch	
Double-Throw Knife Switch	
Ground	
Battery	
Contactor	C
Photoelectric Cell	PE
Voltage Cycles, Phase	Ex: 480/60/3
Relay	R
Equipment Connection (As Noted)	▲

*Note: Use heavy weight line to identify service and leaders. Indicate empty conduit by notation CO (conduit only).
‡ Note: any circuit without further identification indicates two-wire circuit for a greater number of wires, indicate with cross lines, e.g.:

3 wires 4 wires

Neutral wire may be shown longer. Unless indicated otherwise, the wire size of the circuit is the minimum size required by the specification. Identify different functions of wiring system, e.g. signaling system by notation or other means.
‡ Identify by notation or schedule.

Remote Control Stations for Motors or Other Equipment

Push Button Station	PB
Float Switch (Mechanical)	F
Limit Switch (Mechanical)	L
Pneumatic Switch (Mechanical)	P
Electric Eye (Beam Source)	
Electric Eye (Relay)	
Temperature Control Relay Connection (3 Denotes Quantity.)	R 3
Solenoid Control Valve Connection	S
Pressure Switch Connection	P
Aquastat Connection	A
Vacuum Switch Connection	V
Gas Solenoid Valve Connection	G
Flow Switch Connection	F
Timer Connection	T
Limit Switch Connection	L

Lighting

	Ceiling	Wall
	TYPE	SWITCH
Surface or Pendant Incandescent Fixture (PC = Pull Chain)	WATTS	PC CIRCUIT
Surface or Pendant Exit Light		
Blanked Outlet	B	B
Junction Box	J	J

Figure 41–4. *(Continued)*

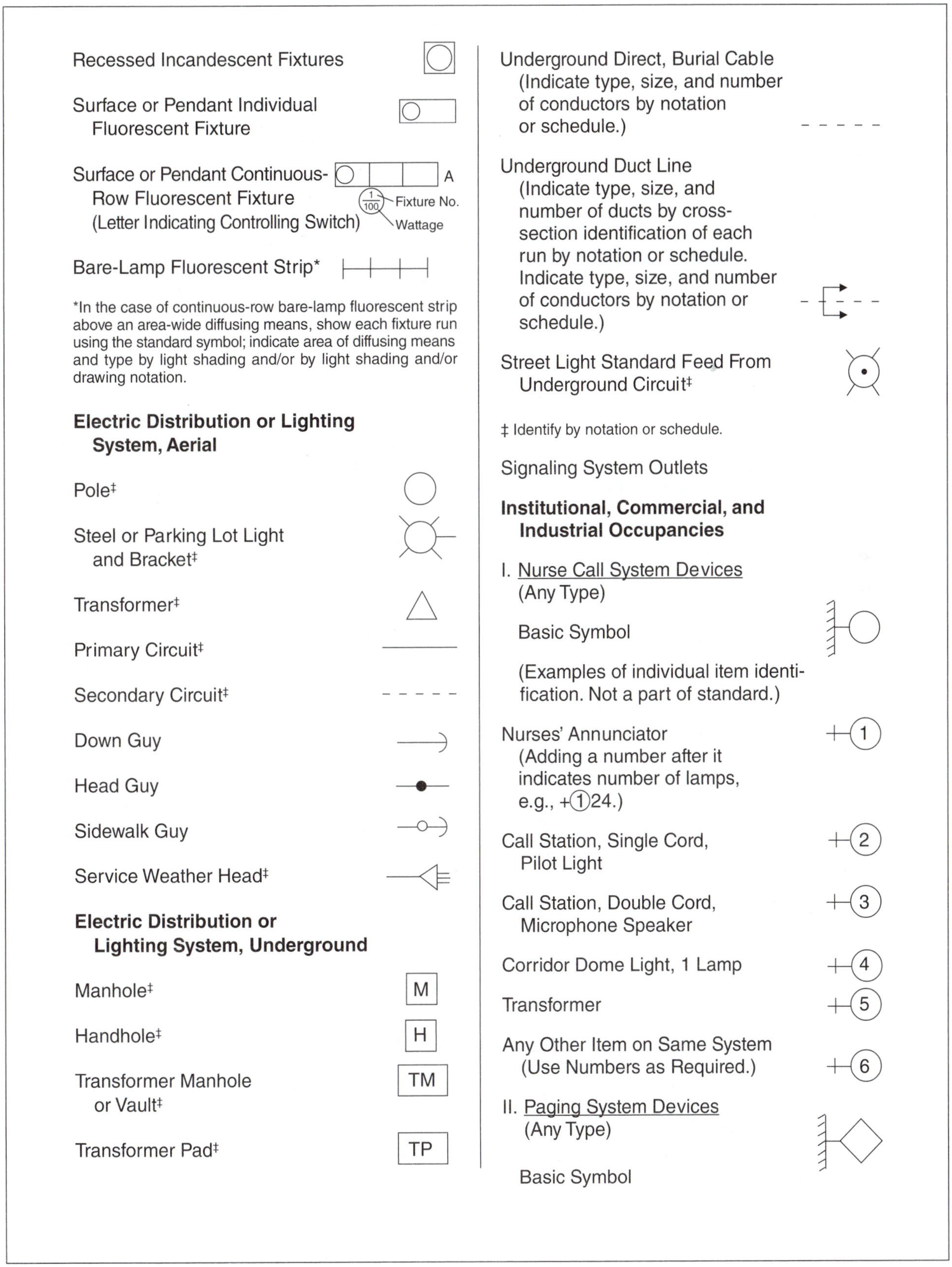

Recessed Incandescent Fixtures

Surface or Pendant Individual Fluorescent Fixture

Surface or Pendant Continuous-Row Fluorescent Fixture (Letter Indicating Controlling Switch)

A

Fixture No.

Wattage

Bare-Lamp Fluorescent Strip*

*In the case of continuous-row bare-lamp fluorescent strip above an area-wide diffusing means, show each fixture run using the standard symbol; indicate area of diffusing means and type by light shading and/or by light shading and/or drawing notation.

Electric Distribution or Lighting System, Aerial

Pole‡

Steel or Parking Lot Light and Bracket‡

Transformer‡

Primary Circuit‡

Secondary Circuit‡

Down Guy

Head Guy

Sidewalk Guy

Service Weather Head‡

Electric Distribution or Lighting System, Underground

Manhole‡ — M

Handhole‡ — H

Transformer Manhole or Vault‡ — TM

Transformer Pad‡ — TP

Underground Direct, Burial Cable (Indicate type, size, and number of conductors by notation or schedule.)

Underground Duct Line (Indicate type, size, and number of ducts by cross-section identification of each run by notation or schedule. Indicate type, size, and number of conductors by notation or schedule.)

Street Light Standard Feed From Underground Circuit‡

‡ Identify by notation or schedule.

Signaling System Outlets

Institutional, Commercial, and Industrial Occupancies

I. Nurse Call System Devices (Any Type)

Basic Symbol

(Examples of individual item identification. Not a part of standard.)

Nurses' Annunciator (Adding a number after it indicates number of lamps, e.g., +①24.) — 1

Call Station, Single Cord, Pilot Light — 2

Call Station, Double Cord, Microphone Speaker — 3

Corridor Dome Light, 1 Lamp — 4

Transformer — 5

Any Other Item on Same System (Use Numbers as Required.) — 6

II. Paging System Devices (Any Type)

Basic Symbol

Figure 41–4. *(Continues)*

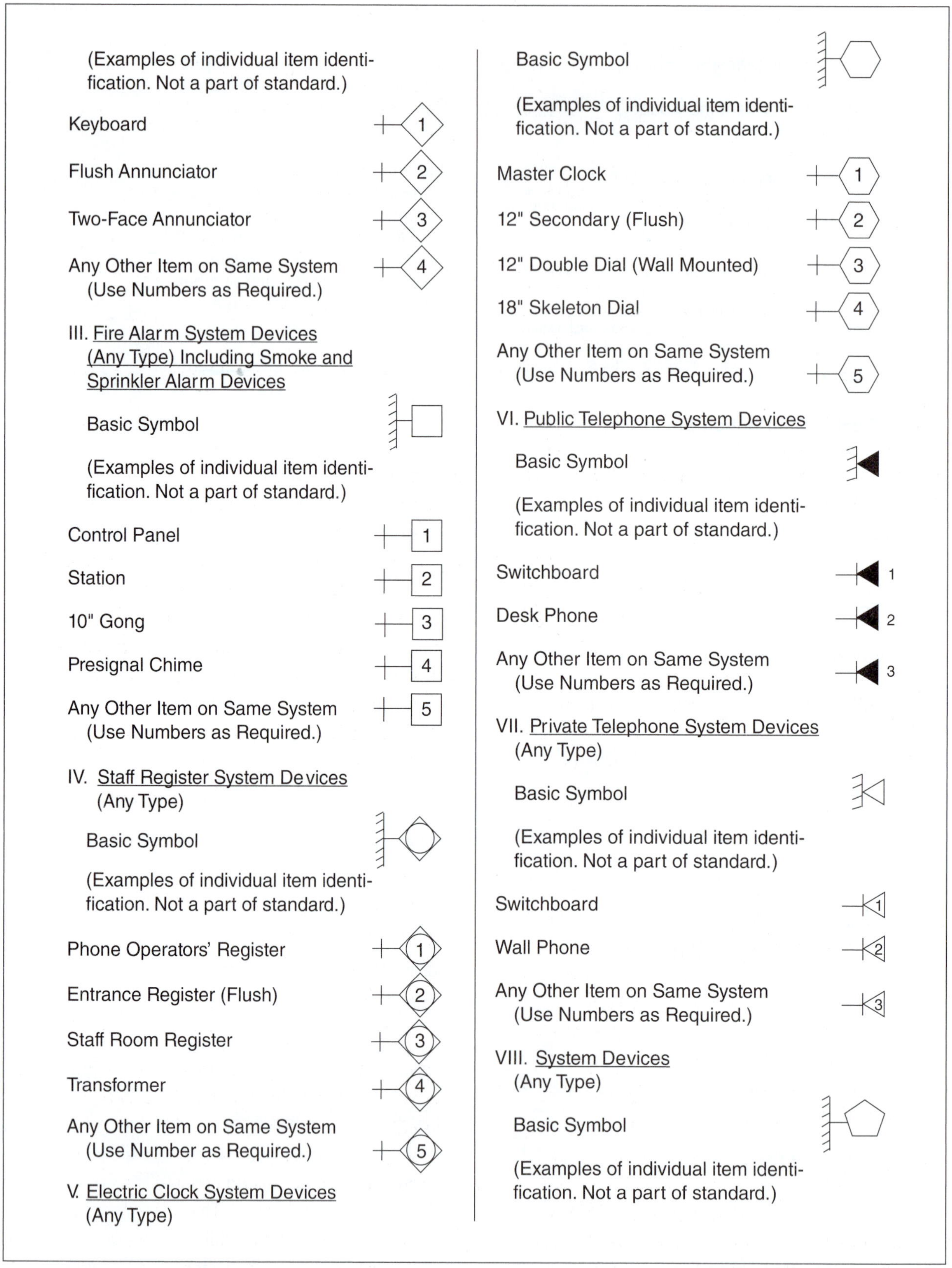

(Examples of individual item identification. Not a part of standard.)

Keyboard

Flush Annunciator

Two-Face Annunciator

Any Other Item on Same System (Use Numbers as Required.)

III. Fire Alarm System Devices (Any Type) Including Smoke and Sprinkler Alarm Devices

Basic Symbol

(Examples of individual item identification. Not a part of standard.)

Control Panel

Station

10" Gong

Presignal Chime

Any Other Item on Same System (Use Numbers as Required.)

IV. Staff Register System Devices (Any Type)

Basic Symbol

(Examples of individual item identification. Not a part of standard.)

Phone Operators' Register

Entrance Register (Flush)

Staff Room Register

Transformer

Any Other Item on Same System (Use Number as Required.)

V. Electric Clock System Devices (Any Type)

Basic Symbol

(Examples of individual item identification. Not a part of standard.)

Master Clock

12" Secondary (Flush)

12" Double Dial (Wall Mounted)

18" Skeleton Dial

Any Other Item on Same System (Use Numbers as Required.)

VI. Public Telephone System Devices

Basic Symbol

(Examples of individual item identification. Not a part of standard.)

Switchboard 1

Desk Phone 2

Any Other Item on Same System (Use Numbers as Required.) 3

VII. Private Telephone System Devices (Any Type)

Basic Symbol

(Examples of individual item identification. Not a part of standard.)

Switchboard

Wall Phone

Any Other Item on Same System (Use Numbers as Required.)

VIII. System Devices (Any Type)

Basic Symbol

(Examples of individual item identification. Not a part of standard.)

Figure 41–4. *(Continued)*

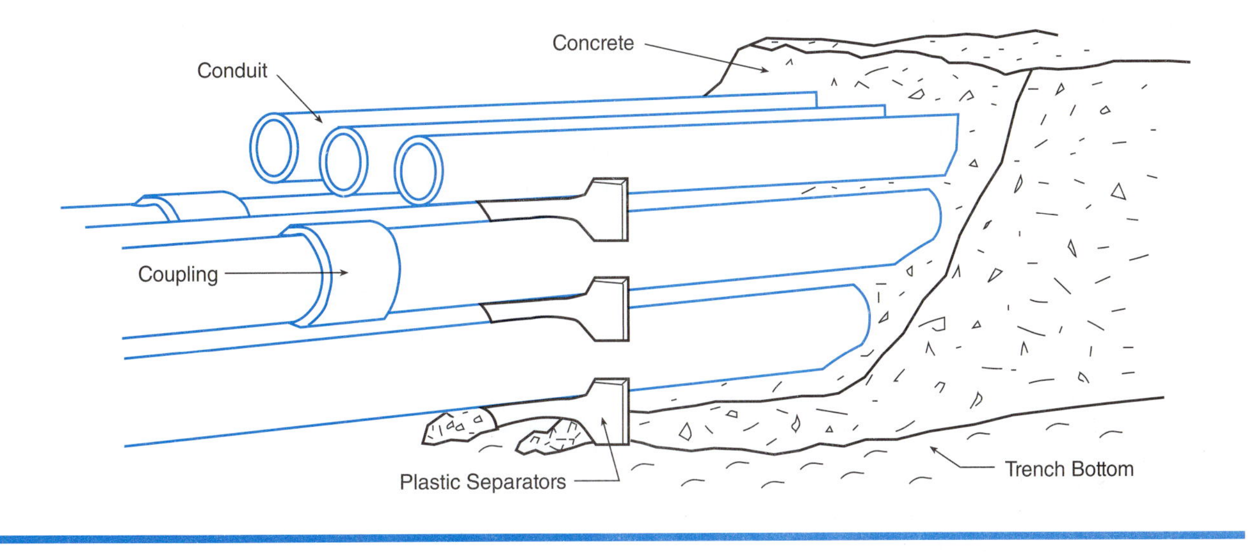

Figure 41–5. Duct bank system.

The feeders for the new elevator, electrical panels L10 and L11, are fed from existing panel MDP. These new electrical loads are:

1. 60-amp circuit breaker with four #4 AWG, one#10 ground, and one #10 isolated ground in a 1¼-inch conduit feeding the elevator
2. 100-amp circuit breaker with four #1 AWG and one #6 ground in a 1½-inch conduit feeding panel L11
3. 100-amp circuit breaker with four #1 AWG and one #6 ground in a 1½-inch conduit feeding panel L10

A complete building electrical floor plan is not a part of the School Addition drawings, and the existing electrical distribution plan on drawing E-2 shows only a portion of the total electrical distribution section. This required notes 6, 7, and 8 to be added to drawing E-2, Single-Line Diagram E-17, giving the contractor the lengths of the feeders for the elevator, panels L10 and L11, which are to be included in this School Addition. In a typical design, a complete building electrical plan is provided indicating the locations of all existing and new electrical equipment. The feeder length would be determined from that drawing.

Riser Diagrams

A *riser diagram* is so named because it usually shows the path of wiring or raceway from one level of a building to another and because the wiring rises from one floor to the next. A riser diagram does not give information about where equipment is to be located in a room or area. Riser diagrams are used because they are particularly easy to understand and do not require much explanation.

A *power riser diagram* (see **Figure 41–6**) shows a typical building's electrical service and related components. This figure is not the same electrical service as the School Addition Single-Line Diagram, E-17, on drawing E-2, but comparing the two diagrams shows how a power riser diagram greatly simplifies the interpretation of an installation drawing.

A *special riser diagram* is used for many systems that include:

1. Fire alarm
2. Security
3. Telephone
4. Clock
5. Signal
   a. Bell
   b. Call (nurse, emergency, etc.)
   c. Water sprinkler

The *fire alarm riser diagram* (see **Figure 41–7**) shows the new School Addition fire alarm system with ¾-inch conduits to ramp area 113. The School Addition drawing E-2 shows three ¾-inch conduits with the note "Three (3) ¾-inch existing conduits from the Fire Alarm Control Panel (150 feet)." These three conduits are to be used for the School Addition fire alarm connection to the existing fire alarm system. The original fire alarm control panel was sized to accommodate this School Addition; however, you should verify that the existing

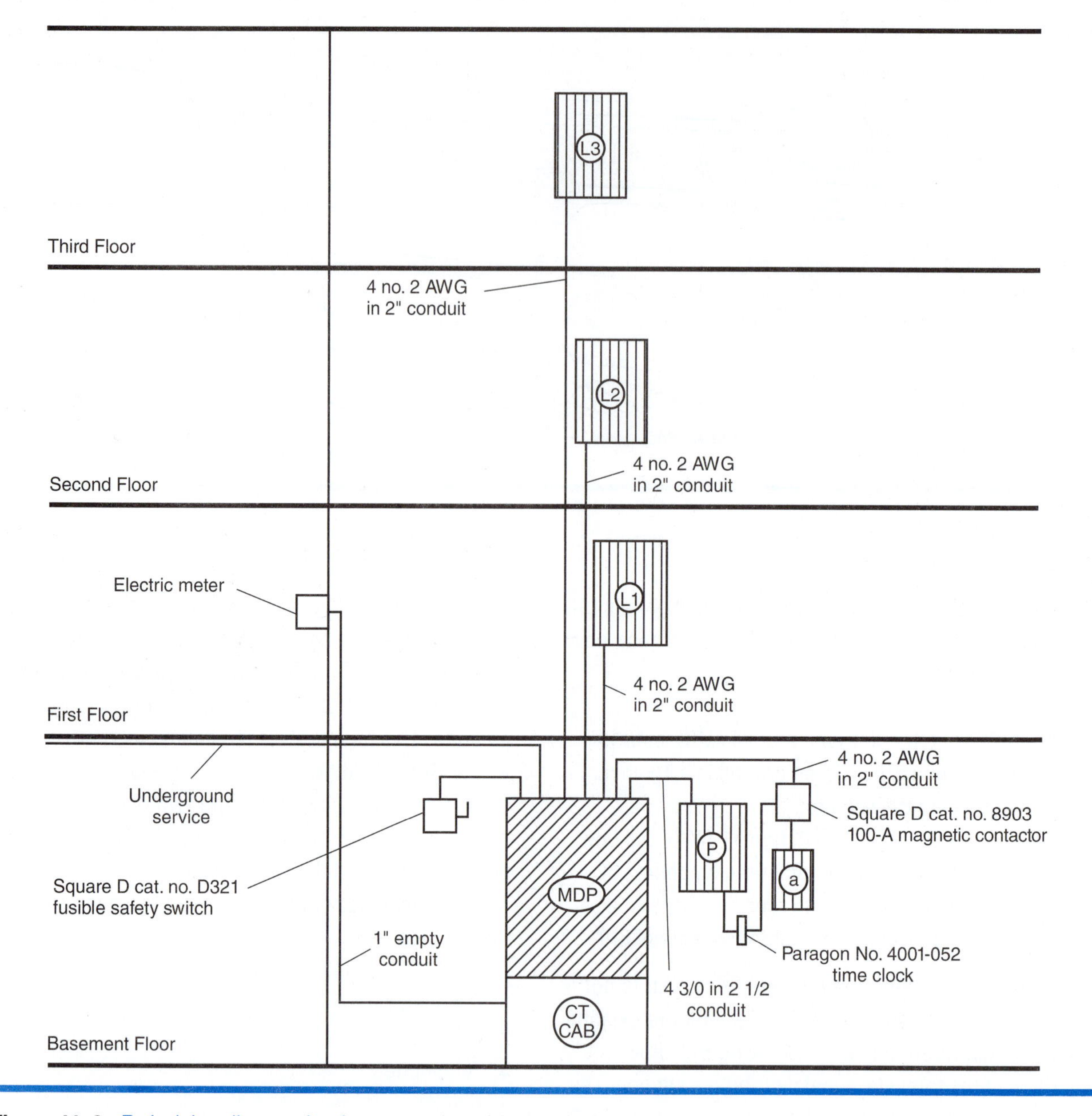

Figure 41–6. Typical riser diagram showing an overview of the building's electrical service and related components.

special systems (fire alarm, security, clock, etc.) will accommodate the additional requirements when adding to or modifying the existing system(s).

The *telephone riser diagram,* **Figure 41–8**, shows an existing 1¼-inch conduit, 150 feet in length, from ramp area 113 to the existing main telephone terminal cabinet. This conduit is to be extended to the telephone terminal board in room 102, First Floor Plan E-7, drawing E-2. The telephone riser diagram shows telephone conduit to be installed in the area above the suspended ceiling from the telephone terminal board to the outlet locations as shown on drawing E-2 and outlet detail A-11.

Schematic Diagrams

A schematic wiring diagram is a drawing that uses symbols and lines to show how the parts of an electrical assembly or unit are connected. A schematic does not necessarily show where parts are actually located, but it does explain how to make electrical connections. Several of the schematics with the School Addition show connections to wires labeled G, N, and H. These stand for *ground, neutral,* and *hot.* Schematics are commonly drawn for electrical equipment that involves internal wiring—everything from washing machines to computers. A basic motor control schematic (see

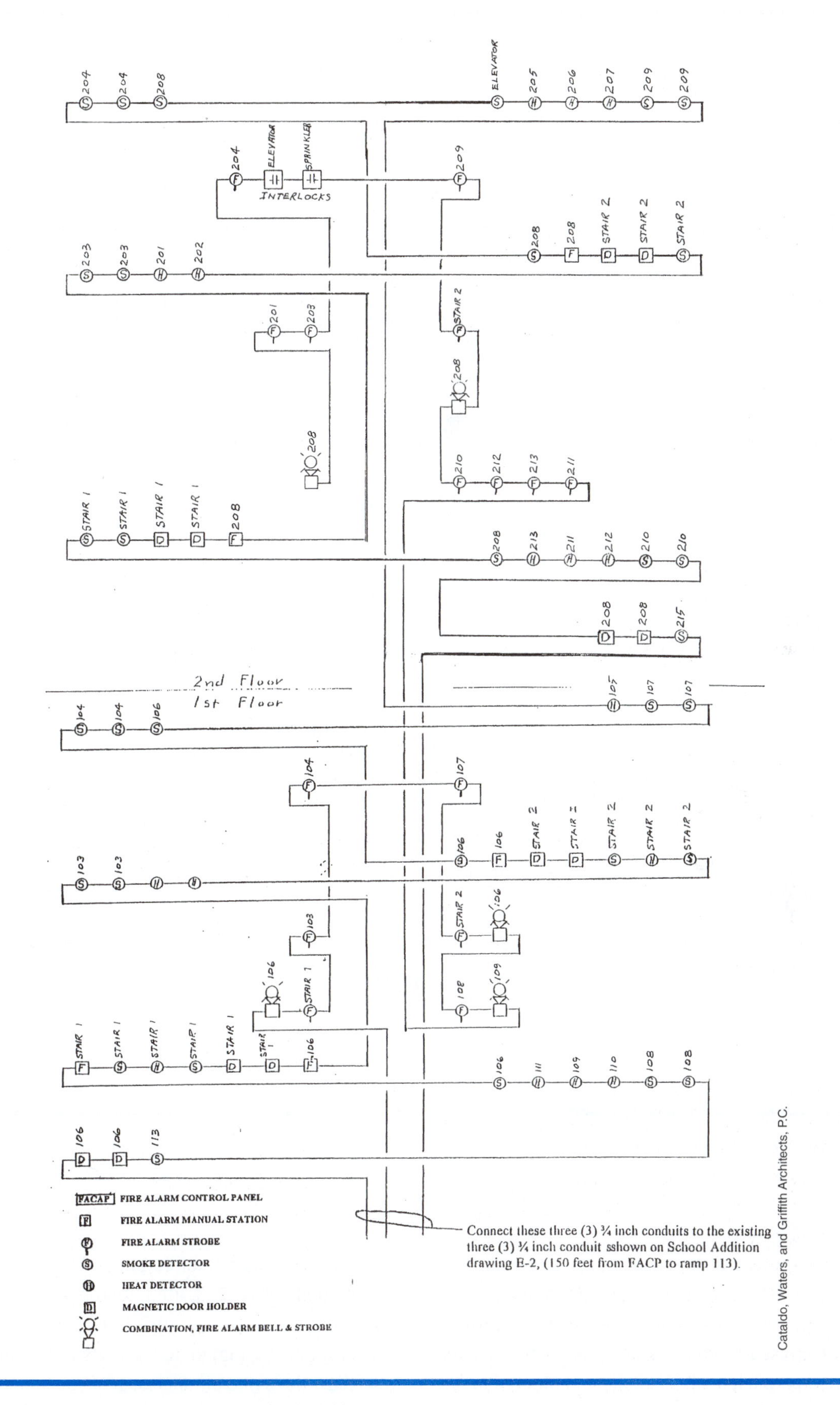

Figure 41–7. Fire alarm riser diagram for the school addition.

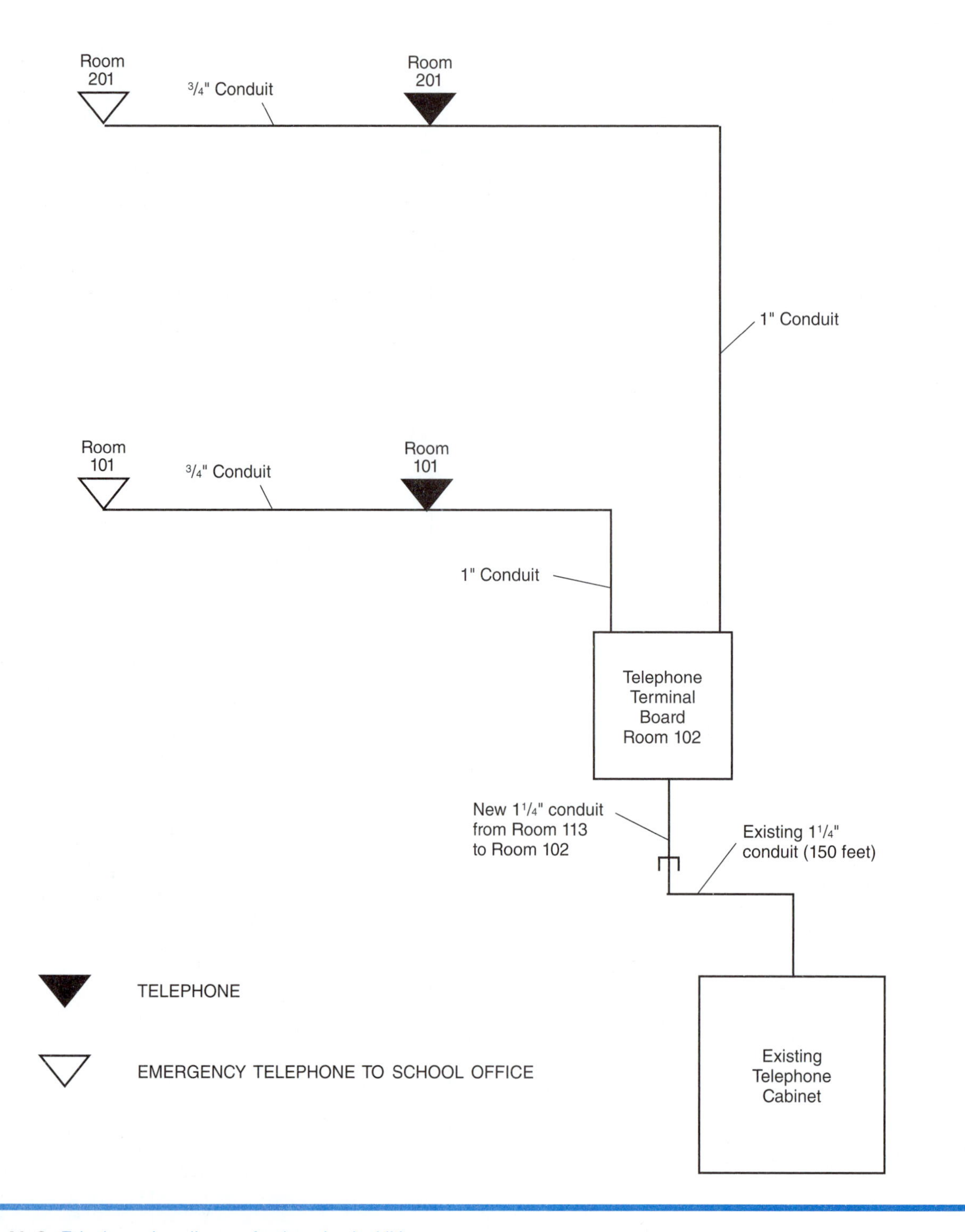

Figure 41–8. Telephone riser diagram for the school addition.

Figure 41–9) shows a three-phase power source (L1, L2, and L3) through the starter contacts (M) and the overloads (OL) powering the motor. The starter control is taken from phase L1 through a stop push button, a start push button paralleled with a latching contact M to the starter coil M, and finally through the normally closed overload contacts back to phase L2. The connection label numbers (1, 2, 3, etc.), which are shown on the schematic, aid the electrician in troubleshooting. Several different labeling methods are used, but they all follow the same principles; so if you understand one method, you can understand the other methods.

Schedules

An electrical schedule is used to systematically list equipment, loads, devices, and information. Schedules organize the information in an easily understood form and can be a valuable method for communicating the design requirements to contractors and their installers.

The *fixture schedule* (see **Figure 41–10**) lists complete information about each fixture type shown on the lighting plan. The following is a list of the kind of information that is usually shown on a fixture schedule:

- *Mark*—The label used to indicate a fixture type. The mark is written on or next to each fixture on the plan.
- *Make*—The identification of the manufacturer being used to establish the specific design requirements

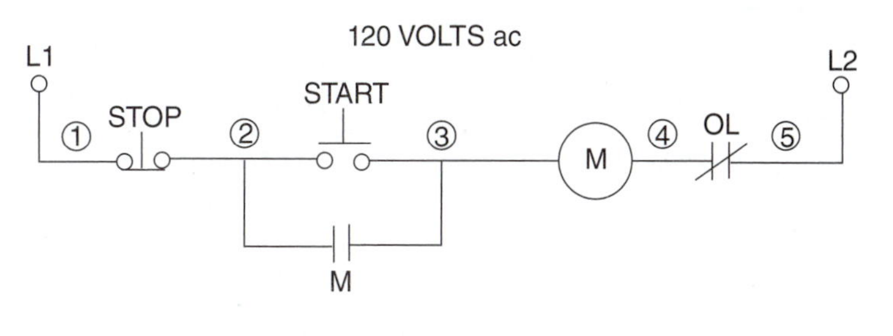

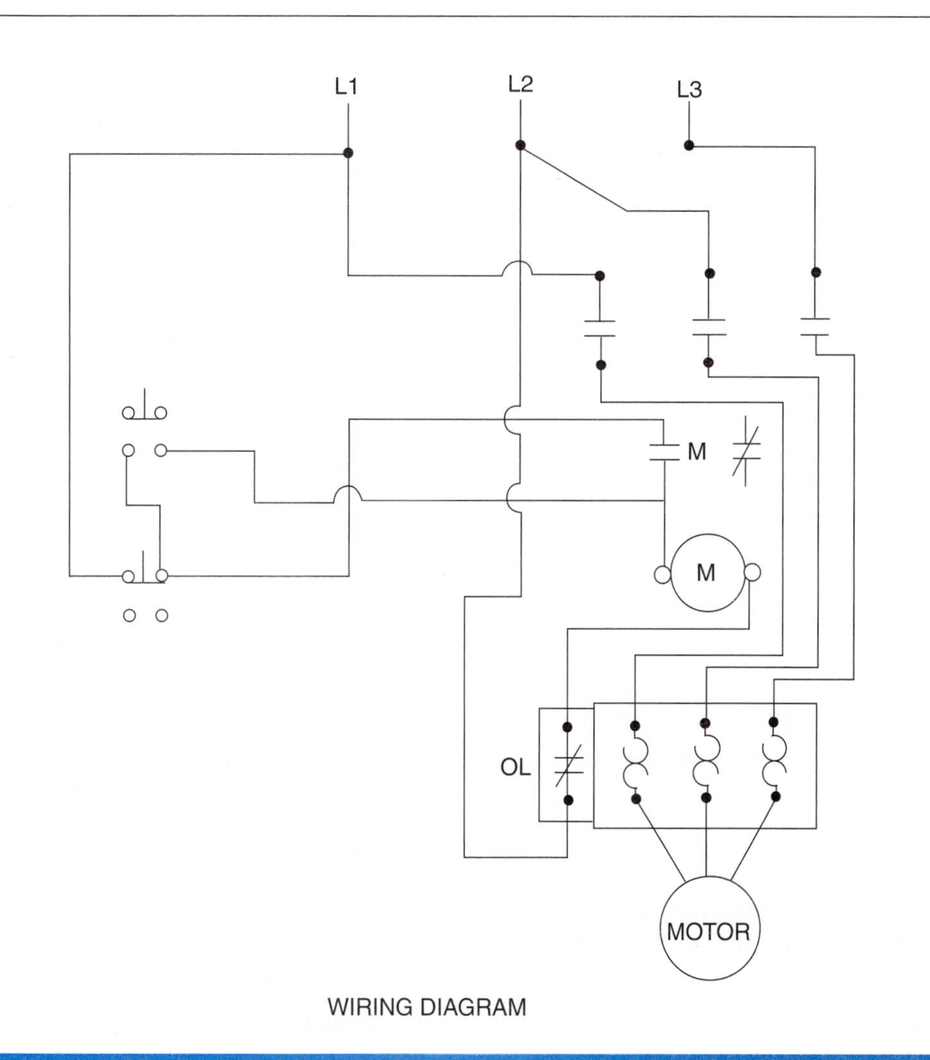

Figure 41–9. Schematic diagram and wiring diagram for a three-phase, AC magnetic, nonreversing motor starter.

MARK	MAKE	MODEL	VOLTS	WATTS	LAMP		REMARKS
					TYPE	NO./FIXT.	
A	STONCO	VWXL11GC	120	100	100WA19	1	CAST GUARD.
FA	WILLIAMS	1222-RWKA125	120	183	F32T8/ SP30	3	SPLIT WIRE IN TANDEM PAIRS – WATTAGE IS FOR 2 FIXTURES W/ TOTAL OF 3 2-LAMP BALLASTS. ELECTRONIC BALLAST.
FA1	WILLIAMS	1222-RWKA125	120	92	F32T8/SP30	3	ELECTRONIC BALLAST.
FB	WILLIAMS	2922-KA	120	61	F32T8/SP30	2	ELECTRONIC BALLAST.
FC	KIRLIN	96617-45-46-61	120	36	PLC13W/27K	2	HPF BALLAST.
FC1	KIRLIN	96617-46-61-SM	120	36	PLC13W/27K	2	HPF BALLAST, SURFACE MOUNT; COLOR AS SELECTED
FD	WILLIAMS	8222	120	61	F32T8/SP30	2	ELECTRONIC BALLAST.
FE	WILLIAMS	1262-RWKA125	120	84	F17T8/SP30	4	ELECTRONIC BALLAST.
FF	TERON	EE26-P-H	120	36	PL13/27	2	HPF BALLAST.
FG	WILLIAMS	2122-IM	120	61	F32T8/SP30	2	ELECTRONIC BALLAST.
FH	WILLIAMS	EPG-R272RWKA-125	120	61	F32T8/SP30	2	TO MATCH EXISTING
M1	KIRLIN	SS-51277-24-43 (35W)-45-46-FR	120	40	35W HPS	1	HPF BALLAST.
M2	STONCO	PAR250LX	120	300	250W HPS	1	HPF BALLAST EQUIP WITH MOUNTING BRACKET ARM.
EXITS	LITHONIA	WLES SERIES	120	65	LED	1	SEE SHEET E-2 FOR LOC. ARROWS & MOUNTING AS IND EQUIP W/ BATTERY BACK-UP
EMERG.LGT. BATT.PACK	EXIDE	B200	120/12	150	H1212		
REMOTE LAMP	EXIDE	H1212	12	24	H1212	2	

Figure 41–10. Fixture schedule.

needed to aesthetically and functionally light the various rooms or areas.

- *Volts*—It is becoming increasingly common for light fixtures to be powered by up to 277/480 volts, while the *switches* might use only 24 volts (low-voltage lighting control) to protect the person using the switch.
- *Watts*—It is necessary to know the wattage of the lamps (bulbs) in each fixture, so that the space will have the amount of illumination intended by the electrical designer.
- *Lamp Type and Quantity*—Lamp manufacturers have similar systems of designating lamp characteristics (see **Figures 41–11**, **41–12**, and **41–13**).
- *Notes or Remarks*—This column is for information that does not clearly belong under the other headings.

The School Addition Fixture Schedule A-9—Floor Plans E-7 and E-14 on drawing E-1—shows type FA light fixture to be the light source for the classrooms. The School Addition fixture schedule, as shown in **Figure 41–10**, shows that the fixture is manufactured by Williams and the catalog number is 1222-RWKA125 with three F32T8/SP30 lamps rated at 120 volts with a 183-watt load. Also, the Remarks column indicates some special wiring and specific wiring and specific ballast-type requirements. The first and second floor Reflected Ceiling Plans, E-15 and E-16 on drawing E-7, indicate light fixture type FA to be a 2′ × 4′ light fixture mounted end to end.

Fifteen other light fixture types are shown on the fixture schedule and on the drawings and are identified by their corresponding mark or label. All are rated at 120 volts except one emergency battery-powered remote lamp, rated at 12 volts. This information is typically given in the specifications but is more readily presented by a schedule.

A *panel schedule* identifies the panels by their mark or label. They are shown by this same designation on the electrical floor plan. The panel schedule for the School Addition lists each of the branch circuits served by a panel, the calculated load for that branch circuit, the voltage of that circuit, and the number of poles and trip rating for each circuit breaker. A panel schedule also might include:

- Type (surface or flush)
- Panel main buss amperes, volts, and phases
- Main circuit breaker/main lugs only
- Breaker frame sizes
- Items fed and/or remarks

On the lighting and power plans, we saw that each device is connected to a branch circuit identified by a panel number (label) and circuit number. Those numbers correspond with the numbers on the panel schedule.

When a commercial project has a kitchen, you should have a *kitchen equipment schedule.* If the drawings do not have one, it is very helpful to make one for the installer. The kitchen equipment schedule should include:

- Equipment number or designation
- Description of each equipment item
- The load in horsepower or kilowatts
- Volts
- Wire size
- Conduit size

How to Read Ordering Guide

Watts	Bulb	Base	One Item No. for each lamp	Ordering Abbreviation	Volts	Package Quantity	Description (Indicates aluminum base; Footnotes)	Filament Code	Average Rated Hours Life	Approx. Lumens	(L.C.L.) Light Center Length	(M.O.L.) Max. Overall Length
25	A-15	Med.	10125	**25A15**	115-125	120	•Inside Frosted, Refrig.	B, C-9	1100	250	2⅜	3½
	A-17	Med.	10417	**25A17/RS**	75	120	I.F. Train, Rough Service	B, C-9	1000	228	2⅜	3⅝
	A-19	Med.	10443	**25A**	6	120	Inside Frosted ⌀	C, C-6	1000	357	2½	3[15]/16

25A17/RS
25 — Lamp Nominal Wattage
A — A shape
17 — 17/8 = 2 1/8" Diameter
RS — Rough Service

Class C - denotes gas filled bulb.
Class B - denotes vacuum lamp.

Incandescent Lamps

A Bulb designation consists of a letter(s) to indicate the shape and a fugure(s) to indicate the approximate major diameter in eights of an inch. Bulbs are measured through their greatest diameter, in eights of an inch. Thus, a F-15 bulb is a flame shape, 15/8 of an inch or 1-7/8 inches in diameter.

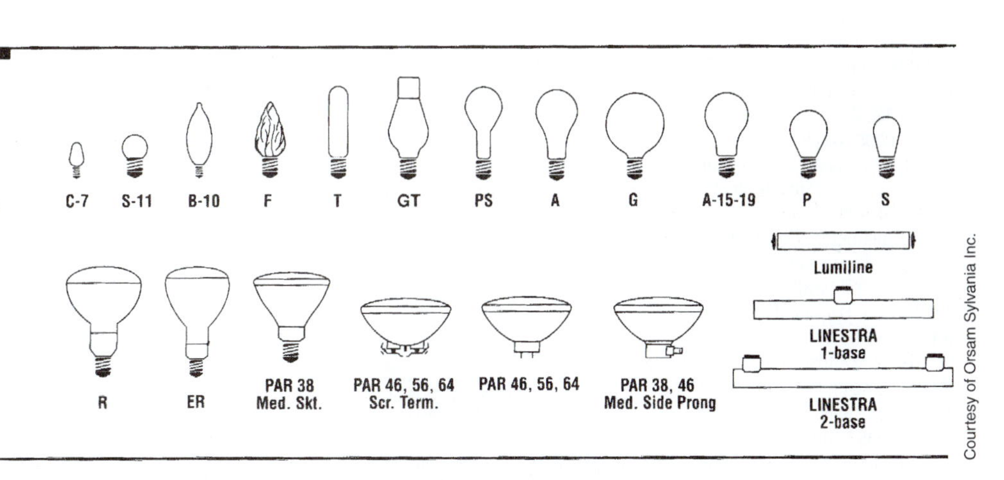

Courtesy of Orsam Sylvania Inc.

Figure 41–11. Incandescent lamp designations.

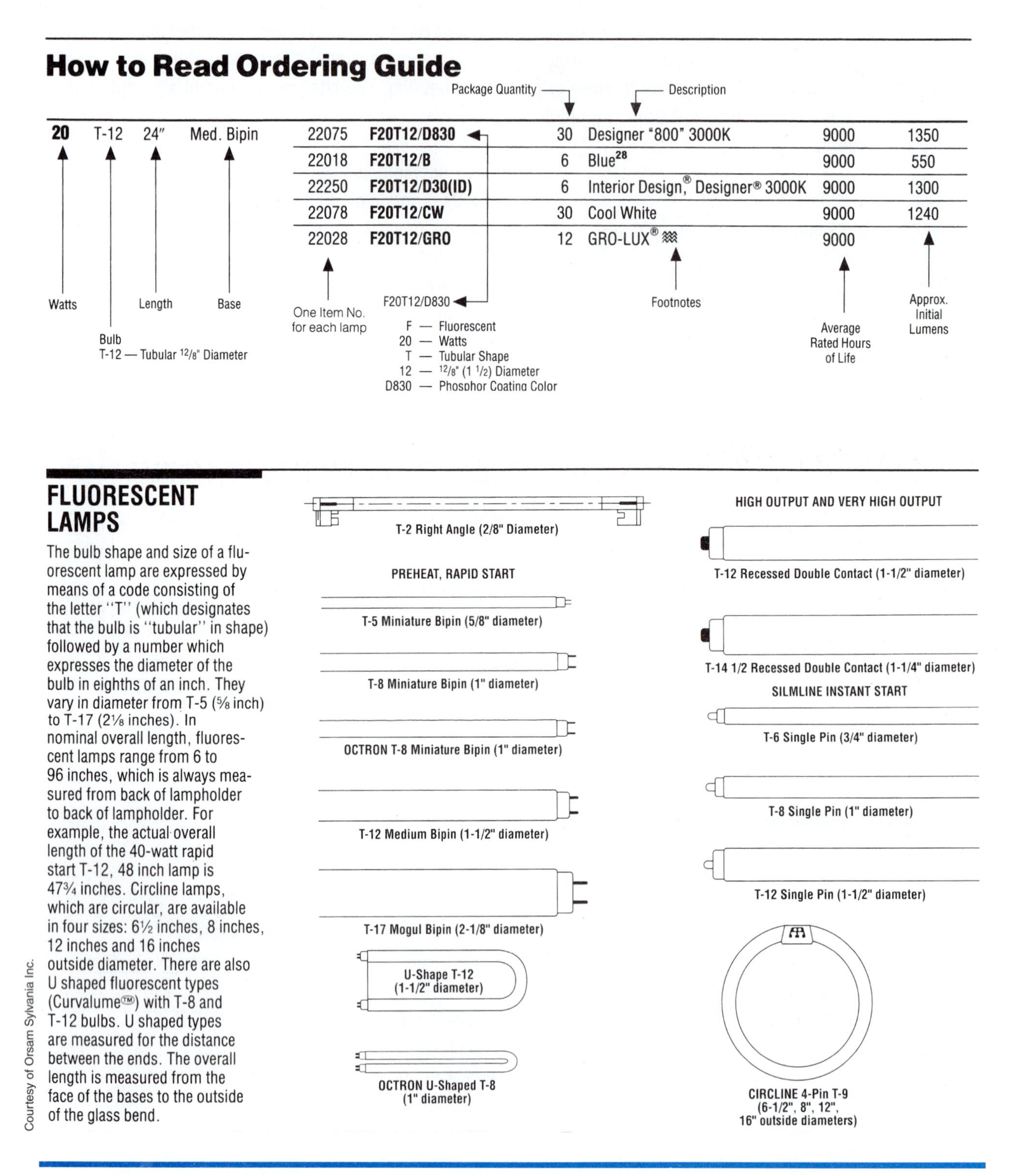

Figure 41–12. Fluorescent lamp designations.

HOW TO READ PRODUCT INFORMATION - COMPACT FLUORESCENT

Nominal Wattage	Bulb	MOL (in)	MOL (mm)	Base	Product Number	Ordering Abbreviation	NEMA Generic Designation	Pkg Qty	Avg Rated Life (hrs)	CCT (K)	CRI	Approx Lumens Initial @25°C/77°F (@35°C/95°F)	Approx Lumens Mean @25°C/77°F (@35°C/95°F)	Symbols & Footnotes
13	Twist	4.6	117	Med	29116	**CF13EL/MINITWIST**		6	8000	3000	82	800	640	2,21,28, 36,63,64
26	D(T4)	6.8	173	G24D-3	20710	**CF26DD/830/ECO**	CFQ26W/G24D/30	50	10000	3000	82	1800	1548	2,21,28, 34,37
32	T(T4)	5.5	140	GX24Q-3	20885	**CF32DT/E/IN/835/ECO**	CFTR32W/GX24Q/35	50	10000	3500	82	2328 2400	2002 2064	2,21,28, 33,35,59
40	L(T5)	22.6	573	2G11	20586	**FT40DL/841/RS/ECO**	FT40W/2G11/RS/41	10	20000	4100	82	3150	2709	2,21,28

Nominal Wattage	Design wattage on reference ballast. Actual wattage dependent on ballast.
Bulb	Describes the shape of the bulb.
Base	Base designations for compact fluorescent lamps are the NEMA designations. Please see page 111 for base illustrations.
MOL	Maximum overall length. The actual length of the lamp measured from the bottom of the base to the top outside edge of the glass. In many cases, the bottom of the base is the bottom of the center post of the base of the lamp.
Symbols & Footnotes	Most symbols and footnotes that apply to a specific product will appear in this space. The explanations of the symbols and footnotes are at the end of the fluorescent section.
Ordering Abbreviation	A text description of the lamp. Please see below for several examples and explanations of some of the codes.
NEMA Generic Designation	Designation assigned by NEMA (National Electrical Manufacturers Association).
CCT	Correlated Color Temperature. The degree of "whiteness" of the light. Expressed in kelvins (K). Please see page 109 for more information.
CRI	Color Rendering Index. A numbering system for rating the relative color rendering quality of a light source compared to a standard. Please see page 109 for more information.
Initial & Mean Lumens	Initial lumens are measured when the lamp has been operating for 100 hours. Mean lumens are typically measured at 40% of the rated life of the lamp. Compact Fluorescent lamp lumens are measured at 25°C (77°F) and 35°C (95°F)

How to Read Ordering Abbreviations

CF26DD/830		CF32DT/E/IN/835/ECO		FT40DL/841/RS/ECO		CF20EL/830/MED/ECO	
CF	Compact Fluorescent	CF	Compact Fluorescent	FT	Fluorescent Twin	CF	Compact Fluorescent
26	Nominal lamp wattage	32	Nominal lamp wattage	40	Nominal lamp wattage	20	Nominal lamp wattage
DD	DULUX® Double	DT	DULUX Triple	DL	DULUX Long	EL	Electronic Lamp
8	82 CRI	E	Electronic or dimming operation	8	82 CRI	8	82 CRI
30	3000K CCT	IN	Amalgam	41	4100K CCT	30	3000K CCT
		8	82 CRI	RS	Rapid Start	MED	Medium screw base
		35	3500K CCT	ECO	ECOLOGIC	ECO	ECOLOGIC
		ECO	ECOLOGIC				

Figure 41–13. Compact fluorescent lamp designations. *(Continues)*

- Protection in amperes
- Who will furnish each equipment item (furnished by others or by contractor)
- Installation requirements
- Remarks column for any detailed specific information required
- Number of wires and poles
- Voltage rating
- NEMA type
- The configuration of the blades or slots
- A manufacturer catalog number reference
- Special information (duplex, single, three phase, etc.)

A *receptacle schedule* is valuable when a number of special or specific receptacle types are found on the electrical drawings. If one is not provided, you should make one to expedite the installation time required and reduce the chance of installation error. A receptacle schedule should include:

- The symbol designation used by the designer
- Amperage rating

Note: Receptacle information may be found on an *equipment schedule.*

The local plans review and the utility company typically require a *connected load schedule.* This type of schedule includes:

- Type of load
- Building or area designation

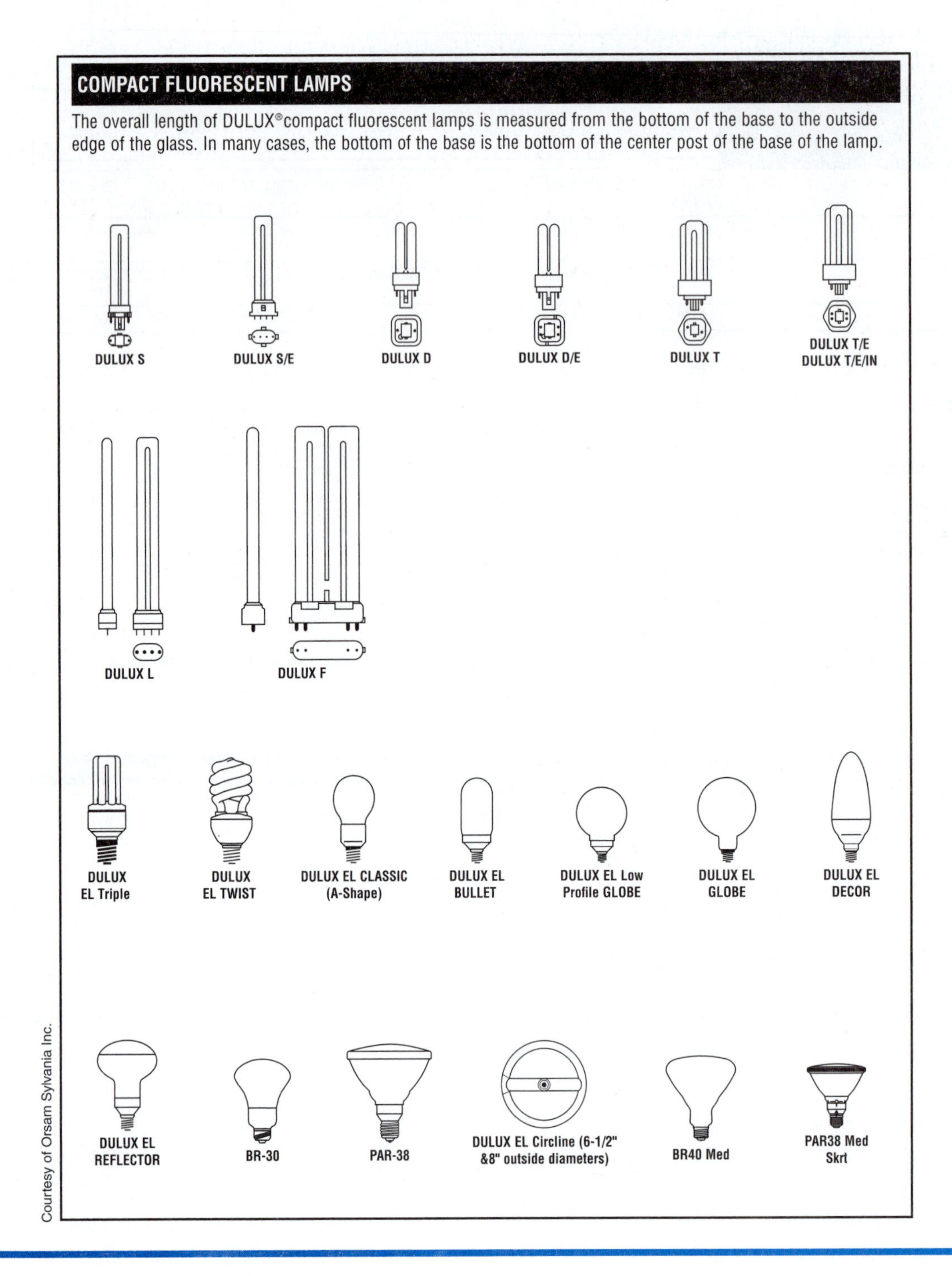

Figure 41–13. *(Continued)*

- Size of load (kilowatts or horsepower)
- Total electrical load by type or area
- Notes explaining any special methods used in the load calculations

Specifications

The drawings and the ***specifications*** are the items that establish the intended design and the construction requirements required by the owner, architect, and engineers. The contractors and their installers must review the ***contract*** documents for conflicts and/or discrepancies between the contract, the specifications, and the drawings. The drawings should be reviewed for conflicts between sections (architectural, mechanical, plumbing, structural, civil, HVAC, electrical, etc.). The drawings in your specific section should be reviewed for conflicts sheet by sheet.

The electrical specifications give the quality of materials intended to be used and the installation and testing requirements. Sometimes a specific manufacturer or catalog number is specified with no substitution or "equal" allowed. This usually inflates the cost of the project. There is no standard specification. The contractors and their installers should read through all project specifications and become knowledgeable of their content prior to starting any installation.

USING WHAT YOU LEARNED

A basic skill for reading electrical drawings is knowing which circuits feed each electrical device and the corresponding panel for that circuit. There are two duplex receptacles shown in the elevator area on the first floor plan. Are those receptacles in the elevator car or elsewhere? From which panel are they fed? What circuit are they on? The receptacles are labeled as being "in pit," so they are not in the elevator car. The numbers 15 and L10 indicate that they are on circuit 15 which is fed from panel L10. This can be verified on the panel schedule E-11 of Sheet E7.

Assignment

Refer to the School Addition drawings in your textbook packet to complete this assignment.

1. What are the four basic methods of showing electrical circuits?
2. How is a home run circuit shown?
3. What is a reflected ceiling plan?
4. In a low-voltage lighting system, what does the light switch activate?
5. Where do you find the industry standardized electrical symbols?
6. What are the four basic types of electrical drawings?
7. What is the name of the method used to show the path of wiring or raceway from one level of a building to another?
8. What is a schematic wiring diagram?
9. What is used to systematically list equipment, loads, devices, and information?
10. Where do you find the quality of material intended to be used on a project?
11. How many F17T8/SP30 lamps are required on the first floor of the addition?
12. Explain why one of the switches in room 109 is listed as S_{3M} and the other is simply S_3.
13. How are the lights turned on and off in the first-floor corridor?
14. What circuit carries the lights for stair #2?
15. What is the approximate total wattage of the lamps in the boys' toilet room on the first floor?
16. What is the total load for the circuit that serves the lights in the boys' toilet room on the first floor?
17. What is indicated by the D in a square near the doors from the existing building into the addition?
18. What is on circuit L10, 24?
19. Where are the devices on circuit L10, 15?
20. Explain what each of the colored terminals on a classroom lighting control is to be connected to:
    a. Green
    b. Orange
    c. Black
    d. White
    e. Blue (inner terminal)
    f. Blue (outer terminal)

7. On what branch circuit are the light fixtures in stair #1?
8. How are the light fixtures in the second floor corridor turned on and off?
9. How many lamps are required for all the fixtures in the second floor corridor?
10. What branch circuit supplies the convenience outlets in the second floor corridor?
11. Which terminal on the corridor lighting control unit is connected to the neutral leg of the supply?
12. What is the diameter of the sanitary drain where it leaves the building?
13. What are the pipe sizes that storm water passes through as it flows from the roof drain on the north canopy to the point where it exits the building?
14. Where is the nearest cleanout to the roof drain at the west end of the addition?
15. What type of unit provides heat in storage room 109?

Appendix A

School Addition Master Keynotes

Master Keynotes

2.11	Demolition, Removals & Relocation
2.20	Site Preparation & Earthwork
2.20A	Select Fill - Bank Run Gravel
2.20B	Select Granular Material
2.20D	Topsoil
2.20F	Crushed Gravel
2.20G	4" perforated P.v.c.
2.20I	Compacted Subgrade
2.22	Structural Excavation, Backfill and Compaction (Building Area)
2.60	Pavement and Walks
2.60A	Vehicular Area Sub-Base Course Granular
2.60B	Asphaltic Concrete - Binder Course
2.60C	Asphaltic Concrete Top - Wearing Course
2.60D	Concrete Walk/Paving
2.60E	Precast Concrete Curb
2.60G	Reinforcing Mesh
2.60H	Expansion Joint Filler
2.60I	Control Joint - 4'-0" O.C. Maximum Saw Cut or Tooled
2.60J	Stabilization Fabric
2.60P	Asphaltic Concrete-Base Course
2.60Q	Detectable Warning Pavers
2.61	Pavement Markings
2.61B	Painted ANSI Handicap Symbol

9.25J Gypsum Sheathing

9.25k Carrying Channel

9.25L 5/8" Exterior Gypsum Ceiling Board

9.25N Metal Angle Runner

9.30 Tile

See Room Finish Schedule*

9.30A Glazed Wall Tile

9.30B Unglazed Ceramic Mosaic Tile

9.30D Glazed Ceramic Tile Wall Base

9.30E Marble Thresholds - See Door Schedule*

9.30F Accent Tile - Continuous Around Room

9.30Q Unglazed Ceramic Mosaic Tile Base

9.40 Terrazzo

9.40A Thin Set Terrazzo

9.40C Terrazzo Cove Base

9.50 Acoustical Treatment

See Room Finish Schedule*

9.50A Suspended Ceiling System

9.65 Resilient Flooring

See Room Finish Schedule*

9.65A Vinyl Composition Tile

9.65B Vinyl Cove Base

9.65D Rubber Stair Treads and Risers

9.65E Molded Rubber Tile

9.65J Rubber Base

9.80 Special Coating System

9.80A Special Coating System

9.80B Finish Coat

9.80C Base Coat

9.80D Reinforcing Mesh

9.80E Insulation

9.80F Gypsum Sheathing

9.80G Metal Studs

9.80H Sealant and Backer Rod

9.80I Waterproof Base Coat

9.80J Routed Joint

9.80K Insulation Board Below Grade

9.80L Below Grade Waterproofing

9.80M Slide Clip

9.80N Bent Galvanized Metal (Size and Gauge as Noted)

9.80O Expansion Joint

9.90 Painting

See Room Finish Schedule*

9.90E-1 Paint E-1

9.90E-2 Paint E-2

9.90P-1 Paint P-1

9.90P-2 Paint P-2

9.90P-3 Paint P-3

9.90P-4 Paint P-4

9.90P-6 Paint P-6

9.90P-8 Paint P-8

10.10 Chalkboards and Tackboards

10.10B Liquid Marker Board

10.10C Tackboard

10.10F Projection Screen

10.25 Firefighting Devices

10.25A Fire Extinguisher Cabinet. Paint P-2. Model * 2409-R 2 By Larson Manufacturing Co., or Approved Equal

10.42 Signage and Graphics

10.42A 12″ X 18″ Aluminum Handicap Sign

10.50 Lockers

10.50A Single Tier Lockers

10.50G Sloping Top

10.50H Metal Base

MATH REVIEW 3 Adding Combinations of Fractions, Mixed Numbers, and Whole Numbers

- To add mixed numbers or combinations of fractions, mixed numbers, and whole numbers, express the fractional parts of the numbers as equivalent fractions having the lowest common denominator. Add the whole numbers. Add the fractions. Combine the whole number and the fraction and express in lowest terms.

Example 1 Add: $3\frac{7}{8} + 5\frac{1}{2} + 9\frac{3}{16}$

Express the fractional parts as equivalent fractions with 16 as the common denominator. Add the whole numbers. Add the fractions. Combine the whole number and the fraction. Express the answer in lowest terms.

Example 2 Add $6\frac{3}{4} + \frac{9}{16} + 7\frac{21}{32} + 15$

Express the fractional parts as equivalent fractions with 32 as the common denominator. Add the whole numbers. Add the fractions. Combine the whole number and the fraction. Express the answer in lowest terms.

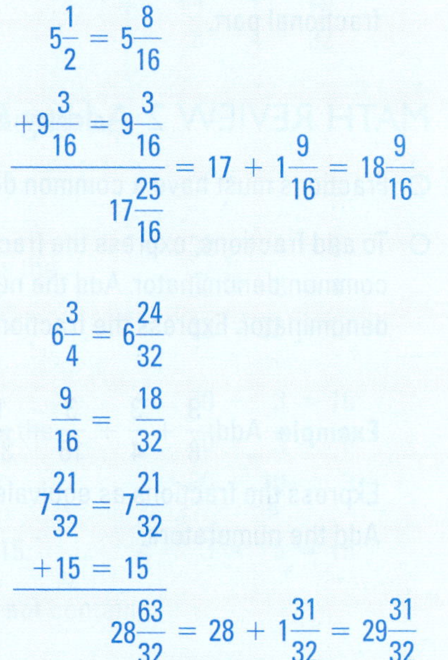

MATH REVIEW 4 Subtracting Fractions from Fractions

- Fractions must have a common denominator in order to be subtracted.
- To subtract a fraction from a fraction, express the fractions as equivalent fractions having the lowest common denominator. Subtract the numerators. Write their difference over the common denominator.

Example Subtract $\frac{3}{4}$ from $\frac{15}{16}$.

Express the fractions as equivalent fractions with 16 as the common denominator. Subtract the numerator 12 from the numerator 15. Write the difference 3 over the common denominator 16.

$$\frac{15}{16} = \frac{15}{16}$$
$$-\frac{3}{4} = -\frac{12}{16}$$
$$\frac{3}{16}$$

MATH REVIEW 5 Subtracting Fractions and Mixed Numbers from Whole Numbers

- To subtract a fraction or a mixed number from a whole number, express the whole number as an equivalent mixed number. The fraction of the mixed number has the same denominator as the denominator of the fraction that is subtracted. Subtract the numerators of the fractions and write their difference over the common denominator. Subtract the whole numbers. Combine the whole number and fraction. Express the answer in lowest terms.

Example 1 Subtract $\frac{3}{8}$ from 7

Express the whole number as an equivalent mixed number with the same denominator as the denominator of the fraction that is subtracted $\left(7 = 6\frac{8}{8}\right)$.

Subtract $\frac{3}{8}$ from $\frac{8}{8}$

Combine whole number and fraction.

$$\begin{array}{r} 7 = 6\frac{8}{8} \\ -\frac{3}{8} = -\frac{3}{8} \\ \hline 6\frac{5}{8} \end{array}$$

Example 2 Subtract $5\frac{15}{32}$ from 12

Express the whole number as an equivalent mixed number with the same denominator as the denominator of fraction that is subtracted $\left(12 = 11\frac{32}{32}\right)$.
Subtract fractions.
Subtract whole numbers.
Combine whole number and fraction.

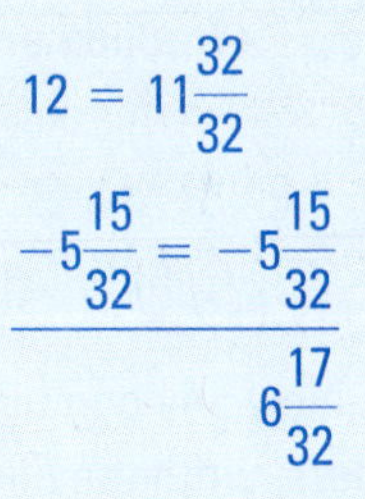

MATH REVIEW 6 Subtracting Fractions and Mixed Numbers from Mixed Numbers

- To subtract a fraction or a mixed number from a mixed number, the fractional part of each number must have the same denominator. Express fractions as equivalent fractions having a common denominator. When the fraction subtracted is larger than the fraction from which it is subtracted, one unit of a whole number is expressed as a fraction with the common denominator. Combine the whole number and fractions. Subtract fractions and subtract whole numbers.

Example 1 Subtract $\frac{7}{8}$ from $4\frac{3}{16}$

Express the fractions as equivalent fractions with the common denominator 16.

Since 14 is larger than 3, express one unit of $4\frac{3}{16}$ as a fraction and combine whole number and fractions.
$\left(4\frac{3}{16} = 3 + \frac{16}{16} + \frac{3}{16} = 3\frac{19}{16}\right)$.
Subtract.

$$\begin{array}{r} 4\frac{3}{16} = 4\frac{3}{16} = 3\frac{19}{16} \\ -\frac{7}{8} = \frac{14}{16} = -\frac{14}{36} \\ \hline 3\frac{5}{16} \end{array}$$

Example 2 Subtract 13 from 20
Express the fractions as equivalent fractions with the common denominator 32.
Subtract fractions.
Subtract whole numbers.

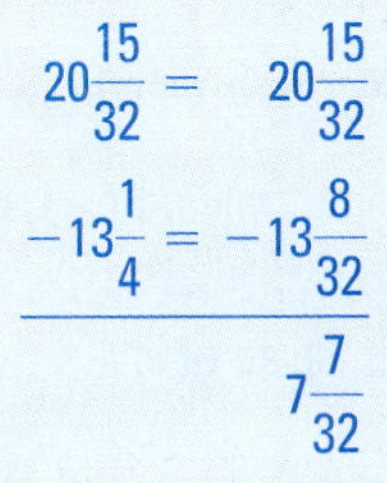

MATH REVIEW 7 Multiplying Fractions

- To multiply two or more fractions, multiply the numerators. Multiply the denominators. Write as a fraction with the product of the numerators over the product of the denominators. Express the answer in lowest terms.

Example 1 Multiply $\frac{3}{4} \times \frac{5}{8}$

Multiply the numerators.
Multiply the denominators.
Write as a fraction.

$$\frac{3}{4} \times \frac{5}{8} = \frac{15}{32}$$

Example 2 Multiply

Multiply the numerators.
Multiply the denominators.
Write as a fraction and express answer in lowest terms.

$$\frac{1}{2} \times \frac{2}{3} \times \frac{4}{5} = \frac{8}{30} = \frac{4}{15}$$

MATH REVIEW 8 Multiplying Any Combination of Fractions, Mixed Numbers, and Whole Numbers

- To multiply any combination of fractions, mixed numbers, and whole numbers, write the mixed numbers as fractions. Write whole numbers over the denominator 1. Multiply numerators. Multiply denominators. Express the answer in lowest terms.

Example 1 Multiply $3\frac{1}{4} \times \frac{3}{8}$

Write the mixed number $3\frac{1}{4}$ as the fraction $\frac{13}{4}$.
Multiply the numerators.
Multiply the denominators.
Express as a mixed number.

$$3\frac{1}{4} \times \frac{3}{8} = \frac{13}{4} \times \frac{3}{8} = \frac{39}{32} = 1\frac{7}{32}$$

Example 2 Multiply $2\frac{1}{3} \times 4 \times \frac{4}{5}$

Write the mixed number $2\frac{1}{3}$ as the fraction $\frac{7}{3}$.
Write the whole number 4 over 1.
Multiply the numerators.
Multiply the denominators.
Express as a mixed number.

$$2\frac{1}{3} \times 4 \times \frac{4}{5} = \frac{7}{3} \times \frac{4}{1} \times \frac{4}{5} = \frac{112}{15}$$

$$\frac{112}{15} = 7\frac{7}{15}$$

MATH REVIEW 9 Dividing Fractions

- Division is the inverse of multiplication. Dividing by 4 is the same as $\frac{1}{4}$. multiplying by *j*. 4 is the inverse of $\frac{1}{4}$. and is the inverse of 4.

  The inverse of is $-\frac{5}{16}$ is $\frac{16}{5}$.

- To divide fractions, invert the divisor, change to the inverse operation and multiply. Express the answer in lowest terms.

Example Divide: $\frac{7}{8} \div \frac{2}{3}$

Invert the divisor $\frac{2}{3}$

$\frac{2}{3}$ inverted is $\frac{3}{2}$.

Change to the inverse operation and multiply.

Express as a mixed number.

$$\frac{7}{8} \div \frac{2}{3} = \frac{7}{8} \times \frac{3}{2} = \frac{21}{16} = 1\frac{5}{16}$$

MATH REVIEW 10 Dividing Any Combination of Fraction, Mixed Numbers, and Whole Numbers

- To divide any combination of fractions, mixed numbers, and whole numbers, write the mixed number as fractions. Write whole numbers over the denominator 1. Invert the divisor. Change to the inverse operation and multiply. Express the answer in lowest terms.

Example 1 Divide: $6 \div \frac{7}{10}$

Write the whole number 6 over the denominator 1.

Invert the divisor $\frac{7}{10}$; $\frac{7}{10}$ inverted is $\frac{10}{7}$.

Change to the inverse operation and multiply.

Express as a mixed number.

$$\frac{6}{1} \div \frac{7}{10} =$$

$$\frac{6}{1} \times \frac{10}{7} = \frac{60}{7} = 8\frac{4}{7}$$

Example 2 Divide: $\frac{3}{4} \div 2\frac{1}{5}$

Write the mixed number divisor $2\frac{1}{5}$ as the fraction $\frac{11}{5}$.

Invert the divisor $\frac{11}{5}$; $\frac{11}{5}$ inverted is $\frac{5}{11}$.

Change to the inverse operation and multiply.

$$\frac{3}{4} \div \frac{11}{5} =$$

$$\frac{3}{4} \times \frac{5}{11} = \frac{15}{44}$$

Example 3 Divide: $4\frac{5}{8} \div 7$

Write the mixed number $4\frac{5}{8}$ as the fraction $\frac{37}{8}$.

Write the whole number divisor over the denominator 1.

Invert the divisor $\frac{7}{1}$; $\frac{7}{1}$ inverted is $\frac{1}{7}$.

Change to the inverse operation and multiply.

$$\frac{37}{8} \div \frac{7}{1} =$$

$$\frac{37}{8} \times \frac{1}{7} = \frac{37}{56}$$

MATH REVIEW 11 Rounding Decimal Fractions

- To round a decimal fraction, locate the digit in the number that gives the desired number of decimal places. Increase that digit by 1 if the digit that directly follows is 5 or more. Do not change the value of the digit if the digit that follows is less than 5. Drop all digits that follow.

- *Equivalent Units of Volume Measure:*
  1 cubic foot (cu ft) =
  12 in × 12 in × 12 in = 1728 cubic inches (cu in)
  1 cubic yard (cu yd) =
  3 ft × 3 ft × 3 ft = 27 cubic feet (cu ft)
- To express a given unit of volume as a larger unit of volume, divide the given volume by the number of cubic units contained in one of the larger units.

  Example 1 Express 6,048 cubic inches as cubic feet
  Since 1728 cu in = 1 cu ft, divide 6048 by 1728.

  6048 ÷ 1728 = 3.5
  6048 cubic inches = 3.5 cubic feet

  Example 2 Express 167.4 cubic feet as cubic yards.
  Since 17 cu ft = 1 cu yd, divide 167.4 by 27.

  167.4 ÷ 27 = 6.2
  167.4 cubic feet = 6.2 yards

- To express a given unit of volume as a smaller unit of volume, multiply the given volume by the number of cubic units contained in one of the larger units.

  Example 1 Express 1.6 cubic feet as cubic inches.
  Since 1728 cu in = 1 cu ft, divide 1.6 by 1728.

  1.6 × 1728 = 2764.8
  1.6 cubic feet = 2764.8 cubic inches

  Example 2 Express 8.1 cubic yards as cubic feet.
  Since 27 cu ft = 1 cu yd, divide 8.1 by 27.

  8.1 × 27 = 218.7
  8.1 cubic yards = 218.7 cubic feet

- *Computing Volumes of Common Solids*
- A prism is a solid that has two identical faces called bases and parallel lateral edges. In a right prism, the lateral edges are perpendicular (at 90°) to the bases. The altitude or height (h) of a prism is the perpendicular distance between its two bases. Prisms are named according to the shapes of their bases.
- The volume of any prism is equal to the product of the area of its base and altitude or height.
  Volume = area of base 3 altitude ($V = AB \times h$)

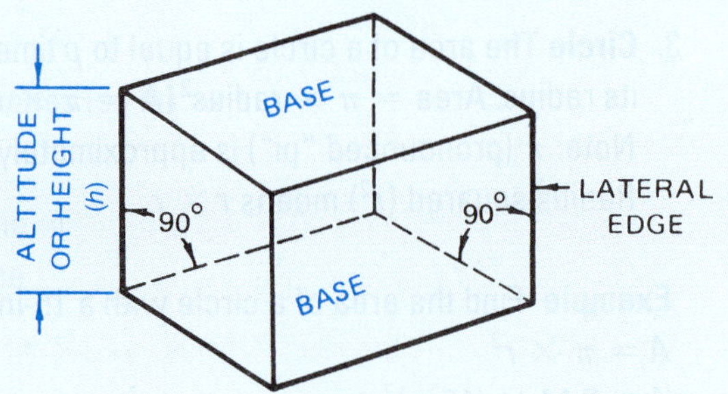

- *Right Rectangular Prism:*
  A right rectangular prism has rectangular bases.
  Volume = area of base 3 altitude
  $V = A_B \times h$

  Example Find the volume of a rectangular prism with a base length of 20 feet, a base width of 14 feet, and a height (altitude) of 8 feet.
  $V = A_B \times h$
  Compute the area of the base (A_B):
  Area of base = length × width
  $A_B = 20 \text{ ft} \times 14 \text{ ft}$
  $A_B = 280 \text{ sq ft}$

Compute the volume of the prism:

$V = A_B \times h$

$V = 280 \text{ sq ft} \times 8 \text{ ft}$

$V = 2{,}240 \text{ cu ft}$

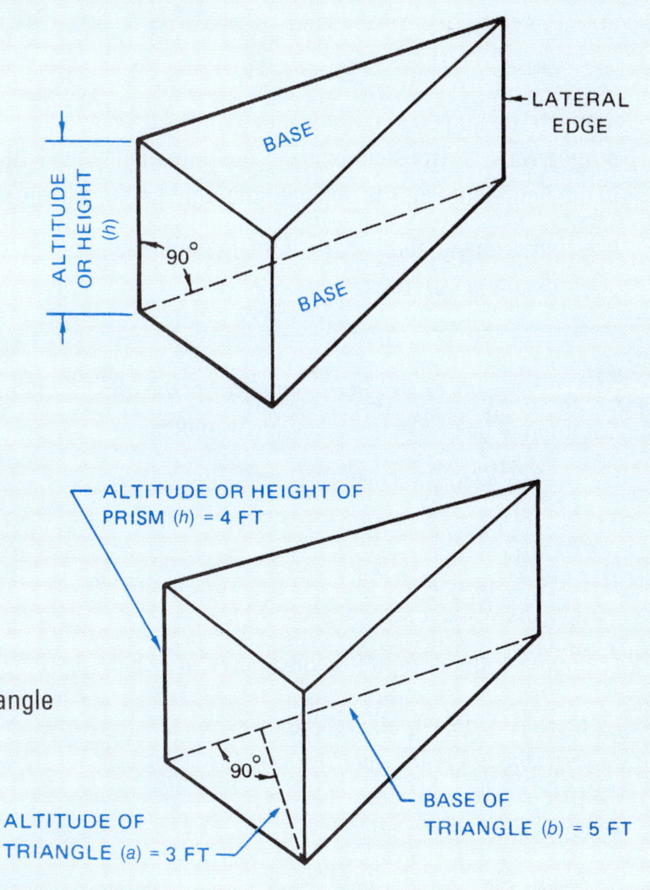

- *Right Triangular Prism:*

  A right triangular prism has triangular bases.

  Volume = area of base × altitude

  $V = A_B \times h$

Example Find the volume of a triangular prism in which the base of the triangle is 5 feet, the altitude of the triangle is 3 feet, and the altitude (height) of the prism is 4 feet. Refer to the accompanying figure.

Volume = area of base × altitude

$V = A_B \times h$

Compute the area of the base:

Area of base = $\frac{1}{2}$base of triangle × altitude of triangle

$A_B = \frac{1}{2} b \times a$

$A_B = \frac{1}{2} \times 5 \text{ ft} \times 3 \text{ ft}$

$A_B = 7.5 \text{ sq ft}$

Compute the volume of the prism:

$V = A_B \times h$

$V = 7.5 \text{ sq ft} \times 4 \text{ ft}$

$V = 30$ cubic feet

- *Right Circular Cylinder:*

  A right circular cylinder has circular bases.

  Volume = area of base × altitude

  $V = A_B \times h$

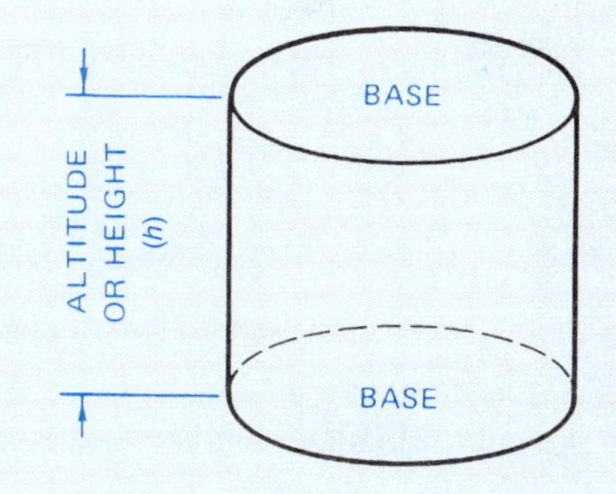

Example Find the volume of a circular cylinder 1 foot in diameter and 10 feet high.

Note: Radius = $\frac{1}{2}$ Diameter; Radius = 1 × 1 ft = 0.5 ft.

$V = A_B \times h$

Compute the area of the base:

Area of base = π × radius squared

$A_B = 3.14 \times (0.5)^2$

$A_B = 3.14 \times 0.5 \text{ ft} \times 0.5 \text{ ft}$

$A_B = 3.14 \times 0.25 \text{ sq ft}$

$A_B = 0.785 \text{ sq ft}$

Compute the volume of the cylinder:

$V = A_B \times h$

$V = 0.785 \text{ sq ft} \times 10 \text{ ft}$

$V = 7.85$ cubic feet

MATH REVIEW 24 Finding an Unknown Side of a Right Triangle, Given Two Sides

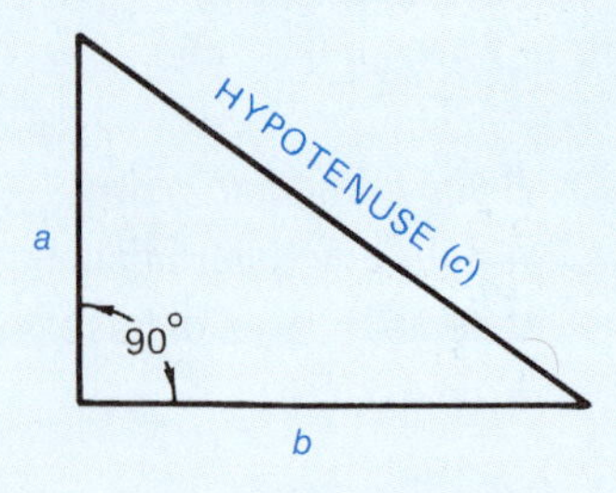

- If one of the angles of a triangle is a right (90°) angle, the figure is called a *right triangle*. The side opposite the right angle is called the *hypotenuse*. In the figure shown, *c* is opposite the right angle; *c* is the hypotenuse.
- In a right triangle, the square of the hypotenuse is equal to the sum of the squares of the other two sides:

$$C^2 = a^2 + b^2$$

If any two sides of a right triangle are known, the length of the third side can be determined by one of the following formulas:

$$c = \sqrt{a^2 + b^2}$$

$$a = \sqrt{c^2 - b^2}$$

$$b = \sqrt{c^2 - a^2}$$

Example 1 In the right triangle shown, $a = 6$ ft, $b = 8$ ft, find c.

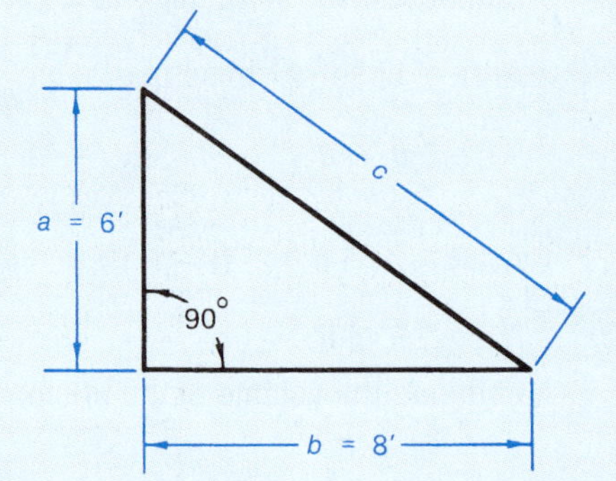

$$c = \sqrt{a^2 + b^2}$$

$$c = \sqrt{6^2 + 8^2}$$

$$c = \sqrt{36 + 64}$$

$$c = \sqrt{100}$$

$$c = 10 \text{ feet}$$

Example 2 In the right triangle shown, $c = 30$ ft, $b = 20$ ft, find a.

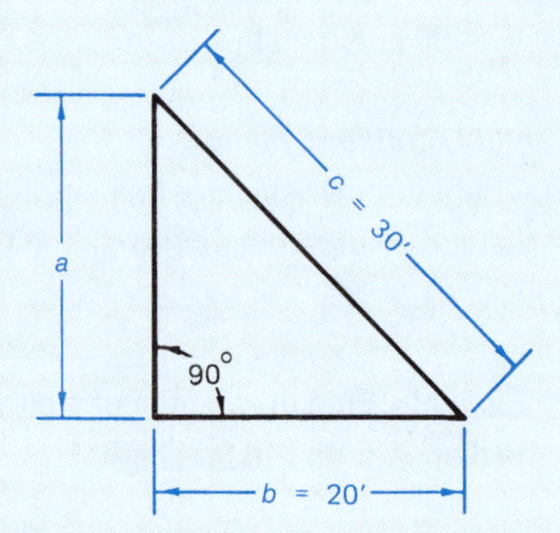

$$a = \sqrt{c^2 - b^2}$$

$$a = \sqrt{30^2 - 20^2}$$

$$a = \sqrt{900 - 400}$$

$$a = \sqrt{500}$$

$$a = 22.36 \text{ feet (to two decimal places)}$$

Example 3 In the right triangle shown, $c = 18$ ft, $a = 6$ ft, find b.

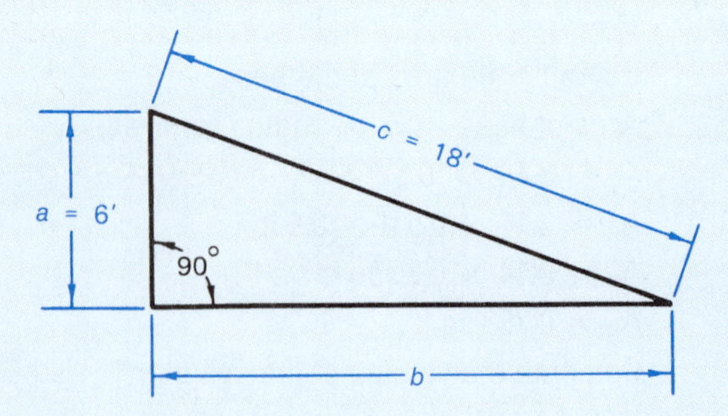

$$b = \sqrt{c^2 - a^2}$$

$$b = \sqrt{18^2 - 6^2}$$

$$b = \sqrt{324 - 36}$$

$$b = \sqrt{288}$$

$$b = 16.97 \text{ feet (to two decimal places)}$$

Appendix C

MATERIAL SYMBOLS IN SECTIONS

EARTH

ROCK

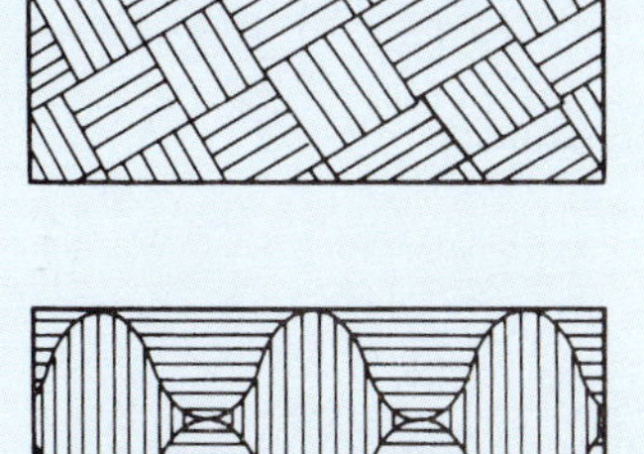

GRAVEL OR CRUSHED STONE

CONCRETE

CONCRETE BLOCK

FACE BRICK OR COMMON BRICK

FIRE BRICK

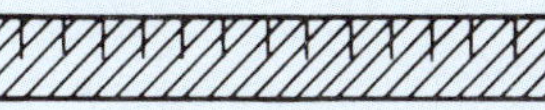

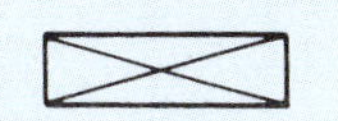

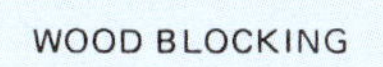

FINISH WOOD

STRUCTURAL STEEL

REINFORCING BARS

GENERAL METAL

BATT INSULATION

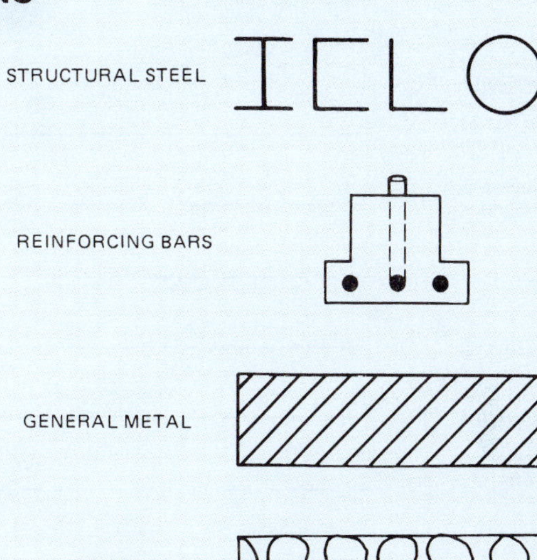

RIGID INSULATION

PLASTER OR GYPSUM BOARD

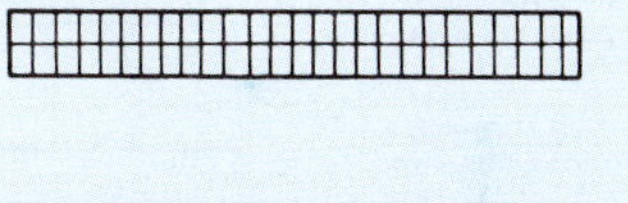

THIN SHEET MATERIALS (PLASTIC FILM, SHEET METAL, PAPER, ETC.)

Appendix D

PLUMBING SYMBOLS

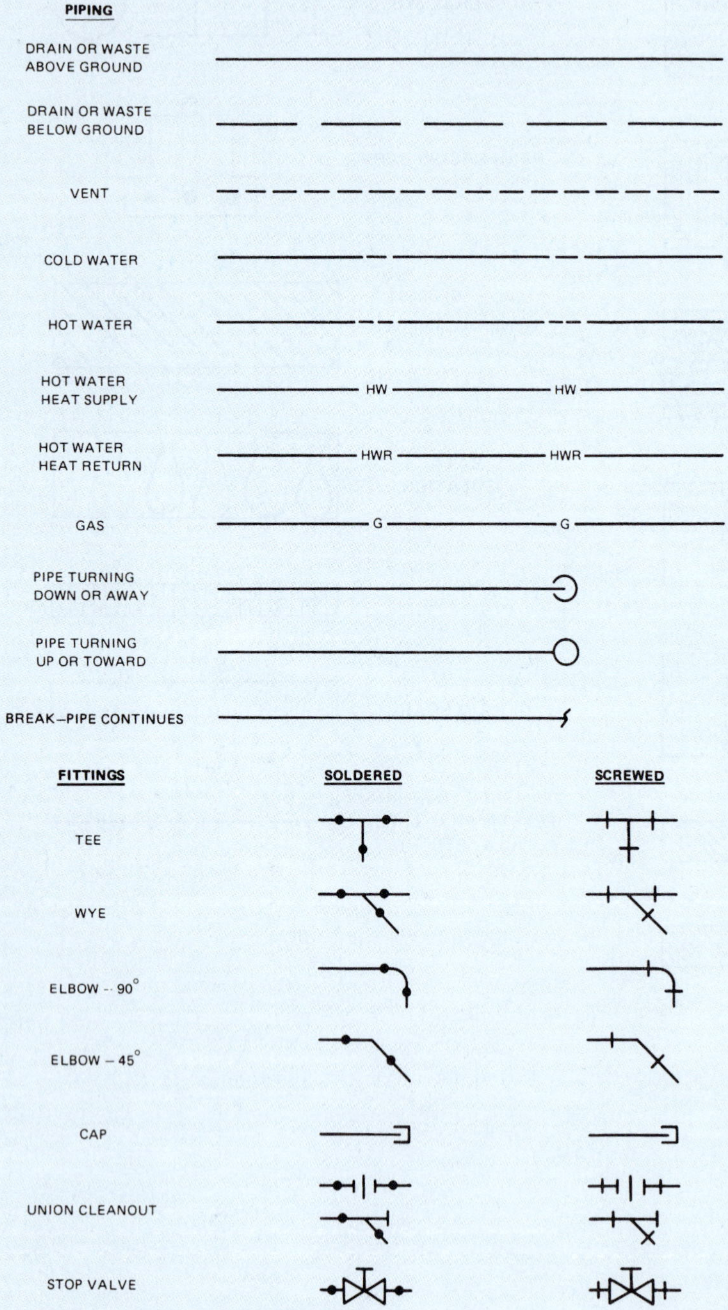

PIPING

DRAIN OR WASTE ABOVE GROUND

DRAIN OR WASTE BELOW GROUND

VENT

COLD WATER

HOT WATER

HOT WATER HEAT SUPPLY — HW — HW —

HOT WATER HEAT RETURN — HWR — HWR —

GAS — G — G —

PIPE TURNING DOWN OR AWAY

PIPE TURNING UP OR TOWARD

BREAK—PIPE CONTINUES

FITTINGS	SOLDERED	SCREWED
TEE		
WYE		
ELBOW – 90°		
ELBOW – 45°		
CAP		
UNION CLEANOUT		
STOP VALVE		

Appendix E

ELECTRICAL SYMBOLS

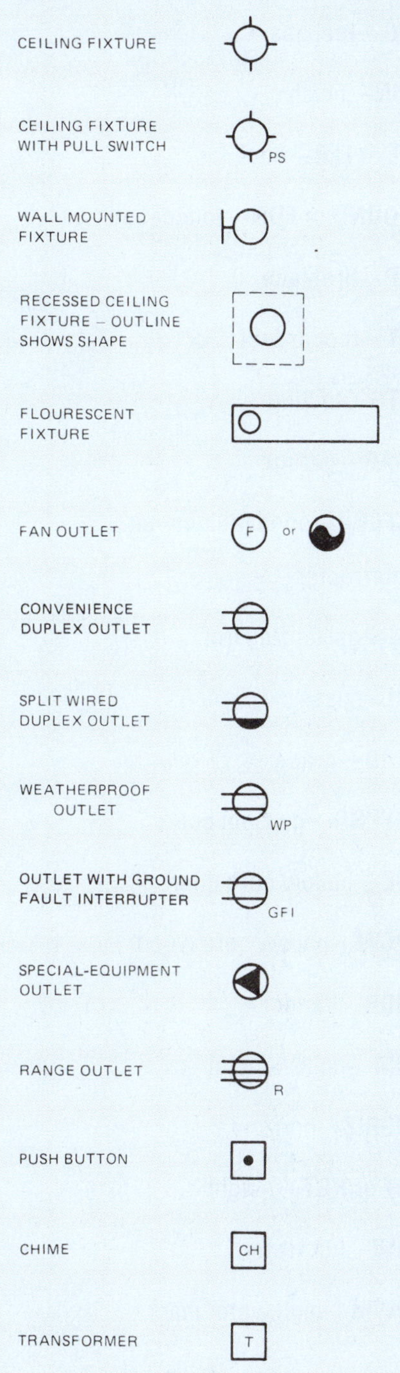

TELEPHONE	
INTERCOM	
TELEVISION ANTENNA	TV
SMOKE DETECTOR	
DISTRIBUTION PANEL	
JUNCTION BOX	J
SINGLE-POLE SWITCH	S
THREE-WAY SWITCH	S_3
SWITCH WITH PILOT LIGHT	S_p
WEATHERPROOF SWITCH	S_{wp}
SWITCH WIRING	or

Appendix F

COMMON ABBREVIATIONS

Note: These common abbreviations can appear on construction drawings and like documentation with or without periods (e.g., AC or A.C.).

AB—anchor bolt

AC—air conditioning

AL or **ALUM**—aluminum

BA—bathroom

BLDG—building

BLK—block

BLKG—blocking

BM—beam

BOTT—bottom

BPL—base plate

BR—bedroom

BRM—broom closet

BSMT—basement

CAB—cabinet

CL—centerline

CLNG or **CLG**—ceiling

CMU—concrete masonry unit (concrete block)

CNTR—center or counter

COL—column

COMP—composition

CONC—concrete

CONST—construction

CONT—continuous

CORRUG—corrugated

CRNRS—corners

CU—copper

d—penny (nail size)

DBL—double

DET—detail

DIA or ∅—diameter

DIM—dimension

DN—down

DO—ditto

DP—deep or depth

DR—door

DW—dishwasher

ELEC—electric

ELEV—elevation

EQ—equal

EXP—exposed or expansion

EXT—exterior

FG—fuel gas

FIN—finish

FL or **FLR**—floor

FOUND or **FDN**—foundation

FP—fireplace

FT—foot or feet

FTG—footing

GAR—garage

GFCI—ground fault circuit interrupter

GI—galvanized iron

GL—glass

GRD—grade

GYPBD—gypsum board

HC—hollow core door

HCW—hollow core wood

HDR—header

HM—hollow metal

HORIZ—horizontal

HT or HGT—height

HW—hot water

HWM—high water mark

IN—inch or inches

INSUL—insulation

INT—interior

JSTS—joists

JT—joint

LAV—lavatory

LH—left hand

LIN—linen closet

LT —light

MANUF—manufacturer

MAS—masonry

MATL—material

MAX—maximum

MIN—minimum

MTL—metal

NAT—natural

N/F—now or formerly

NIC—not in contract

o/—overhead or over

OC—on centers

OH Door—overhead door

OSB—Oriented strand board

PERF—perforated

PL—plate

PLYWD—plywood

PT—pressure-treated lumber

R—risers

REF—refrigerator

REINF—reinforcement

REQ—requirement

RH—right hand

RM—room

ROB—run of bank (gravel)

ROW—right of way

SCRND—screened

SHT—sheet

SHTG—sheathing

SHWR—shower

SIM—similar

SL—sliding

S&P—shelf and pole

SPEC—specifications

SQ or ⧄—square

STD—standard

STL—steel

STY—story

T&G—tongue and groove

THK—thick

T'HOLD—threshold

TYP—typical

VB—vapor barrier

w/—with

WARD—wardrobe

WC—water closet

WD—wood

WDW—window

WH—water heater

WI—wrought iron

Glossary

A

Addendum—A change or modification to the bid documents, plans, and specifications that is made prior to the contractor's bid date

Aggregate—Hard materials such as sand and crushed stone used to make concrete

Ampere (amp)—Unit of measure of electric current

Anchor Bolt—A bolt placed in the surface of concrete for attaching wood framing members

Apron—Concrete slab at the approach to a garage door—also the wood trim below a window stool

Architect's Scale—A flat or triangular scale used to measure scale drawings

Ash Dump—A small metal door in the bottom of a fireplace

Awning Window—A window that is hinged near the top so the bottom opens outward

B

Backfill—Earth placed against a building wall after the foundation is in place

Backsplash—The raised lip on the back edge of a countertop to prevent water from running down the backs of the cabinets

Balloon Framing—Type of construction in which the studs are continuous from the sill to the top of the wall—upper floor joists are supported by a let-in ribbon

Balusters—Vertical pieces that support a handrail

Balustrade—The complete assembly of railings, balusters, and newel posts around a stair or balcony

Batt Insulation—Flexible, blanket-like pieces, usually of fiber-glass, used for thermal or sound insulation

Batten—Narrow strip of wood used to cover joints between boards of sheet materials

Batter Boards—An arrangement of stakes and horizontal pieces used to attach lines for laying out a building

Beam—Any major horizontal structural member

Beam Pocket—A recessed area to hold the end of a beam in a concrete or masonry wall

Bird's Mouth—A notch cut in a rafter to fit around the top plate of the wall. The bird's mouth consists of a level cut that rests on top of the wall plate and a plumb cut that fits against the vertical surface of the wall.

Board Foot—144 cubic inches of wood or the amount contained in a piece measuring 12″ × 12″ × 1″

Bottom Chord—The bottom horizontal member in a truss

Box Sill—The header joist nailed across the ends of floor joists at the sill

Branch Circuit—The electrical circuit that carries current from the distribution panel to the various parts of the building

British Thermal Unit (BTU)—The amount of heat required to raise the temperature of 1 pound of water 1° Fahrenheit

Building Lines—The outside edge of the exterior walls of a building

C

Casement Window—A window that is hinged at one side so the opposite side opens outward

Casing—The trim around a door or window

Centerline—An actual or imaginary line through the exact center of any object

Change Order—A change or modification to the contract documents, plans, and specifications that is made after the contract has been awarded to the selected trade contractor

Cleanout—A pipe fitting with a removable plug that allows for cleaning the run of piping in which it is installed, or an access door at the bottom of a chimney

Collar Beam—Horizontal members that tie opposing rafters together, usually installed about halfway up the rafters

Collar Ties—A horizontal board between two opposing rafters, usually about one-third the vertical height of the roof from peak. Collar ties strengthen the roof frame and help prevent the rafters from forcing the walls outward.

Column—A metal post to support an object above

Common Rafter—A rafter extending from the top of the wall to the ridge

Concrete—Building material consisting of fine and coarse aggregates bonded together by Portland cement

Conductor—Electrical wire—a cable may contain several conductors

Contour Lines—Lines on a topographic map or site plan to describe the contour of the land

Contract—Any agreement in writing for one party to perform certain work and the other party to pay for the work

Convenience Outlet—Electrical outlet provided for convenient use of lamps, appliances, and other electrical equipment

Cornice—The construction that encloses the ends of the rafters at the top of the wall

Cornice Return—The construction where the level cornice meets the sloping rake cornice

Course—A single row of building units such as concrete blocks or shingles

Cove Mold—Concave molding used to trim an inside corner

CSI Format—The standard format for specifications developed by the Construction Specifications Institute.

D

Damper—A door installed in the throat of a fireplace to regulate the draft

Dampproofing—Vapor barrier or coating on foundation walls or under concrete slabs to prevent moisture from entering the house

Datum—A reference point from which elevations are measured

Delivery Sheet (Trusses)—A summary sheet included with a packet of truss drawings to show how many of each type of truss is required

Detail—A drawing showing special information about a particular part of the construction—details are usually drawn to a larger scale than other drawings and are sometimes section views

Dormer—A raised section in a roof to provide extra headroom below

Double-hung Window—A window consisting of two sashes that slide up and down past one another

Drip Cap—A wood ledge over wall openings to prevent water from running back under the frame or trim around the opening

Drip Edge—Metal trim installed at the edge of a roof to stop water from running back under the edge of the roof deck

Drywall—Interior wall construction using gypsum wallboard

E

Elevation—A drawing that shows vertical dimensions—it may also be the height of a point, usually in feet, above sea level

F

Fascia—The part of a cornice that covers the ends of the rafters

Firestop—Blocking or noncombustible material between wall studs to prevent vertical draft and flame spread

Flashing—Sheet metal used to cover openings and joints in walls and roofs

Float—To level concrete before it begins to cure—floating is done with a tool called a *float*

Floor Plan—A drawing showing the arrangement of rooms, the locations of windows and doors, and complete dimensions—a floor plan is actually a horizontal section through the entire building

Flue—The opening inside a chimney—the flue is usually formed by a terra cotta flue liner

Flush Door—A door having flat surfaces

Footing—The concrete base upon which the foundation walls are built

Footing Drain—(also called *perimeter drain*) An underground drain pipe around the footings to carry groundwater away from the building

Frieze—A horizontal board beneath the cornice and against the wall above the siding

Frostline—The maximum depth to which frost penetrates the earth

Furring—Narrow strips of wood attached to a surface for the purpose of creating a plumb or level surface for attaching the wall, ceiling, or floor surface

G

Gable—The triangular area between the roof and the top plate walls at the ends of a gable roof

Gable Studs—The studs placed between the end rafters and the top plates of the end walls

Gauge—A standard unit of measurement for the diameter of wire or the thickness of sheet metal

Girder—A beam that supports floor joists

Grout—A thin mixture of high-strength concrete or mortar

Gusset—A reinforcing piece of metal or plywood fastened to a truss where the members meet.

Gypsum Wallboard—Drywall material made of gypsum encased in paper to form boards

H

Header—A joist fastened across the ends of regular joists in an opening, or the framing member above a window or door opening

Hearth—Concrete or masonry apron in front of a fireplace

Hip—Outside corner formed by intersecting roofs

Hip Rafter—The rafter extending from the corner of a building to the ridge at a hip

Hopper—A type of window that is hinged at the bottom and opens inward at the top.

Hose Bibb—An outside faucet to which a hose can be attached

I

I-joist—An engineered wood joist with a vertical web of plywood or OSB and solid wood upper and lower chords

Insulated Glazing—Two or more pieces of glass in a single sash with an air space between them for the purpose of insulation

Invert Elevation—The elevation at the lowest point inside a pipe

Isometric Drawing—A kind of drawing in which horizontal lines are 30° from true horizontal and vertical lines are vertical

J

Jack Rafter—A rafter between the outside wall and a hip rafter or the ridge and a valley rafter

Jamb—Side members of a door or window frame

Joists—Horizontal framing members that support a floor or ceiling

L

Lintel—Steel or concrete member that spans a clear opening—usually found over doors, windows, and fireplace openings

M

Masonry Cement—Cement that is specially prepared for making mortar

Measuring Line—An imaginary line running the length of a rafter and passing through the deepest part of the bird's mouth. This is the line from which all rafter measurements are made.

Mil—A unit of measure for the thickness of very thin sheets—1 mil equals .001″

Miter—A 45° cut so that two pieces will form a 90° corner

Mortar—Cement and aggregate mixture for bonding masonry units together

Mullion—The vertical piece between two windows that are installed side by side—window units that include a mullion are called *mullion windows*

Muntins—Small vertical and horizontal strips that separate the individual panes of glass in a window sash

N

Nailer—A piece of wood used in any of several places to provide a nailing surface for other framing members

Nominal Size—The size by which a material is specified—the actual size is often slightly smaller

Nosing—The portion of a stair tread that projects beyond the riser

O

Oblique Drawing—A drawing type in which one side (usually the one with the most detail) is drawn in proportion as though it is flat against the paper. Other sides are shown with parallel lines to show depth.

Orthographic Projection—A method of drawing that shows separate views of an object

Overhang—The horizontal distance covered by a roof outside the walls of the structure.

P

Panel Door—A door made up of panels held in place by rails and stiles

Penny Size—The length of nails

Perimeter Drain—(See *Footing Drain*)

Pilaster—A masonry or concrete pier built as an integral part of a wall

Pitch—Refers to the steepness of a roof—the pitch is written as a fraction with the rise over the span

Plan View—A drawing that shows the layout of an object as viewed from above

Plate—The horizontal framing members at the top and bottom of the wall studs

Platform Framing—(also called *Western framing*) A method of framing in which each level is framed separately—the subfloor is laid for each floor before the walls above it are formed

Plenum—A chamber within a forced-air heating system that is pressurized with warm air

Plumb—Truly vertical or true according to a plumb bob

Ply (trusses)—If trusses are plied, they are joined together face-to-face to make a stronger unit

Portland Cement—Finely powdered limestone material used to bond the aggregates together in concrete and mortar

Potable Water—Water that is safe for drinking

R

R value—The ability of a material to resist the flow of heat

Rafter—The framing members in a roof

Rail—Also the horizontal members in a balustrade, such as the railing around a stairway

Rake—The sloping cornice at the end of a gable roof

Resilient Flooring—Vinyl, vinyl-asbestos, and other man-made floor coverings that are flexible yet produce a smooth surface

Ridge Board—The framing member between the tops of rafters that runs the length of the ridge of a roof

Rise—The vertical dimension of a roof or stair

Riser—The vertical dimension of one step in a stair—the board enclosing the space between two treads is called a riser

Rowlock—Position of bricks in which the bricks are laid on edge

Run—The horizontal distance covered by an inclined surface such as a rafter or stair

S

Sash—The frame holding the glass in a window

Saturated Felt—Paper-like felt that has been treated with asphalt to make it water resistant

Screed—A straight board used to level concrete immediately after it is placed

Section View—A drawing showing what would be seen by cutting through a building or part

Setback—The distance from a street or front property line to the front of a building

Sheathing—The rough exterior covering over the framing members of a building

Shim—Thin material, typically wood shingle, used to adjust a small space

Sill—The framing member in contact with a masonry or concrete foundation

Sill Sealer—Compressible material used under the sill to seal any gaps

Site Constructed—Built on the job

Site Plan—The drawing that shows the boundaries of the building, its location, and site utilities

Sliding Window—A window with two or more sashes that slide horizontally past one another

Slope—The steepness of a roof, usually indicated in inches of rise per foot of run.

Soffit—The bottom surface of any part of a building, such as the underside of a cornice or lowered portion of a ceiling over wall cabinets

Soldier—Brick position in which the bricks are stood on end

Span—The horizontal dimension between vertical supports—the span of a beam is the distance between the posts that support it

Specifications—Specific written instructions for materials to be used and methods of construction or application

Square—The amount of siding or roofing materials required to cover 100 square feet

Stack—The main vertical pipe into which plumbing fixtures drain

Stair Carriage—The supporting framework under a stair

Stile—The vertical members in a sash, door, or other panel construction

Stool—Trim piece that forms the finished window sill

Stop—Molding that stops a door from swinging through the opening as it is closed—also used to hold the sash in place in a window frame

Stud—Vertical framing member in a wall

Subfloor—The first layer of rough flooring applied to the floor joists

Sweat—Method of soldering used in plumbing

T

Tail—The portion of a truss top chord or a rafter that extends beyond the walls of the structure.

Termite Shield—Sheet metal shield installed at the top of a foundation to prevent termites from entering the wood superstructure

Thermal-break Window—Window with a metal frame that has the interior and exterior separated by a material with a higher R value than the metal itself

Thermostat—An electrical switch that is activated by changes in temperature

Top Chord—The top horizontal or sloped member of a truss

Trap—A plumbing fitting that holds enough water to prevent sewer gas from entering the building

Tread—The surface of a step in stair construction

Trimmers—The double framing members at the sides of an opening

Truss—A manufactured assembly used to support a load over a long span

Truss Detail—An engineering drawing giving detailed specification for the manufacture of one particular truss; a truss drawing packet will usually include a truss detail for each type of truss to be used

Truss Drawings—Any of several types of drawings provided by an engineer to give detailed information about the construction and placement of trusses

Truss Layout Plan—A plan drawing giving detailed information about where each truss is to be placed, including spacing between trusses

U

Underlayment—Any material installed over the subfloor to provide a smooth surface over which floor covering will be installed

V

Valley—The inside corner formed by intersecting roofs

Valley Rafter—The rafter extending from an inside corner in the walls to the ridge at a valley

Vapor Barrier—Sheet material used to prevent water vapor from passing through a building surface

Veneer—A thin covering; in masonry, a single wythe of finished masonry over a wall; in woodwork, a thin layer of wood

Vent Pipe—A pipe, usually through the roof, that allows atmospheric pressure into the drainage system

Vertical Contour Interval—The difference in elevation between adjacent contour lines on a topographic map or site plan

Volt—The unit of measurement for electrical force

W

Water Closet—A plumbing fixture commonly called a *toilet*

Watt—The unit of measurement of electrical power—1 watt is the amount of power from 1 ampere of current with 1 volt of force

Web—The bracing members of a truss between the top chord(s) and the bottom chord.

Weep Hole—A small hole through a masonry wall to allow water to pass

Western Framing—(See *Platform Framing*)

Wythe—A single thickness of masonry construction

Index

Page numbers followed by "f" indicate figure.

D

E

F

R

S

T

U

V

W

Z